Studies in Computational Intelligence

Volume 565

Series editor

Janusz Kacprzyk, Polish Academy of Sciences, Warsaw, Poland
e-mail: kacprzyk@ibspan.waw.pl

About this Series

The series "Studies in Computational Intelligence" (SCI) publishes new developments and advances in the various areas of computational intelligence—quickly and with a high quality. The intent is to cover the theory, applications, and design methods of computational intelligence, as embedded in the fields of engineering, computer science, physics and life sciences, as well as the methodologies behind them. The series contains monographs, lecture notes and edited volumes in computational intelligence spanning the areas of neural networks, connectionist systems, genetic algorithms, evolutionary computation, artificial intelligence, cellular automata, self-organizing systems, soft computing, fuzzy systems, and hybrid intelligent systems. Of particular value to both the contributors and the readership are the short publication timeframe and the world-wide distribution, which enable both wide and rapid dissemination of research output.

More information about this series at http://www.springer.com/series/7092

Elias Kyriakides · Marios Polycarpou
Editors

Intelligent Monitoring, Control, and Security of Critical Infrastructure Systems

Editors
Elias Kyriakides
Marios Polycarpou
KIOS Research Center for Intelligent Systems and Networks
Department of Electrical and Computer Engineering
University of Cyprus
Nicosia
Cyprus

ISSN 1860-949X ISSN 1860-9503 (electronic)
ISBN 978-3-662-44159-6 ISBN 978-3-662-44160-2 (eBook)
DOI 10.1007/978-3-662-44160-2

Library of Congress Control Number: 2014948739

Springer Heidelberg New York Dordrecht London

Printed on acid-free paper

Springer is part of Springer Science+Business Media (www.springer.com)

COST Information

COST—the acronym for European Cooperation in Science and Technology—is the oldest and widest European intergovernmental network for cooperation in research. Established by the Ministerial Conference in November 1971, COST is presently used by the scientific communities of 35 European countries to cooperate in common research projects supported by national funds.

The funds provided by COST—less than 1% of the total value of the projects—support the COST cooperation networks (COST Actions) through which, with €30 million per year, more than 30,000 European scientists are involved in research having a total value which exceeds €2 billion per year. This is the financial worth of the European added value which COST achieves.

A "bottom up approach" (the initiative of launching a COST Action comes from the European scientists themselves), "à la carte participation" (only countries interested in the Action participate), "equality of access" (participation is open also to the scientific communities of countries not belonging to the European Union) and "flexible structure" (easy implementation and light management of the research initiatives) are the main characteristics of COST.

As a precursor of advanced multidisciplinary research COST has a very important role for the realization of the European Research Area (ERA) anticipating and complementing the activities of the Framework Programs, constituting a "bridge" toward the scientific communities of emerging countries, increasing the mobility of researchers across Europe, and fostering the establishment of "Networks of Excellence" in many key scientific domains such as: Biomedicine and Molecular Biosciences; Food and Agriculture; Forests, their Products and Services; Materials, Physical and Nanosciences; Chemistry and Molecular Sciences and Technologies; Earth System Science and Environmental Management; Information and Communication Technologies; Transport and Urban Development; Individuals, Societies, Cultures and Health. It covers basic and more applied research and also addresses issues of pre-normative nature or of societal importance.

Web: http://www.cost.eu

Preface

Modern society relies on the availability and smooth operation of a variety of complex engineering systems. These systems are termed *Critical Infrastructure Systems*. Some of the most prominent examples of critical infrastructure systems are electric power systems, telecommunication networks, water distribution systems, transportation systems, wastewater and sanitation systems, financial and banking systems, food production and distribution, health, security services, and oil/natural gas pipelines. Our everyday life and well-being depend heavily on the reliable operation and efficient management of these critical infrastructures.

The citizens expect that critical infrastructure systems will always be available, 24 hours a day, 7 days a week, and at the same time, they will be managed efficiently so that the services are provided at a low cost. Experience has shown that this is most often true. Nevertheless, critical infrastructure systems fail occasionally. Their failure may be due to natural disasters (e.g., earthquakes and floods), accidental failures (e.g., equipment failures, software bugs, and human errors), or malicious attacks (either direct or remote). When critical infrastructures fail, the consequences are tremendous. These consequences may be classified into societal, health, and economic. For example, if a large geographical area experiences a blackout for an extended period of time, that may result in huge economic costs, as well as societal costs. In November 2006, a local fault in Germany's power grid cascaded through large areas of Europe, resulting in 10 million people left in the dark in Germany, France, Austria, Italy, Belgium, and Spain. Severe cascading blackouts have taken place in North America as well. Similarly, there may be significant health hazards when there is a serious fault in the water supply, especially if it is not detected and accommodated quickly. When the telecommunication networks are down, many businesses can no longer operate. In the case of faults or unexpected events in transportation systems, we witness the effect of traffic congestion quite often in metropolitan areas of Europe and around the world. In general, failures in critical infrastructure systems are low probability events, which however may have a huge impact on everyday life and well-being.

Technological advances in sensing devices, real-time computation and the development of intelligent systems, have instigated the need to improve the performance of critical infrastructure systems in terms of security, accuracy, efficiency, reliability, and fault tolerance. Consequently, there is a strong effort in developing new algorithms for monitoring, control, and security of critical infrastructure systems, typically based on computational intelligence techniques and the real time processing of data received by networked embedded systems and sensor/actuator networks, dispersed throughout the system. Depending on the application, such data may have different characteristics: multidimensional, multiscale, spatially distributed, time series, event-triggered, etc. Furthermore, the data values may be influenced by controlled variables, as well as by external environmental factors. However, in some cases the collected data may be incomplete, or it may be faulty due to sensing or communication problems, thus compromising the sensor-environment interaction and possibly affecting the ability to manage and control key variables of the environment.

Despite the technological advances in sensing/actuation design and data processing techniques, there is still an urgent need to intensify the efforts towards intelligent monitoring, control, and security of critical infrastructure systems. The problem of managing critical infrastructure systems is becoming more complicated since they were not designed to be so large in size and geographical distribution; instead, they evolved due to the growing demand, while new technologies have been combined with outdated infrastructures in a single system that is required to perform new and more complex tasks. Furthermore, deregulation and the new market structure in several of these infrastructures has resulted in more heterogeneous and distributed infrastructures, which make them more vulnerable to failures and attacks. The introduction of renewable energy sources and environmental issues have incorporated new objectives to be met and new challenges in the operation and economics of some of these infrastructures (for example, power systems, telecommunication, water distribution networks, and transportation).

Two important notions that captivate the attention of researchers and of the industry are the concepts of *cyber-physical systems* and *system of systems*. Cyber-physical systems are the result of the interconnection and interaction of the cyber (computation) and the physical elements in a system. Embedded sensors, computers, and networks monitor and collect data from the physical processes; in turn, it is possible to control the physical processes through the analysis and use of the data collected to take appropriate actions for retaining the stability and security of the system.

The system of systems concept arose from the interconnection of independent systems in a larger, more complex system. These formerly independent systems may now be interacting or be interdependent. There are dependencies between infrastructures (e.g., a fault in the power system removes the supply to a water pump and thus, the water supply to an area), or in some cases interdependencies (e.g., a fault in the power system causes the oil/natural gas pipeline pumping stations to stop working, and as a consequence the supply of fuel to the power station is interrupted). Critical infrastructure dependencies and interdependencies pose an even higher degree of complexity, particularly on the appropriate modeling and simulation of the effects that one infrastructure has on another infrastructure. The fact that fewer people nowadays understand how these networks operate and the interactions between all the components, creates a necessity for further research and in-depth analysis of the various infrastructures.

Given the current challenges faced by critical infrastructures and given that it is not realistic to consider rebuilding them from scratch, it is necessary to derive approaches and develop methods to transform and optimize these infrastructures through the use of instrumentation, embedded software, and "smart" algorithms. In the global effort towards developing a more systematic and efficient approach for all critical infrastructures, it is useful to consider that these systems have some common characteristics. Critical infrastructure systems are safety critical systems that are complex in operation, spatially distributed, dynamic, time-varying, and uncertain. There is a wealth of data that can be obtained from various parts of these systems. Their dynamics have significant similarities in their analysis, while the effects of faults or disturbances can be modeled in similar ways.

This book was motivated by the European Science Foundation COST Action *Intelligent Monitoring, Control, and Security of Critical Infrastructure Systems* IC0806 (IntelliCIS) and is supported by the COST (European Cooperation in Science and Technology) Office. The book aims at presenting the basic principles as well as new research directions for intelligent monitoring, control, and security of critical infrastructure systems. Several critical infrastructure application domains are presented, while discussing the key challenges that each is facing. Appropriate state-of-the-art algorithms and tools are described that allow the monitoring and control of these infrastructures, based on computational intelligence and learning techniques. Some of the book chapters describe key terminology in the field of critical infrastructure systems: risk evaluation, intelligent control, interdependencies, fault diagnosis, and system of systems.

The Chapter "Critical Infrastructure Systems—Basic Principles of Monitoring, Control, and Security" provides an overview of critical infrastructure systems. It describes the basic principles of monitoring, control, and security and sets the stage for the rest of the book chapters. Chapters "Electric Power Systems", "Telecommunication Networks", "Water Distribution Networks", and "Transportation Systems: Monitoring, Control, and Security" concentrate on four key critical infrastructure systems: electric power systems, telecommunication networks, water distribution networks, and transportation systems. Their basic principles and key challenges are described.

Chapters "Algorithms and Tools for Intelligent Monitoring of Critical Infrastructure Systems", "Algorithms and Tools for Intelligent Control of Critical Infrastructure Systems", and "Algorithms and Tools for Risk/Impact Evaluation in Critical Infrastructures" focus on algorithms and associated tools for intelligent monitoring, control, and security of critical infrastructure systems, as well as risk/impact evaluation. The chapters provide the necessary theory, but also provide real life examples in the application of these tools and methodologies.

The Chapter "Infrastructure Interdependencies—Modeling and Analysis" presents several approaches for modeling critical infrastructure interdependencies. The Chapter "Fault Diagnosis and Fault Tolerant Control in Critical Infrastructure

Systems" provides a theory-based overview of fault diagnosis and fault tolerant control in critical infrastructure systems and illustrates the application of these methodologies in the case of water distribution networks.

The Chapter "Wireless Sensor Network Based Technologies for Critical Infrastructure Systems" examines the role of telecommunication networks in supporting the monitoring and control of critical infrastructure applications. The physical network is examined, as well as reliability and security issues. The Chapter "System-of-Systems Approach" concentrates on the reliability, security, risk, and smart self-healing issues in critical infrastructures, viewed from a system of systems perspective. The interdependencies between systems are examined with a focus on the electric power grid. Finally, the Chapter "Conclusions" discusses the main attributes that a future critical infrastructure system is expected to have and provides some potential future research directions in the areas of intelligent monitoring, control, and security of critical infrastructure systems.

Contents

Critical Infrastructure Systems: Basic Principles of Monitoring, Control, and Security

Georgios Ellinas, Christos Panayiotou, Elias Kyriakides and Marios Polycarpou

Abstract Critical Infrastructures have become an essential asset in modern societies and our everyday tasks are heavily depended on their reliable and secure operation. Critical Infrastructures are systems and assets, whether physical or virtual, so vital to the countries that their incapacity or destruction would have a debilitating impact on security, national economy, national public health or safety, or any combination of these matters. Thus, monitoring, control, and security of these infrastructures are extremely important in order to avoid the disruption of their normal operation (either due to attacks, component faults, or natural disasters) or to ensure that the infrastructure continues to function after a failure event. This chapter aims at presenting the basic principles and new research directions for the intelligent monitoring, control, and security of critical infrastructure systems.

Keywords Control systems · Critical infrastructure systems · Electric power systems · Fault diagnosis · Monitoring · Security · Telecommunication networks · Transportation systems · Water distribution networks

G. Ellinas (✉) · C. Panayiotou · E. Kyriakides · M. Polycarpou
KIOS Research Center for Intelligent Systems and Networks, Department of Electrical and Computer Engineering, University of Cyprus, 1678 Nicosia, Cyprus
e-mail: gellinas@ucy.ac.cy

C. Panayiotou
e-mail: christosp@ucy.ac.cy

E. Kyriakides
e-mail: elias@ucy.ac.cy

M. Polycarpou
e-mail: mpolycar@ucy.ac.cy

E. Kyriakides and M. Polycarpou (eds.), *Intelligent Monitoring, Control, and Security of Critical Infrastructure Systems*, Studies in Computational Intelligence 565,
DOI 10.1007/978-3-662-44160-2_1

1 Introduction

Everyday life relies heavily on the reliable and secure operation and intelligent management of large-scale critical infrastructures and any destruction or disruption of these infrastructures would cause tremendous consequences and will have a debilitating impact on security, national economy, national public health or safety, or any combination of these matters [1]. Specifically, critical infrastructures are defined as "*assets, systems, or parts thereof, essential for the maintenance of vital societal functions, health, safety, security, economic, or social well-being*" [2]. Thus, citizens nowadays expect that critical infrastructures will always be available and that they will be managed efficiently (with low cost).

Examples of Critical Infrastructures (CIs) include, amongst others, the electrical power plants and the national electrical grid, oil and natural gas systems, telecommunication and information networks, transportation networks, water distribution systems, banking and financial systems, healthcare services, and security services. Figure 1 shows the South East Asia—Middle East—Western Europe 4 (SEA-ME-WE 4) undersea fiber-optic transport network as an example of a (telecommunications) critical infrastructure. SEA-ME-WE 4 is a fiber-optic cable system approximately 19,000 km long that is used to carry information (providing the primary Internet backbone) between South East Asia, the Indian subcontinent, the Middle East, and Europe. This cable connects a large number of countries and is used to carry telephone, Internet, multimedia, and various broadband data applications utilizing a data transmission rate of 1.28 Tbps.

The monitoring, control, and security of critical infrastructure systems are becoming increasingly more challenging as their size, complexity, and interactions are steadily growing. Moreover, these critical infrastructures are susceptible to natural disasters (such as earthquakes, fires, and flooding), frequent faults (e.g., equipment faults, human error, software errors), as well as malicious attacks (directly or remotely) (Figure 2 shows possible threats to critical infrastructures).

There is thus an urgent need to develop a common framework for modeling the behavior of critical infrastructure systems and for designing algorithms for intelligent monitoring, control, and security of such systems. This chapter aims at presenting the basic principles and new research directions for the intelligent monitoring, control, and security of critical infrastructure systems. Subsequent chapters in this book provide more specific information on the monitoring and control of particular infrastructures such as Electric Power Systems (Chapter "Electric Power Systems"), Telecommunication Networks (Chapter "Telecommunication Networks"), Water Distribution Networks (Chapter "Water Distribution Networks"), and Transportation Systems (Chapter "Transportation Systems: Monitoring, Control, and Security"). Additional information on algorithms and tools for CI monitoring and control, as well as on critical infrastructure interdependencies are included in later chapters. In particular, Chapters "Algorithms and Tools for Intelligent Monitoring of Critical Infrastructure Systems," "Algorithms and Tools for Intelligent Control of Critical Infrastructure Systems," and "Algorithms and Tools for Risk/Impact Evaluation in Critical

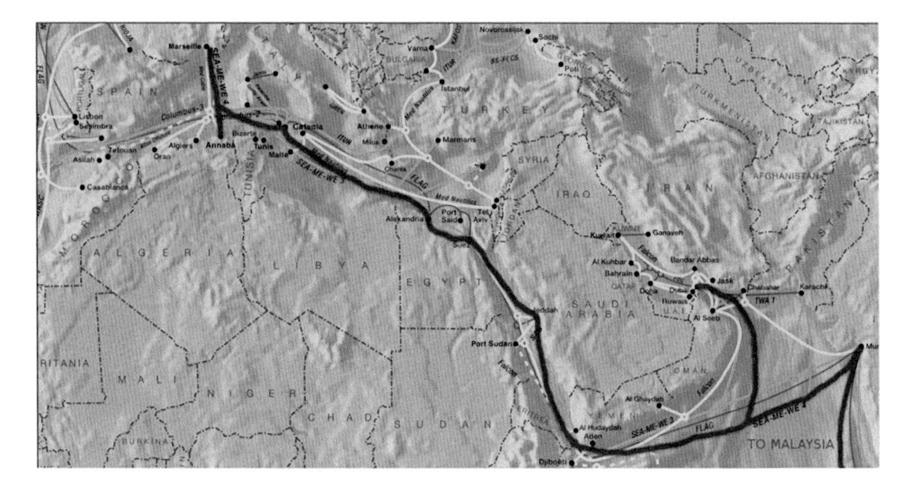

Fig. 1 SEA-ME-WE 4 undersea fiber-optic cable system (in *bold*) as an example of a (telecommunications) critical infrastructure

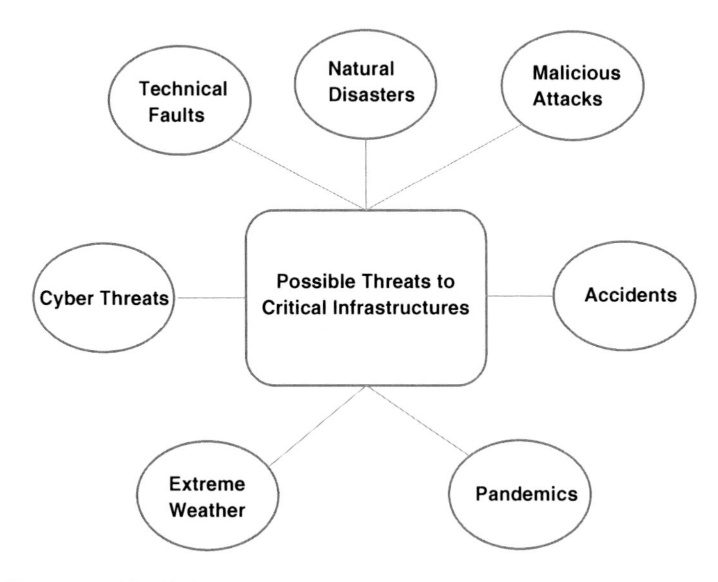

Fig. 2 Threats to critical infrastructures

Infrastructures" describe different algorithms and tools for intelligent monitoring and control and for risk/impact evaluation of CIs, while Chapter "Fault Diagnosis and Fault Tolerant Control in Critical Infrastructure Systems" addresses specifically the problems of fault diagnosis and fault tolerance control in critical infrastructure systems. In addition, Chapter "Infrastructure Interdependencies: Modeling and Analysis" discusses modeling and analysis of infrastructure interdependencies, Chapter "Wireless Sensor Network Based Technologies for Critical Infrastructure Systems" describes wireless sensor network based technologies for CIs and Chapter "System-of-Systems Approach" presents a system-of-systems approach for the intelligent monitoring, control, and security of critical infrastructures.

2 Disruption of Critical Infrastructures

A *failure event* in CIs (accidental or intentional) is defined as a negative event which influences the *inoperability* of infrastructures and subsystems, where the inoperability of an infrastructure or subsystem is defined as the inability to perform its intended function. An example of a failure event in the CI shown in Fig. 1 was the January 2008 accidental destruction of the submarine system's fiber-optic link (speculated to have happened by a ship's anchor outside Alexandria's port). As a result of this accident, Internet services were widely disrupted in the Middle East and in the Indian subcontinent, including more than 50 % disruption on Internet services in some of the countries affected.

Note that throughout this chapter the term "failure" denotes infrastructure failures while the term "fault" denotes component faults that even though cannot be avoided they can be dealt with by utilizing redundancy, such that, even under some component faults, the infrastructure continues to operate. Thus, an overall system requirement is that the system should continue to operate (perhaps suboptimally) even when one or more of its constituent components have failed. However, when a system operates at a suboptimal point it could potentially waste energy and other resources, or operate at a high risk region. This creates a false sense of security for the entire system. Thus, autonomous ways for quickly detecting, isolating, and recovering the faults are needed for the successful deployment of critical infrastructures as it is discussed in detail in the sections that follow. This is because a fault that goes unattended for a long period of time can cause both tangible and intangible losses for the company/organization that provides the service, as well as for its clients. Therefore, the current trend is for more and more CIs to provide services that are virtually uninterruptible.

Table 1, for example, shows examples of monetary losses (per hour) incurred by various industries when even simple IT outages occur in their networks [3]. On average, businesses lose between $84,000 and $108,000 (US) for every hour of IT system downtime, according to estimates from studies and surveys performed by IT industry analyst firms. Losses in these industries can be tangible or intangible. Tangible/direct financial losses include such things as lost transaction revenue, lost wages, lost inventory, and legal penalties from not delivering on service level agreements. Conversely, intangible/indirect financial losses may include lost business opportunities, loss of customer/partner goodwill, brand damage, and bad publicity/press.

One important aspect of CIs related to their management is that they consist of various autonomous systems due to deregulation. This makes coordination and protection more difficult, since each autonomous system may have its own

Table 1 Projected losses/hour for various industries during IT outages [3]

Industry	Typical hourly cost of downtime (in US dollars)
Brokerage service	6.48 million
Energy	2.8 million
Telecom	2.0 million
Manufacturing	1.6 million
Retail	1.1 million
Health care	636,000
Media	90,000

objectives. A failure event can also be propagated or propagate its effects to other interdependent infrastructures, according to specific concepts of proximity (e.g., geographical, physical, cyber, etc.). This is due to the interdependencies that exist between infrastructures. These interdependencies are mostly highlighted when infrastructures are experiencing catastrophic natural disasters or are under terrorist attacks and there is an attempt to respond and recover from severe disruptions in the infrastructures [4]. Because of the interdependencies between critical infrastructures, potential failures in one infrastructure may lead to unexpected cascade failures to other infrastructures that may have severe consequences.

Specifically, over the last few years, several efforts have been undertaken in the literature on how to take measures aimed at preventing/reducing risks, preparing for, and protecting citizens and critical infrastructures from accidental faults, terrorist attacks, and other security related incidents. These measures entail (i) identifying critical infrastructures and interdependencies and developing risk assessment tools and methodological models for the critical infrastructures, (ii) developing monitoring, control, and security strategies for these infrastructures based on the risk assessment tools, (iii) developing contingency planning, stress tests, awareness raising, training, incident reporting, etc., as part of the prevention and preparedness strategy, and (iv) developing protection mechanisms, as part of the response strategy, that can enable the infrastructure to recover from the failure/attack.

3 Modeling of Critical Infrastructures

The problem of controlling and managing critical infrastructures is becoming more and more difficult as CIs are becoming very large due mainly to the growing demand for the services they provide. Furthermore, deregulation has resulted in more heterogeneous and distributed infrastructures, and has created more interdependencies between them, which make them more vulnerable to failures and attacks. As these infrastructures become larger and more complex, fewer people understand how these networks work and the interactions between all the components. Thus, models are created in order to represent these infrastructures and try to predict their behavior under failure/attack scenarios.

It is important to note that even though critical infrastructures apply to different sectors and provide different types of services, they share certain characteristics, which enable the creation of a common modeling framework that can be applied to any type of critical infrastructure. For example, CIs are safety-critical systems with similarities in the impact of failures, they are dynamic, complex, and large scale, and they are data rich environments that are mostly spatially distributed and time-varying.

Because of the large scale and the great importance of the various critical infrastructures, real practical solutions, such as introducing scenarios and triggering events for analyzing the effect on critical infrastructures, are not feasible. Thus, the modeling and analysis of critical infrastructures are the best and safest methods that can be used to understand the best way to monitor and control these infrastructures, as well as to understand how the infrastructures function in the presence of a failure event, and how they can recover from such an event.

There are a number of ways that can be utilized to model critical infrastructures including network flow models, system dynamics models, agent-based models, as well as combinations of these models. Clearly, there is a large body of work in the literature on the modeling of critical infrastructures that cannot be elaborated on in this chapter. However, as an indicative example, individual critical infrastructures can be described as complex adaptive systems (CASs), since they are complex collections of interacting components in which change often occurs as a result of a learning process (influenced by past experiences and adapting to future expectations) [5]. The CAS methodology does not only model the infrastructure as a collection of individual components, but also considers the synergies and interactions between components (adding an additional complexity exhibited by the infrastructure as a whole), an attribute termed as the "emergent behavior of the system". Figure 3 shows a representation of a critical infrastructure as a CAS.

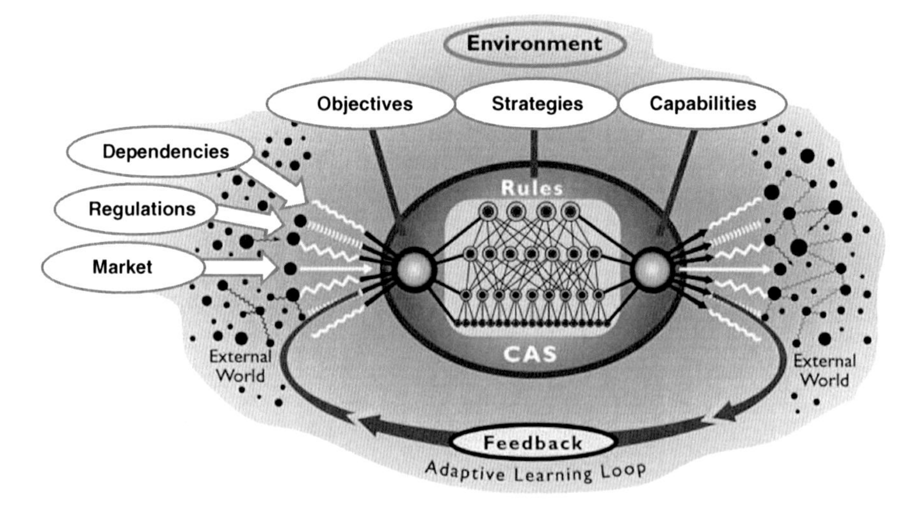

Fig. 3 Modeling of a critical infrastructure as a complex adaptive system (CAS) [adopted from 7]

An efficient way to model critical infrastructures (under the CAS methodology) is to consider each component of the critical infrastructure as an agent entitythat has a specific location (e.g., geographical, logical, cyber location, etc.), certain capabilities, and memory in terms of past experiences [6]. As an example, consider an agent identity that models a fiber-optic link that is part of a telecommunications infrastructure (example of Fig. 1). The fiber's physical location can then be the location attribute of the agent; the capabilities of the fiber (agent) can include its ability to transmit information (e.g., the number of wavelengths multiplexed on the fiber, the bit-rate for each wavelength, etc.), as well as how it responds to failure events (e.g., redirecting the flow of information to other fibers). The memory attribute of the agent in this case can include such information as the average and peak traffic flow through that fiber, the number of wavelengths utilized, etc. It is also important to note that each infrastructure component (agent) communicates with other agents (receives information from some agents and sends information to other agents) so as to model the synergies and interdependencies between components in a critical infrastructure as well as to model the state of the agents.

4 Monitoring and Control of Critical Infrastructures

Based on the CI modeling framework, there is a need to develop techniques for efficient monitoring and control of CIs. Monitoring and control of CIs can be realized utilizing networked intelligent agent systems that have sensing (for monitoring) and actuator (for control) capabilities, as well as communication, computing, and data processing capabilities. An example of such an agent is shown in Fig. 4. These agents can also potentially communicate amongst themselves and cooperate to effect fault accommodation and self-healing during a fault/attack event.

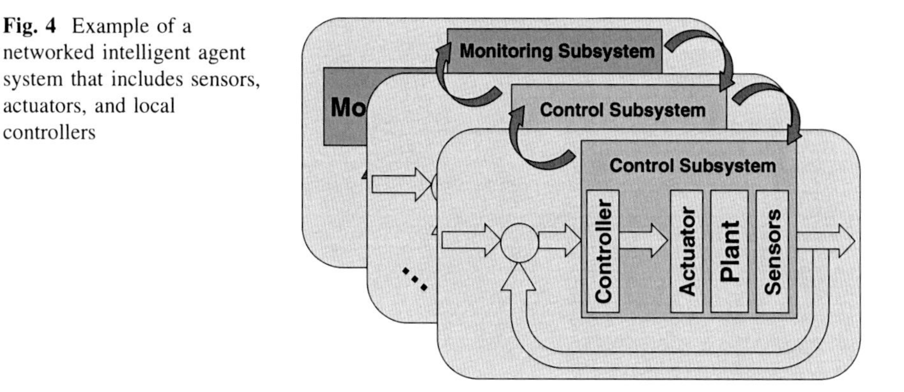

Fig. 4 Example of a networked intelligent agent system that includes sensors, actuators, and local controllers

4.1 Monitoring of CIs

With the advancement in technology over the years, remote monitoring of critical infrastructures has become easier, providing an effective tool for the control of these infrastructures. However, it is important to note that even though technology has become more capable, infrastructures have also become more complex, thus having as a net result the increased difficulty in efficient monitoring of these infrastructures. Nevertheless, the ability of network operators to better prepare for and respond to failures has greatly improved through the enhanced capabilities that are possible nowadays for remotely monitoring and controlling critical infrastructures.

One of the problems that received considerable attention in the literature concerning the monitoring problem of CIs is the question of the placement of these intelligent agents (as well as how many should be used) with sensing and actuator capabilities. This placement will then facilitate various objectives of the CI under investigation, including the control, fault diagnosis, and security objectives. This is achieved in one of several ways; for example, the data gathered via the sensing capabilities can provide a central control module with information on the state of the CI that can be utilized for effective system control. In another example, the sensing information can be utilized by the intelligent agents (possibly in cooperation with other agents in a decentralized manner) to effectively address fault diagnosis and security. The different architectures that can be used to control the network are described in detail in Sect. 4.2 that follows, whereas this section deals specifically with the problem of monitoring the CIs.

Efficient monitoring and control in any distributed system requires the use of sensors and actuators which are installed at optimal (or near optimal) locations, in accordance to certain objectives. In general, the problem of where to place facilities, sensors or actuators, in order to keep certain objectives and constraints satisfied within a network has been examined in various research disciplines. This section outlines several sensor and actuator placement techniques using as case studies power and water distribution networks. It should be noted, though, that most of the sensor and actuator placement approaches can in general be applied to many different CIs (with variations on the quantities measured and controlled depending on the CI examined).

In general, the sensor and actuator placement problem in networks can be considered as a combinatorial problem of selecting certain locations out of all the possible locations in the network for the installations. However, in practice, due to the large-scale nature of these networks and the large number of possible installation locations, the solution space may be non-tractable. The general solution methodology is to formulate the sensor and actuator placement problem as a mathematical program with one or more objective functions and constraints, and to utilize an optimization methodology such as linear, quadratic, non-linear, or evolutionary optimization algorithms considering either one, or multiple objectives. Furthermore, in the case of multiple objective formulations, a single optimal solution may not be available, and as a result, a decision maker should make a selection based on the solutions which reside on a Pareto front, through expert reasoning.

In relation to the two critical infrastructures utilized as use cases in this section, it is observed that both electric power systems and water distribution networks share certain similarities with respect to the sensor placement solution methods used in the literature. For the case of power systems, various approaches have been outlined in the literature addressing the problem of sensor placement for power system observability as well as approaches for sensor placement for minimizing power system state estimator accuracy. For the case of water distribution networks, most of the approaches deal with sensor and actuator placement for water quality monitoring, control, and security.

Specifically, for the case of electrical networks, the increase in electricity demand and the need to decrease the cost of electricity production has forced electric utilities to operate their system close to their physical limits. This fact makes the electric utilities prone to various contingencies and/or complex interactions that may lead to islanding or blackout; thus, the monitoring of power systems is an essential need for their normal operation. Monitoring information in power systems is provided in terms of measurements utilizing various sensors (measurement units), placed in strategic locations of the power system network. The sensors used nowadays in the measurement system of the electric utilities are separated in two categories, namely the conventional and the synchronized measurement units. The conventional measurement units usually provide measured electric quantities to the control center every 3–5 s. The category of the conventional measurements includes measurements of real and reactive power flow of a transmission line, real and reactive net power injection in a bus, voltage magnitude of a bus, current magnitude flow of a transmission line, and transformer tap changer position. Synchronized measurement units (called Phasor Measurement Units (PMUs)) that use the Global Positioning System (GPS) for synchronization are also used to provide extremely accurate voltage and current phasor measurements in a very fast reporting rate (up to 60 phasors per second).

In order to have a unique and reliable solution by the state estimator (located at the Supervisory Control and Data Acquisition (SCADA) controller as it will be described in Sect. 4.2 that follows), proper sensor placement for rendering the system observable is needed. Therefore, the development of methodologies for sensor placement for making the power system fully observable has been investigated by several researchers. For example, with the deployment of PMUs in the power system, many researchers have dealt with the placement of the minimum number of PMUs to make the system observable (either by utilizing PMUs only or by utilizing both PMUs and conventional measurement units). Besides the observability issue, many approaches aim at improving state estimator accuracy by determining the optimal placement of sensors. For example, in [8], a methodology which determines which measurements should be added in the measurement system for making the system observable was proposed. In addition, in [9], a methodology for finding the minimal PMU placement for making the system (topologically) observable was proposed with the utilization of a dual search technique which is based on both a modified bisecting search and a simulated annealing method. The modified bisecting search is executed first for fixing the number of PMUs that will be placed, and then the simulated annealing based method determines the placement set that will make the power system topologically observable. Furthermore, other techniques aim to determine the optimal PMU locations in case of complete or incomplete observability using a graph theoretic procedure [10] and a non-dominated sorting genetic algorithm [11], without taking into consideration the presence of conventional measurements.

It should be noted that the conventional measurement units are already installed in the measurement system of the electric utilities constituting an irreplaceable source of measurements providing valuable information to the control centers. Further, the trend of the various stakeholders is to install PMUs incrementally since the high cost of rendering a power system observable only by PMUs is unaffordable. Therefore, significant research activity has also been undertaken for the development of methodologies for determining the optimal PMU locations in the presence of both synchronized and conventional measurements [12–14]. Finally, as it is desired by the electric utilities to accommodate accurate and reliable state estimators at their control centers (so as to provide power flow and voltage stability analysis), the accuracy of the state estimator is also very important. This accuracy is however heavily dependent on the quality of the measurements, the measurement redundancy, and the measurement unit configuration. Hence, another issue that is extensively investigated by many researchers is the determination of the optimal sensor placement for improving the accuracy of the state estimator [15–18].

The second use case of water distribution networks also utilizes sensors to measure relevant quantities that can be utilized for the efficient control of the network during normal operation and operation under failure conditions. Since the sensors measuring quality characteristics, as well as the actuators responsible for water disinfection, are quite expensive, and not feasible to deploy everywhere in the network, various methodologies have again been proposed to optimize (using some objective function) the placement of a limited number of sensors and actuators in water distribution networks. In addition to solving the quality observability problem, e.g., detecting a contamination event given a certain sensor placement, the problem has also been examined from the control perspective (see Sect. 4.2) as well as from the security perspective, for computing sensor placements which minimize the risk of severe impact (as outlined in Sect. 5.3).

The "Set Covering" method was one of the first mathematical formulations of the placement problem and it has since been applied in various fields, such as in facility location [19]. According to this approach, an integer optimization program is formulated in order to determine a set of nodes from a topological graph to install facilities, so that all the remaining nodes are next to at least one facility. A related approach is the "Maximal Covering" formulation described in [20] for the calculation of a set of nodes which maximize the population served in an area within a certain distance. The mathematical program tries to maximize the number of people served, constrained by a fixed number of facilities and a limiting service distance. The authors in [20] examined heuristic search and linear programming as possible methods for solving this problem. A similar formulation to the Maximal Covering was considered in [21], which selected the locations to install water quality sensors in drinking water distribution systems, so that the largest volume of water consumed was examined. Following this formulation, aiming at solving bigger networks, other works utilized heuristics [22] and genetic algorithms [23]. Furthermore, a mixed-integer problem formulation was presented in [24] and a multi-objective weighted-sum extension of [21] was presented in [25] which considered certain physical network characteristics and time-delays.

Clearly, there are a number of methods and techniques for placing the sensors (and actuators) depending on the measurable quantities and on the critical infrastructure monitored. However, as previously mentioned, most of these techniques can be applied to different CIs as these infrastructures have a number of common characteristics and control/security objectives.

4.2 Control of CIs

Based on the data received by the sensors and communicated to the controllers, there is a need to develop intelligent data processing methods for extracting meaning and knowledge out of the data, analyzing, and interpreting the data, such that systems are controlled and faults are detected, isolated, and identified as soon as possible, and accommodated in future decisions or actuator actions. Processing this information is becoming more and more complex for large-scale critical infrastructure systems, due mainly to the large volume of data obtained and due to the fact that the monitored data may have different characteristics (multi-dimensional, multi-scale, spatially distributed, etc.). Further, as infrastructures get quite large and complex, it is also difficult to anticipate the data that will be received (combinatorial explosion).

Subsequently, the monitored data received and the knowledge extrapolated will be utilized to design software, hardware, and embedded systems that can operate autonomously within the CIs in some intelligent manner. The ultimate goal would be to create smarter infrastructure systems that can provide real-time decisions in the management of large-scale, complex, and safety-critical systems, including functionalities such as system identification, prediction/forecasting, optimization, scheduling and coordination, as well as fault monitoring/isolation/accommodation (including different types of abrupt and incipient faults). Figure 5 illustrates a generic figure of a controller module and its communication links. This module receives information from the CI system, processes this information and outputs system information, warnings, and alarms, and also communicates with the agents placed at the CI system remote locations in order to control the system via specific control/actuator actions.

Clearly, the architecture used for monitoring and control of the CI will dictate how the information obtained by the sensors can be used to meet the objectives of

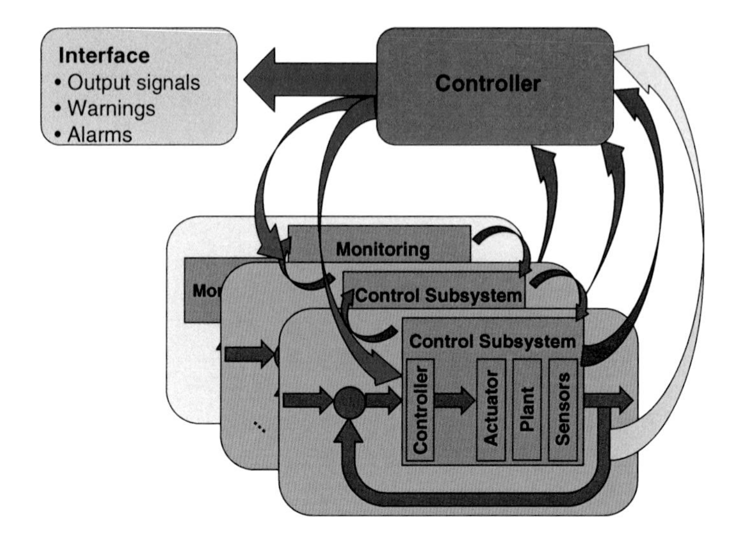

Fig. 5 A generic figure of a controller module and its communication links

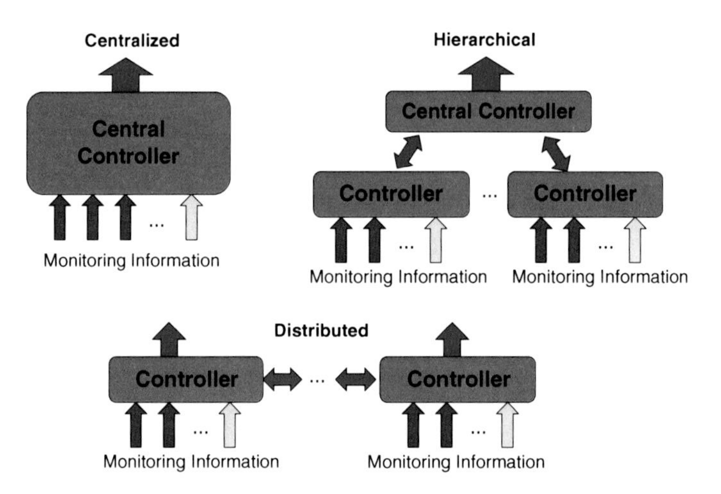

Fig. 6 Different types of control architectures

the CI. There are several control architectures that can be used in CI systems including centralized, distributed, and hierarchical as shown in Fig. 6.

In the case of the centralized control architecture, all monitored information is processed at a single central controller entity, whereas at the opposite spectrum in the distributed approach there is no central controller and local controllers (networked agent systems) process the information themselves and communicate/cooperate with neighboring agents in order to formulate a control action (including fault detection, isolation, and accommodation). An architecture that contains aspects of both the centralized and distributed approach is the hierarchical control architecture where (local) controllers communicate with a central (global) controller (and with each other via this central controller) and some of the data processing is performed (and control action decisions taken) at the local controllers while others take place at the central controller.

Clearly, each one of these architectures has its advantages and drawbacks and the question of which one to use depends on several characteristics of the infrastructure as well as metrics/objectives that are of interest. For example, in terms of scalability, the distributed control approach is the best, as it allows the network infrastructure and the controlled elements to grow to large numbers without any effect on the architecture. On the contrary, a centralized approach cannot scale to large numbers of controlled elements, as limits in database sizes, memory, processing speed, etc., start to appear as the infrastructure grows.

In general, centralized techniques have complete information of the state of the network and can provide optimal solutions for some of the network functionalities (in contrast to distributed approaches who may not have complete and updated information on the network state that may result in sub-optimal solutions). However, centralized approaches do not scale, they require large databases, and they may have to process a very large amount of data. The hierarchical control architecture model can be seen as a (hybrid) technique that shares advantages from both the centralized and distributed approaches, as some of the control/fault management functionalities are undertaken at the local controllers and others at the central (global) controller.

Most of the control architectures for critical infrastructure systems that are currently deployed typically utilize a supervisory control and data acquisition (SCADA) system to monitor and control these infrastructures. These industrial control systems typically incorporate sensors, actuators, and control software that are deployed at various local or remote locations throughout the critical infrastructure, wherever equipment needs to be monitored or controlled. SCADA systems are used for gathering real-time data, monitoring equipment, and controlling processes in most of the public utilities. A SCADA network can cover large geographical areas, especially in the case of public utilities, such as water distribution networks and power systems. An example to demonstrate how geographically dispersed these systems can be is the SCADA system at the ETSA electric utility company in South Australia. Because of the vastness of the country and the remoteness of many of the utility plants and field stations, this system covers more than 70,000 square miles of terrain, comprising of more than 25,000 physical input/output points monitoring a large number of network parameters such as current, temperature, power variables, load shedding systems, and fire alarms [26].

Figure 7 illustrates a generic SCADA network architecture, comprising of a control center (located at the main plant) and the infrastructure it controls (shown as remote and local sites). Local sites are locations of the infrastructure that are collocated with (or very close to) the control center, while remote sites are sites that are far away from the control center and they are connected to it via a communications infrastructure. For example, in the figure it is shown that these sites can be connected to the control center via satellite links, leased lines, or via public access (e.g., the Internet).

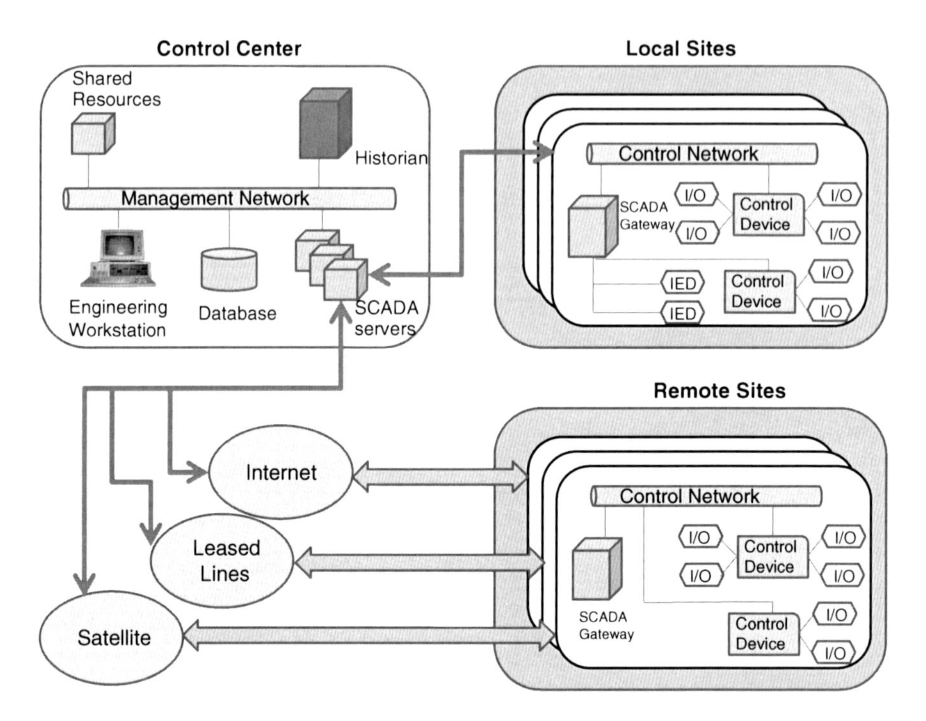

Fig. 7 A generic SCADA architecture [adopted from 27]

The control center of the SCADA system is the one responsible for processing the information received by the remote and local sites and issuing control commands through its communication with the infrastructure local and remote controllers. All main databases (including databases with historic system data), relevant processors, servers, and controllers are located at the control center of the SCADA system. The local/remote site controllers (called network controllers or field devices) include control devices (such as programmable logic controllers (PLCs), remote telemetry units (RTUs), input/output controllers (IOCs) and intelligent electronic devices (IEDs)) that are used to implement the control logic, input/output (I/O) devices (sensors and actuators) that interface with the control devices, and a SCADA gateway for interfacing with the control center [27].

Specifically, sensors are used to collect monitored data and actuators/controllers are used to perform actions based on the received sensor data and predefined control algorithms. At the control center dedicated processors (servers) are used to collect and analyze the various inputs received from all of the field devices. When the processing is completed, actions are undertaken in the form of alarms and warnings (when a fault has occurred), or in the form of a control action (e.g., start and stop processes, automate processes, control of a specific system component, etc.).

For example, the SCADA system installed in most of the electric utilities control centers nowadays is responsible for providing the operating condition of the power system. Some of the tools included in these SCADA systems are responsible for contingency analysis, power flow analysis, corrective real and reactive power dispatch, and bad data detection and correction. Although the aforementioned SCADA tools are of great importance for assessing power system conditions, most of them depend heavily on the results of the state estimator. The state estimator is the main tool in the SCADA system which provides snapshots of the power system operating condition in consecutive time intervals of 1–5 min. Specifically, the state estimator provides the state of the power system (i.e., bus voltage magnitudes and bus voltage angles) by processing redundant measurements provided by various sensors (conventional and synchronized measurement units) that, as previously explained in Sect. 4.1, are placed in strategic locations of the power system network.

Another example is the use of a SCADA system in water distribution networks. This control system is responsible for the delivery of adequate and safe water to the consumers by administrating the tasks of monitoring and control of water delivery. The management of the operation of water distribution systems, however, is a complex task as it involves a large number of subsystems, which need to constantly operate within the specifications set by regulations, while reducing the distribution costs. In general, discrete flow measurements are available for some parts of the network. However, due to the time-varying water demand consumption at each network node, large uncertainties can be observed in the measurements throughout the day. In addition, water distribution systems are prone to hydraulic faults, e.g., due to leakages or pipe bursts, as well as to quality faults, e.g., due to accidental or intentional water contamination. Thus, these uncertainties and possible faults are clearly making the control and management of these networks a very difficult task. Figure 8 shows an example of a graphical representation of a water distribution network (Fig. 8a) and an example of a controller for such a network (Fig. 8b).

A third example of a SCADA controlled infrastructure is the combined sewage system (CSS) that combines sanitary and storm water flows within the same network. The criticality of these infrastructures stems from the fact that during heavy rain these networks may overload, leading to the release of the excess water into the environment. However, as this excess water may contain biological and chemical contaminants, a major environmental and public health hazard may be created. SCADA systems are thus utilized in these infrastructures for highly sophisticated real-time control of the system that will ensure the performance of the network under adverse meteorological conditions (e.g., minimize flooding and combined sewer overflow to the environment and maximize waste water treatment plant utilization) [28]. Figure 9 illustrates an example of the control architecture for a CCS that in this case uses a wireless network and wireless sensor and actuator devices for monitoring and control.

As previously mentioned, sensors and actuators are deployed at various locations throughout the critical infrastructure, wherever equipment needs to be monitored or controlled. Section 4.1 discussed several methods for the intelligent placement of sensors. Even though in most cases the sensors and actuators are

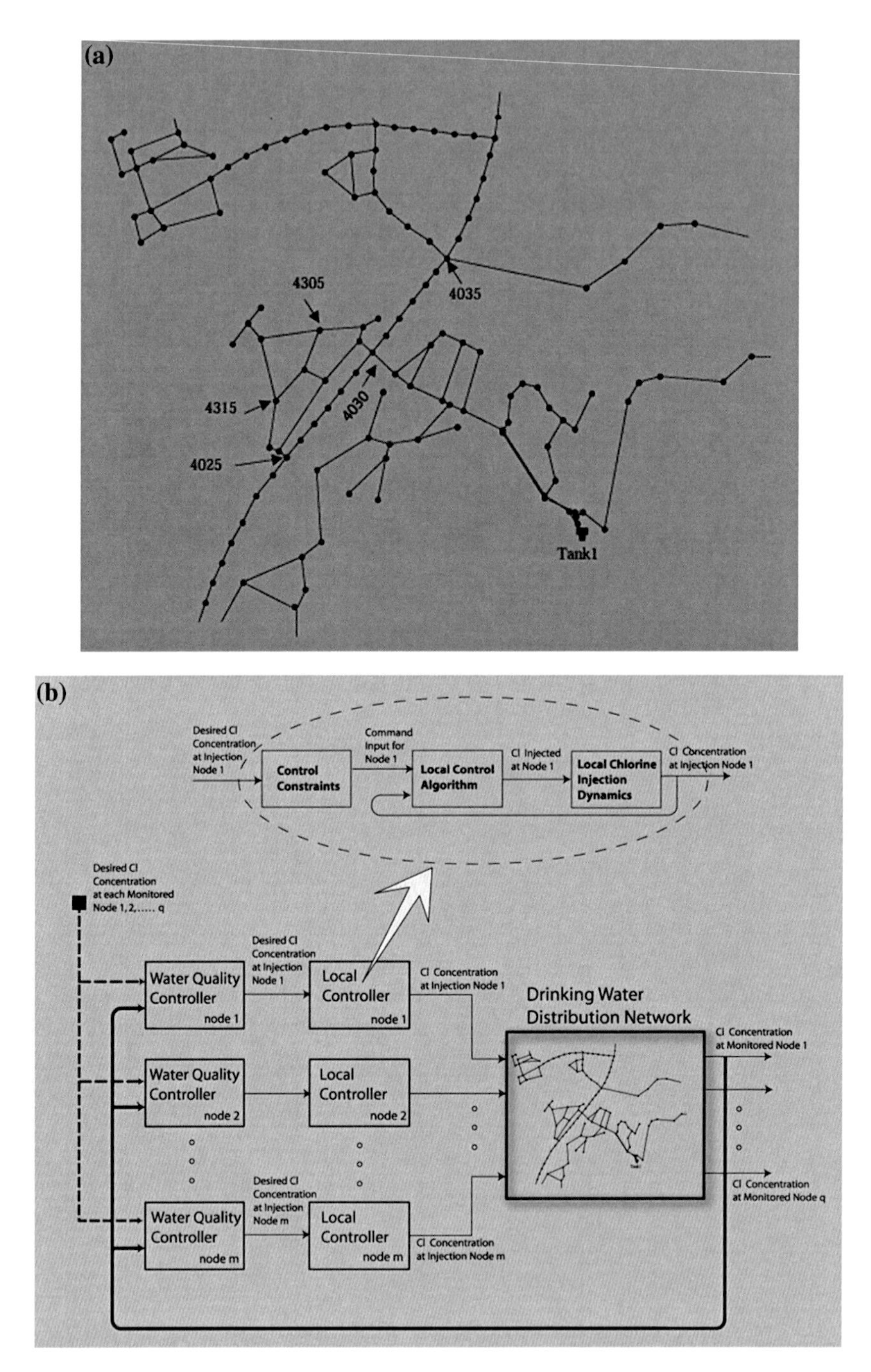

Fig. 8 **a** Graphical representation of a drinking water distribution network. **b** Example of a controller for a water distribution network

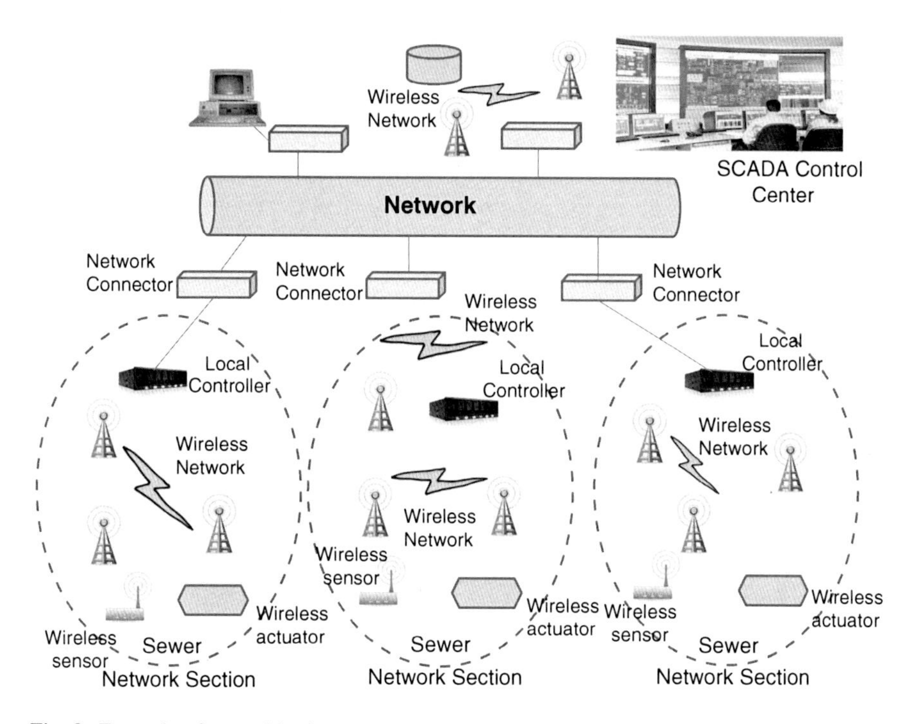

Fig. 9 Example of a combined sewage system control architecture [adopted from 28]

collocated, similar placement approaches can be used when the actuator placement problem is dealt independently of the sensor placement problem.

Let us consider as a use case the problem of actuator placement in water distribution networks for quality control. Disinfection booster stations are sometimes used by water utilities to reduce the total injected chlorine mass while keeping the concentration of chlorine throughout the network above a certain safety level but well below high concentrations which may have harmful effects on the consumers. The objective in these networks is to control the spatio-temporal distribution of drinking water disinfectant throughout the network by the injection of appropriate amounts of disinfectant at appropriately chosen actuator locations.

In general, the problem of selecting locations for installing quality actuators, although it is related with the sensor placement, is sometimes considered as a separate problem. In previous works, various optimization formulations were proposed to solve the problem, while taking into consideration the various constraints. A common assumption is that chlorine dynamics are linear, and that a water demand model which describes consumption at each node is available. For example, in [29] the problem was formulated as a mixed integer optimization program, for minimizing the disinfectant dosage which is required to maintain residuals throughout the water distribution network. A modification of that work was presented in [30] to reduce the computational burden by adding certain disinfectant concentration constraints. Further, a mixed integer quadratic program was formulated in [31] for locating booster disinfectants and computing the injection signals. Multi-objective problem formulations were also presented for this problem. For example, in [32] the booster actuator placement problem considers two objective functions, the total disinfectant dose and the volumetric demand; the problem is solved using an evolutionary multi-objective optimization algorithm. Furthermore, in [33] a multi-objective optimization formulation was presented using a grid implementation for computational efficiency, while considering multiple demand scenarios. The previous formulations considered known demands, however in practice these demands are subject to uncertainties. Thus, in [34] a stochastic optimization approach was presented, which minimized the installation cost and disinfectant mass required to satisfy health regulations, while taking into account demand and chemical reaction uncertainties.

It is evident that the problem of actuator placement in CIs depends on the controlled quantities and on the type of critical infrastructures controlled. However, similar to the monitor placement problem, most of the techniques for actuator placement can be applied to different CIs that share similar characteristics and control objectives.

5 Security of CIs

5.1 SCADA Security

Clearly, there are fault/security risks to the critical infrastructure (plant) itself (e.g., component plant faults), as well as to the control infrastructure (sensors, actuators, communication links, controllers, etc.). Considering specifically the security risks due to malicious attacks, it is easily recognizable that the SCADA system controlling the infrastructure is particularly susceptible to attacks. Attacks/failures on the critical infrastructure can potentially have enormous consequences (monetary and societal). For example, during the power blackout in North America in 2003, more than 100 power plants were shut down, affecting 50 million people in the U.S. and Canada [26]. This emphasizes the need to protect critical infrastructures, and particularly SCADA systems, especially from targeted attacks [35].

SCADA security was a main focus of the U.S. President's Information Technology Advisory Committee in their 2005 report [36]. High probability/high consequence risks for SCADA systems may include cases where a SCADA system crashes due to unexpected or unauthorized traffic from ICT systems (denial of service attacks) [37] or due to virus/worm attacks. In the latter case, a virus or worm causes unpredictable behavior or shuts down key SCADA components and disrupts production processes. Examples include the Stuxnet, Zotob.E, and Slammer attacks that targeted industrial control systems/public utilities in 2010, 2005, and 2003 respectively [38, 39].

One of the main security issues identified is that a SCADA system is not on a separate closed network, but rather it is either interconnected with the corporate network and/or utilizes shared communication medium (e.g., the Internet). Thus, possible risks for the SCADA systems may include attacks on the TCP/IP carrier (the protocol is used to facilitate interconnections between components of the SCADA system), as well as attacks on the SCADA system via the corporate information network that is usually interconnected to the SCADA network [40–42].

It is evident that comprehensive solutions are required to address security issues in SCADA systems, especially due to the fact that these systems are large and geographically diverse, they comprise proprietary hardware and software, and are interconnected with shared (public) computing and network equipment. These solutions must be cost-efficient, scalable, and must also not negatively impact the operations of the infrastructure.

Initially, in order to address these security issues, various organizations attempted to document the probabilities and consequences of likely incidents for individual CIs in a risk matrix. For example, an industry best practice called ISBR (Information Security Baseline Requirements for Process Control, Safety and Support ICT Systems) [43] was developed by the Norwegian Oil Industry Association to accommodate this procedure. Based on the risk matrices created, standards were created to provide information technology security solutions in SCADA systems [44–54]. Clearly, based on the aforementioned risks, there is increased need to more securely integrate SCADA and ICT systems. This is not an easy task, as the two technologies have inherent differences and complexities. For example, SCADA systems have long lifecycles and changes in these systems do not happen often. On the other hand, ICT systems have much shorter lifecycles and changes in these systems are frequent. However, despite their differences, systematic testing of integrated SCADA-ICT systems is required to ensure that both systems are secure and that the ICT system is not used as a vehicle in order to attack the SCADA infrastructure.

A recent work on this issue that appears in [55] discusses two strategies for securing SCADA networks. The first strategy deploys a security services suite that responds to known risk factors and protocol weaknesses by providing risk mitigation facilities. The second strategy analyzes SCADA network traffic using a forensic tool that can monitor process behavior and trends, investigate security events, and optimize performance. Both approaches try to provide security for the SCADA system while having minimal impact on the real-time operations of the infrastructure.

5.2 *Fault Management*

Apart from attacks to the SCADA infrastructure, security is also related to fault management of CIs. Large scale systems in general consist of a large number of components (sensors, actuators, controllers, communication links, etc.). At any point in time, one or more of these components are bound to fail either due to component degradation, human error, natural disaster, etc., or due to a malicious attack. Thus, as previously pointed out, an overall requirement is that the system should continue to operate (perhaps not at an optimal point) even when one or more of its constituent components have failed.

When a fault or attack occurs, such faults/reconfigurations may remain undetected for long periods of time having a negative impact on the overall system performance and security. It is important to manage the fault by diagnosing and accommodating it quickly and efficiently. In particular, the fault diagnosis problem is subdivided into two problems (a) fault detection and (b) fault isolation/identification. The first problem determines whether a system has a faulty component or not, while the second determines the location, the effect, and the impact of the fault. A key characteristic that makes fault diagnosis feasible is the availability of "redundant" information, either hardware redundancy (e.g., using redundant (extra) sensors and software) or analytical redundancy (either from explicit mathematical models or models built from collected data) that combines measurements from other sensors or from the same sensor in past instances [56–59]. Furthermore, fault diagnosis can be addressed from different perspectives e.g., from a worse-case analysis point of view or from a stochastic system point of view. It is thus important to develop algorithms for automatically detecting and identifying such faults, especially in large-scale critical infrastructures.

In some critical systems (such as space aircraft, airplanes, etc.), hardware redundancy is preferred. In large-scale critical infrastructures, however, mainly due to cost considerations, analytical models are preferred over hardware redundancy. Most of the model-based analytical methods use centralized state-estimation techniques (all sensing information is centrally processed). There are, however, also decentralized fault detection and isolation (FDI) implementations, such as the work that was proposed in [60], that nevertheless assumed full-mesh connectivity between all the sensors. Additionally, in [61] a distributed estimation technique was introduced that only required a sparse but connected network of sensors, thus reducing the communication complexity. That work was extended in [62] that defined for each node of the sensor network a bank of micro-Kalman filters, each utilizing a certain model (the fault models and the nominal one) and generating a residual sequence (each node then had all different residual sequences). Each data collection was then analyzed by computing its conditional probabilities (beliefs), given that the corresponding model (hypothesis) of the system was valid. In order to perform distributed hypothesis testing on all different hypotheses and reach a common decision, the belief consensus algorithm in [63] was utilized that allowed computation of conditional probabilities corresponding to different models of faults. The most likely model was then chosen as the one with the maximum conditional probability.

Furthermore, as previously discussed, monitoring and control of critical infrastructures is achieved with the use of a wide deployment of distributed sensing devices, which provide temporal and spatial information through wired and wireless links [64]. The information provided by sensors is used for safety-critical tasks and decisions; therefore, if one or more of the sensors provide erroneous information, then the efficiency of the system may be degraded, or even worse, the system may become unstable, thus jeopardizing the safety of humans and/or expensive equipment.

Again, one approach to the problem of the sensor fault detection and isolation (SFDI) is the physical redundancy method. However, in most applications, the physical redundancy is not feasible due to the high cost of installation and maintenance, and space restrictions. Therefore, again, analytical redundancy approaches for SFDI are widely used [56, 65]. Among them are the observer-based SFDI schemes, which have been extensively developed for linear dynamic systems. The general framework of an observer-based SFDI method consists of the generation of residuals, which are compared with fixed or adaptive thresholds. An initial classification of the existing observer-based architectures for SFDI in nonlinear systems is based on the different ways of modeling the inherent system nonlinearity and the sensor faults. A usual approach is the linearization of the monitored nonlinear system either at a finite number of operating points [66] or operating zones [67]. Then, the well-established observer-based SFDI methods for linear systems can be applied. However, the linear approximation of the system introduces additional errors in the residuals affecting the fault detectability. Also, if the early detection of a fault is not achieved, it may cause the trajectory to move away from the operating point or zone, thus making the linearization even less effective.

Recent research work has focused on the detection and isolation of a single sensor fault in nonlinear uncertain systems [68, 69]. However, the isolation of multiple sensor faults in nonlinear dynamic systems has become a challenging problem. Several researchers, using nonlinear models of the monitored systems, have developed diagnostic schemes based on a single nonlinear observer, capable of isolating multiple sensor faults with a specific time profile or magnitude, or assuming a maximum number of their multiplicity. Exciting research results have been obtained using: (i) a diagonal residual feedback nonlinear observer, for simultaneous sensor faults, whose number is less than the number of states minus the number of disturbances [70], and (ii) a sliding mode observer (SMO) [71] and a neural-network (NN) based observer [72] for bounded sensor faults. Other approaches are based on a linear matrix inequalities-based observer for isolating at most two simultaneous sensor faults [73] and a diagnostic adaptive observer for sensor faults with constant magnitude [74].

On the other hand, there has also been some research activity in addressing the problem of detecting and isolating multiple sensor faults using a bank of observers. For concurrent but not simultaneous sensor faults, a Generalized Observer Scheme (GOS: a bank of observers, in which each observer is driven by all sensor outputs except for one, e.g., the ith sensor output, and the ith residual is sensitive to all faults but the ith sensor fault) using NN-based observers has been developed in [75], while for non-concurrent faults work in [76] proposed a bank of nonlinear observers designed using differential geometry [77]. Some additional studies have attempted to apply the Dedicated Observer Scheme (DOS: a bank of observers, in which each observer of the bank is driven by one sensor output, e.g., the ith sensor output, and the ith residual is sensitive only to the ith sensor fault) in nonlinear systems [78–80]. Furthermore, a bank or banks of observers is also utilized in FDI techniques in which the sensor faults are formulated as actuator faults [81, 82].

Another approach developed deals with the problem of detecting and isolating multiple sensor faults in a distributed framework for a class of nonlinear uncertain systems. To tackle this problem, local SFDI modules were designed, each of which is tailored to monitor the healthy operation of a set of sensors and capture the occurrence of faults in this set. The latter is realized by checking the set of structured residuals that are designed to be sensitive to the faults of the local subsystems, while being insensitive to faults in other subsystems [83]. The fault isolation aims to initially localize the set of sensors containing the faulty sensors through the local SFDI modules and then to localize the faulty sensors themselves by processing the information acquired from the local SFDI modules using a combinatorial decision logic. For analyzing the performance of the proposed distributed SFDI architecture, its robustness is checked with respect to the modeling uncertainties and conditions are established for fault detectability and isolability [84]. From a practical point of view, the distributed architecture described implies lower communication requirements, since the sensor measurements are not transmitted to a single location, and increased architecture reliability with respect to security threats.

Finally, [28] presents another FDI method using a timed discrete-event approach based on interval observers that improves the integration of fault detection and isolation tasks. In this case, the fault isolation module is implemented using a timed discrete-event approach that recognizes the occurrence of a fault by identifying a unique sequence of observable events (fault signals). The states and transitions that characterize such a system can directly be inferred from the relation between fault signals and faults.

Examples of possible controller architectures specifically utilized for fault management (fault detection and isolation), based on the generic controller architectures shown in Fig. 5, are depicted in Figs. 10 and 11. In Fig. 10 it is demonstrated how such a controller uses the monitored information together with a nominal and a fault model to decide whether a fault was detected and isolated respectively.

Figure 11 shows the same controller module architectures when a learning process is also incorporated in relation to the incoming information and the nominal/fault models. A learning process at the input can be utilized to predict sensor values as well as reduce the input size, whereas incorporating a learning process at the nominal/fault models is essential in improving the model parameters as well as learning new models. The need to develop cognitive fault diagnosis approaches that can learn characteristics or system dynamics of the monitored environment is essential in adapting their behavior and predicting missing or inconsistent data to achieve fault tolerant monitoring and control. The main motivation for such approaches is to exploit spatial and temporal correlations between measured variables and to develop algorithms and design methodologies that will prevent situations where a relatively "small" fault event in one or more of the components (e.g., sensor, actuator, communication link) may escalate into a larger fault.

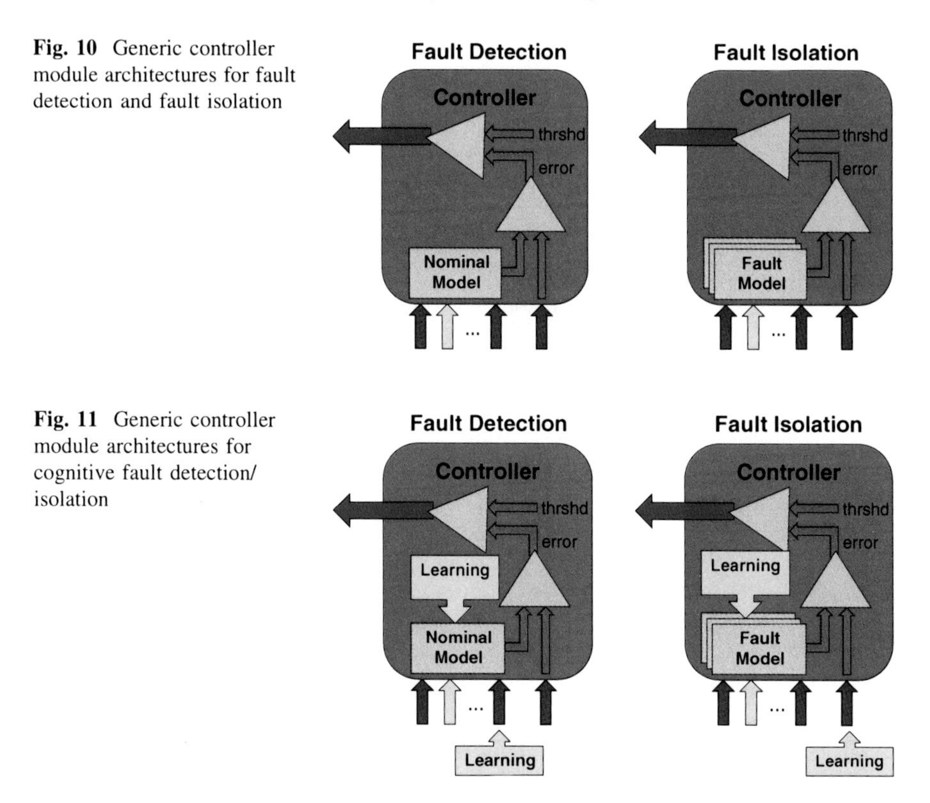

Fig. 10 Generic controller module architectures for fault detection and fault isolation

Fig. 11 Generic controller module architectures for cognitive fault detection/ isolation

5.3 Sensor Placement for Security

In Sect. 4 there was an extensive discussion on the sensor placement problem for monitoring and controlling critical infrastructures. This section presents a variant of that problem, namely sensor placement but now with security being the objective. Again, the use case is the water distribution network. The task is to find sets of locations in the network to install quality sensors, so that a number of objectives are optimized under various fault scenarios.

A mathematical formulation suitable for the security issues related to the location selection is the "p-median" [85], with the objective of minimizing the "maximum distance" of a facility. A similar formulation was examined in [86, 87] for water distribution systems. By considering a number of contamination scenarios and their impacts, the authors formulated a mathematical program to minimize the average "contamination impact". The formulation was extended to take into consideration imperfect sensors in [88] and the "p-median" formulation was further examined in a stochastic framework in [89]. A multi-objective extension was also examined in [90] and in [91] the authors proposed a methodology based on this formulation, to determine locations for monitoring disinfection byproducts.

In general, the problem of sensor placement for security can be formulated as a multi-objective integer optimization program [92–94]. A methodology was proposed in [95] to measure the contamination risk, considering contamination detection faults and its consequences. Some additional single and multiple objective methodologies for sensor placement have also recently been proposed [96–98], where several of the issues related to sensor placement strategies were reviewed.

It is also important to note that in the case of water distribution networks, when online quality sensors are not available, or do not cover part of the distribution network, the standard approach is for the water utility to perform manual water sampling for quality analysis. Water quality may be checked at a few nodes within a network, and for a few times during the day. Sampling locations may be selected by water utility personnel in an arbitrary fashion, based on their own experience, which could be subjective; it may be otherwise chosen using certain regulatory requirements, depending on consumer distributions and historical data.

In practice, due to the large-scale nature of water distribution networks, as well as the partial knowledge of the time-varying, hydraulic, and quality dynamics, it is difficult to optimally identify the best locations and times to conduct manual sampling, or install on-line quality sensors. In addition, each node in the network has certain characteristics, such as the outflow water pattern and the number of customers, which makes the selection problem non-trivial. The problem of manual sampling was discussed in [99], where the authors examined the problem of scheduling manual sampling for contaminant detection. They proposed a mixed integer program for the calculation of the sampling route; i.e., the location and time to take samples, while considering certain real conditions, such as utility working hours, the time required for sampling and the traveling time between nodes. Furthermore, in [100], the TEVA-SPOT software is described for determining where to install quality sensors for increasing system security.

Again, even though the problem of sensor placement for security depends on the type of the critical infrastructure monitored and controlled, most of these techniques can be applied to different CIs with similar characteristics.

6 Conclusions

As pointed out in the preceding sections of this chapter, a large number of problems associated with the monitoring, control, and security of critical infrastructures have been extensively studied in the literature. However, there are still a number of key research issues that are currently under investigation. These include: (a) Development of suitable control architectures (e.g., distributed, decentralized, hierarchical, etc.), (b) Development of communication and cooperation techniques between intelligent agents, (c) Development of software and hardware embedded devices to facilitate the monitoring, control, and security functionalities, taking into account cost, size, reliability, and energy efficiency considerations, (d) Development of efficient algorithms for real-time information processing when the controller receives a very large amount of data, (e) Development of intelligent monitoring and control techniques (including cost-efficient and effective sensor and actuator placement), (f) Development of a common system-theoretic framework for fault diagnosis and security of critical infrastructure systems, (g) Cross-fertilization of ideas between the various applications and between industry and academia, (h) Development of analytical results on detectability, isolability, fault isolation time, fault recovery, etc. and application of these analytical results to different infrastructures, and (i) Analysis of the interdependencies between different infrastructures and creation of efficient interdependency models.

Clearly, the aforementioned list is not exhaustive, but rather it provides a glimpse of some of the issues that need to be resolved efficiently (e.g., taking into account such metrics as cost, size, reliability, energy efficiency, combinatorial complexity, etc.) in order to have effective monitoring, control, and security of CIs capable of quickly and efficiently accommodating changes in the state of the network infrastructure or other interdependent infrastructures (due to normal operating conditions or fault/attack scenarios).

The rest of the chapters in this book address some of these issues. Specifically, Chapters "Electric Power Systems, Telecommunication Networks, Water Distribution Networks and Transportation Systems: Monitoring, Control, and Security" address monitoring, control, and security approaches in four of the main critical infrastructure systems (namely electric power systems, telecommunication networks, water distribution networks, and transportation systems respectively) and Chapters "Algorithms and Tools for Intelligent Monitoring of Critical Infrastructure Systems," "Algorithms and Tools for Intelligent Control of Critical Infrastructure Systems" and "Algorithms and Tools for Risk/Impact Evaluation in Critical Infrastructures" describe different algorithms and tools for intelligent monitoring, control, and risk/impact evaluation of critical infrastructures. Furthermore, Chapter "Infrastructure Interdependencies: Modeling and Analysis" discusses possible approaches for modeling and analysis of infrastructure interdependencies, Chapter "Fault Diagnosis and Fault Tolerant Control in Critical Infrastructure Systems" presents fault diagnosis and fault tolerant control in CIs, and Chapter "Wireless Sensor Network Based Technologies for Critical Infrastructure Systems" deals with the development of wireless sensor network-based technologies for CIs. Finally, Chapter "System-of-Systems Approach" presents a system-of-systems approach for intelligent monitoring, control, and security of interdependent critical infrastructure systems.

From the description of the work in these chapters, it is clear that there are several challenging problems and future directions for the management of these systems. Since this book cannot address the large number of questions that arise, as well as all the new emerging trends, some of the main ones are noted in Chapter Conclusions that concludes the book as topics for future exploration.

References

1. USA Patriot Act, Public Law 107-56, 2001. Available online via http://epic.org/privacy/terrorism/hr3162.html. Accessed 22 Dec 2012
2. EU Directive 2008/114/EC, Identification and designation of European critical infrastructures (2008)
3. Vision Solutions.: Assessing the financial impact of downtime: understand the factors that contribute to the cost of downtime and accurately calculate its total cost in your organization. White Paper (2008)
4. Rinaldi, S.: Modeling and simulating critical infrastructures and their interdependencies. In: Proceedings of the 37th International Conference on System Sciences, 00(C):1–8 (2004)
5. Rinaldi, S., Peerenboom, J., Kelly, T.: Identifying, understanding, and analyzing critical infrastructure interdependencies. IEEE Control Syst. **21**(6), 11–25 (2001)
6. Axelrod, R., Cohen, M.D.: Harnessing Complexity: Organizational Implications of a Scientific Frontier, pp. 32–61. Free Press, New York (1999)
7. Bagheri, E., Ghorbani, A.A.: The state of the art in critical infrastructure protection: a framework for convergence. Int. J. Crit. Infrastruct. **4**(3), 215–244 (2008)
8. Gou, B., Abur, A.: An improved measurement placement algorithm for network observability. IEEE Trans. Power Syst. **16**(4), 819–824 (2001)
9. Baldwin, T.L., Mili, L., Boisen Jr, M.B., Adapa, R.: Power system observability with minimal phasor measurement. IEEE Trans. Power Syst. **8**(2), 707–715 (1993)
10. Nuqui, R.F., Phadke, A.G.: Phasor measurement unit placement techniques for complete and incomplete observability. IEEE Trans. Power Deliv. **20**(4), 2381–2388 (2005)
11. Milosevic, B., Begovic, M.: Nondominated sorting genetic algorithm for optimal phasor measurement placement. IEEE Trans. Power Syst. **18**(1), 69–75 (2003)
12. Xu, B., Abur, A.: Optimal placement of phasor measurement units for state estimation. Final Project Report. PSERC, Ithaca, Oct 2005
13. Valverde, G., Chakrabarti, S., Kyriakides, E., Terzija, V.: A constrained formulation for hybrid state estimation. IEEE Trans. Power Syst. **26**(3), 1102–1109 (2011)
14. Chakrabarti, S., Kyriakides, E., Eliades, D.G.: Placement of synchronized measurements for power system observability. IEEE Trans. Power Deliv. **24**(1), 12–19 (2009)
15. Celik, M.K., Liu, W.H.E.: An incremental measurement placement algorithm for state estimation. IEEE Trans. Power Syst. **10**(3), 1698–1703 (1995)
16. Baran, M.E., Zhu, J., Zhu, H., Garren, K.E.: A meter placement method for state estimation. IEEE Trans. Power Syst. **10**(3), 1704–1710 (1995)
17. Asprou, M., Kyriakides, E.: Optimal PMU placement for improving hybrid state estimation accuracy. In: Proceedings of the IEEE, PowerTech 2011, Trondheim, pp. 1–7, June 2011
18. Rice, M., Heydt, G.T.: Enhanced state estimator: part III: sensor location strategies. Final Project Report. PSERC, Tempe, Nov 2006
19. Toregas, C., ReVelle, C.: Optimal location under time or distance constraints. Pap. Reg. Sci. **28**(1), 131–143 (1972)
20. Church, R., ReVelle, C.: The maximal covering location problem. Pap. Reg. Sci. **32**(1), 101–118 (1974)
21. Lee, B., Deininger, R.: Optimal locations of monitoring stations in water distribution system. ASCE J. Environ. Eng. **118**(1), 4–16 (1992)
22. Kumar, A., Kansal, M.L., Arora, G.: Identification of monitoring stations in water distribution system. ASCE J. Environ. Eng. **123**(8), 746–752 (1997)
23. Al-Zahrani, M.A., Moeid, K.: Locating optimum water quality monitoring stations in water distribution system. In: Proceedings of ASCE World Water and Environmental Resources, pp. 393–402 (2001)
24. Berry, J.W., Fleischer, L., Hart, W.E., Phillips, C.A., Watson, J.-P.: Sensor placement in municipal water networks. ASCE J. Water Resour. Plan. Manage. **131**(3), 237–243 (2005)

25. Harmant, P., Nace, A., Kiene, L., Fotoohi, F.: Optimal supervision of drinking water distribution network. In: Proceedings of ASCE Water Resources Planning and Management, pp. 52–60 (1999)
26. Slay, J., Miller, M.: Lessons learned from the Maroochy water breach. In: Goetz, E., Shenoi, S. (eds.) Critical Infrastructure Protection. Springer, Boston (2008)
27. Goetz, E., Shenoi, S. (eds) Critical Infrastructure Protection. Springer, Heidelberg (2008)
28. Meseguer, J., Puig, V., Escobet, T.: Fault diagnosis using a timed discrete-event approach based on interval observers: application to sewer networks. IEEE Trans. Syst. Man Cybern. Part A Syst. Hum. **40**(5), 900–916 (2010)
29. Tryby, M.E., Boccelli, D.L., Uber, J.G., Rossman, L.A.: Facility location model for booster disinfection of water supply networks. ASCE J. Water Resour. Plan. Manage. **128**(5), 322–333 (2002)
30. Lansey, K., Pasha, F., Pool, S., Elshorbagy, W., Uber, J.: Locating satellite booster disinfectant stations. ASCE J. Water Resour. Plan. Manage. **133**(4), 372–376 (2007)
31. Propato, M., Uber, J.G.: Booster system design using mixed-integer quadratic programming. ASCE J. Water Resour. Plan. Manage. **130**(4), 348–352 (2004)
32. Prasad, T.D., Walters, G.A., Savic, D.A.: Booster disinfection of water supply networks: multiobjective approach. ASCE J. Water Resour. Plan. Manage. **130**(5), 367–376 (2004)
33. Ewald, G., Kurek, W., Brdys, M.A.: Grid implementation of a parallel multiobjective genetic algorithm for optimized allocation of chlorination stations in drinking water distribution systems: Chojnice case study. IEEE Trans. Syst. Man Cybern. Part C Appl. Rev. **38**(4):497–509 (2008)
34. Hernandez-Castro, S.: Two-stage stochastic approach to the optimal location of booster disinfection stations. Ind. Eng. Chem. Res. **46**(19), 6284–6292 (2007)
35. Oman, P., Schweitzer, E., Frincke, D.: Concerns about intrusions into remotely accessible substation controllers and SCADA systems. In: Proceedings of the 27th Annual Western Protective Relay Conference (2000)
36. President's Information Technology Advisory Committee.: Cyber security: a crisis of prioritization. Report to the President, National Coordination Office for Information Technology Research and Development, Arlington (2005)
37. Luders, S.: CERN tests reveal security flaws with industrial networked devices. The Industrial Ethernet Book, GGH Marketing Communications, Titchfield, pp. 12–23, Nov 2006
38. Petroleum Safety Authority. www.ptil.no/English/Frontpage.htm
39. Cherry, S., Langner, R.: How Stuxnet is rewriting the cyberterrorism playbook. In: Proceedings of the IEEE Spectrum, Oct 2010
40. Berg, M., Stamp, J.: A reference model for control and automation systems in electric power. Technical Report, SAND2005-1000C. Sandia National Laboratories, Albuquerque (2005)
41. Byres, E., Carter, J., Elramly, A., Hoffman, D.: Worlds in collision: Ethernet on the plant floor. In: Proceedings of the ISA Emerging Technologies Conference (2002)
42. National Communications System.: Supervisory control and data acquisition (SCADA) systems. Technical Information Bulletin NCS TIB 04-1, Arlington (2004)
43. Ask, R., et al.: Information security baseline requirements for process control, safety and support ICT systems. OLF Guideline No. 104. Norwegian Oil Industry Association (OLF), Stavanger (2006)
44. American Gas Association.: Cryptographic protection of SCADA communications; part 1: background, policies and test plan. AGA Report No. 12 (Part 1), Draft 5, Washington DC. www.gtiservices.org/security/AGA12Draft5r3.pdf (2005)
45. American Gas Association.: Cryptographic protection of SCADA communications; part 2: retrofit link encryption for asynchronous serial communications. AGA Report No. 12 (Part 2), Draft, Washington DC. www.gtiservices.org/security/aga-12p2-draft-0512.pdf (2005)
46. American Petroleum Institute.: API 1164: SCADA security, Washington DC (2004)

47. British Columbia Institute of Technology.: Good practice guide on firewall deployment for SCADA and process control networks. National Infrastructure Security Co-ordination Centre, London (2005)
48. Byres, E., Franz, M., Miller, D.: The use of attack trees in assessing vulnerabilities in SCADA systems. In: Proceedings of the International Infrastructure Survivability Workshop (2004)
49. Graham, J., Patel, S.: Security considerations in SCADA communication protocols. Technical Report TR-ISRL-04-01. Intelligent System Research Laboratory, Department of Computer Engineering and Computer Science, University of Louisville, Louisville (2004)
50. Instrumentation Systems and Automation Society.: Security technologies for manufacturing and control systems (ANSI/ISA-TR99.00.01-2004). Research Triangle Park, North Carolina (2004)
51. Instrumentation Systems and Automation Society.: Integrating electronic security into the manufacturing and control systems environment (ANSI/ISA-TR99.00.02-2004). Research Triangle Park, North Carolina (2004)
52. Kilman, D., Stamp, J.: Framework for SCADA security policy. Technical Report SAND2005-1002C. Sandia National Laboratories, Albuquerque (2005)
53. National Institute of Standards and Technology.: System protection profile: industrial control systems v1.0. Gaithersburg, Maryland (2004)
54. Stouffer, K., Falco, J., Kent, K.: Guide to supervisory control and data acquisition (SCADA) and industrial control systems security—Initial public draft. National Institute of Standards and Technology, Gaithersburg (2006)
55. Chandia, R., Gonzalez, J., Kilpatrick, T., Papa, M., Shenoi, S.: Security strategies for SCADA networks. In: Goetz, E., Shenoi, S. (eds.) Critical Infrastructure Protection. Springer, Heidelberg (2008)
56. Chen, J., Patton, R.J.: Robust Model-Based Fault Diagnosis for Dynamic Systems. Kluwer Academic Publishers, London (1999)
57. Frank, P.M.: Fault diagnosis in dynamic systems using analytical and knowledge–based redundancy: a survey and some new results. Automatica **26**(3), 459–474 (1990)
58. Gertler, J.J.: Fault Detection and Diagnosis in Engineering Systems. Marcel Dekker, New York (1998)
59. Patton, R.J., Frank, P.M., Clark, R.N.: Issues of Fault Diagnosis for Dynamics Systems. Springer, Berlin (2000)
60. Chung, W.H., Speyer, J.L., Chen, R.H.: A decentralized fault detection filter. ASME J. Dyn. Syst. Meas. Control **123**(2), 237–247 (2001)
61. Olfati-Saber, R.: Distributed Kalman filter with embedded consensus filters. In: Proceedings of the Joint CDC and ECC Conference, Sevilla, Dec 2005
62. Franco, E., Olfati–Saber, R., Parisini, T., Polycarpou, M.M.: Distributed fault diagnosis using sensor networks and consensus-based filters. In: Proceedings of the 45th IEEE Conference on Decision and Control, pp. 386–391 (2006)
63. Olfati-Saber, R., Franco, E., Frazzoli, E., Shamma, J.S.: Belief consensus and distributed hypothesis testing in sensor networks. In: Proceedings of the Workshop on Networked Embedded Sensing and Control, University of Notre Dame, IN, Oct 2005
64. Ding, Y., Elsayed, E., Kumara, S., Lu, J., Niu, F., Shi, J.: Distributed sensing for quality and productivity improvements. IEEE Trans. Autom. Sci. Eng. **3**(4), 344–359 (2006)
65. Isermann, R.: Fault-diagnosis Systems: An Introduction from Fault Detection to Fault Tolerance. Springer, Heidelberg (2006)
66. Montes de Oca, S., Puig, V., Blesa, J.: Robust fault detection based on adaptive threshold generation using interval LPV observers. Int. J. Adapt. Control Sig. Process. **26**(3), 258–283 (2011)
67. Orjuela, R., Marx, B., Ragot, J., Maquin, D.: Fault diagnosis for nonlinear systems represented by heterogeneous multiple models. In: Proceedings of Conference on Control and Fault-Tolerant Systems (SysTol), pp. 600–605 (2010)

68. Zhang, X., Parisini, T., Polycarpou, M.: Sensor bias fault isolation in a class of nonlinear systems. IEEE Trans. Autom. Control **50**(3), 370–376 (2005)
69. Zhang, X.: Sensor bias fault detection and isolation in a class of nonlinear uncertain systems using adaptive estimation. IEEE Trans. Autom. Control **56**(5), 1220–1226 (2011)
70. Narasimhan, S., Vachhani, P., Rengaswamy, R.: New nonlinear residual feedback observer for fault diagnosis in nonlinear systems. Automatica **44**(9), 2222–2229 (2008)
71. Yan, X., Edwards, C.: Sensor fault detection and isolation for nonlinear systems based on a sliding mode observer. Int. J. Adapt. Control Sig. Process. **21**, 657–673 (2007)
72. Talebi, H., Khorasani, K., Tafazoli, S.: A recurrent neural-network-based sensor and actuator fault detection and isolation for nonlinear systems with application to the satellite's attitude control subsystem. IEEE Trans. Neural Netw. **20**(1), 45–60 (2009)
73. Rajamani, R., Ganguli, A.: Sensor fault diagnostics for a class of non-linear systems using linear matrix inequalities. Int. J. Control **77**(10), 920–930 (2004)
74. Vemuri, A.: Sensor bias fault diagnosis in a class of nonlinear systems. IEEE Trans. Autom. Control **46**(6), 949–954 (2001)
75. Samy, I., Postlethwaite, I., Gu, D.: Survey and application of sensor fault detection and isolation schemes. Control Eng. Pract. **19**(7), 658–674 (2011)
76. Mattone, R., De Luca, A.: Nonlinear fault detection and isolation in a three-tank heating system. IEEE Trans. Control Syst. Technol. **14**(6), 1158–1166 (2006)
77. De Persis, C., Isidori, A.: A geometric approach to nonlinear fault detection and isolation. IEEE Trans. Autom. Control **46**(6), 853–865 (2001)
78. Adjallah, K., Maquin, D., Ragot, J.: Non-linear observer-based fault detection. In: Proceedings of the 3rd IEEE Conference on Control Applications, pp. 1115–1120 (1994)
79. Rajaraman, S., Hahn, J., Mannan, M.: A methodology for fault detection, isolation, and identification for nonlinear processes with parametric uncertainties. Ind. Eng. Chem. Res. **43**(21), 6774–6786 (2004)
80. Rajaraman, S., Hahn, J., Mannan, M.: Sensor fault diagnosis for nonlinear processes with parametric uncertainties. J. Hazard. Mater. **130**(1–2), 1–8 (2006)
81. Chen, W., Saif, M.: A sliding mode observer-based strategy for fault detection, isolation, and estimation in a class of lipschitz nonlinear systems. Int. J. Syst. Sci. **38**(12), 943–955 (2007)
82. Fragkoulis, D., Roux, G., Dahhou, B.: Detection, isolation and identification of multiple actuator and sensor faults in nonlinear dynamic systems: application to a waste water treatment process. Appl. Math. Model. **35**(1), 522–543 (2010)
83. Gertler, J.: Fault Detection and Diagnosis in Engineering Systems. CRC, Boca Raton (1998)
84. Polycarpou, M., Trunov, A.: Learning approach to nonlinear fault diagnosis: detectability analysis. IEEE Trans. Autom. Control **45**(4), 806–812 (2000)
85. Hakimi, S.: Optimum locations of switching centers and the absolute centers and medians of a graph. Oper. Res. **12**(3), 450–459 (1964)
86. Berry, J., Fleischer, L., Hart, W., Phillips, C.: Sensor placement in municipal water networks. In: Proceedings of the ASCE World Water and Environmental Resources, pp. 40–49 (2003)
87. Berry, J., Hart, W., Phillips, C., Uber, J., Watson, J.: Sensor placement in municipal water networks with temporal integer programming models. ASCE J. Water Resour. Plan. Manage. **132**(4), 218–224 (2006)
88. Berry, J., et al.: Designing contamination warning systems for municipal water networks using imperfect sensors. ACSE J. Water Resour. Plan. Manage. **135**(4), 253–263 (2009)
89. Shastri, Y., Diwekar, U.: Sensor placement in water networks: a stochastic programming approach. ASCE J. Water Resour. Plan. Manage. **132**(3), 192–203 (2006)
90. Watson, J.-P., Greenberg, H.J., Hart, W.E.: A multiple-objective analysis of sensor placement optimization in water networks. In: Proceedings of ASCE World Water and Environmental Resources, pp. 456–465 (2004)

91. Boccelli, D.L., Hart, W.E.: Optimal monitoring location selection for water quality issues. In: Proceedings of the ASCE World Environmental and Water Resources, pp. 522–527 (2007)
92. Krause, A., Leskovec, J., Guestrin, C., VanBriesen, J., Faloutsos, C.: Efficient sensor placement optimization for securing large water distribution networks. ASCE J. Water Resour. Plan. Manage. **134**(6), 516–526 (2008)
93. Ostfeld, A., et al.: The battle of the water sensor networks (BWSN): A design challenge for engineers and algorithms. ASCE J. Water Resour. Plan. Manage. **134**(6), 556–568 (2008)
94. Preis, A., Ostfeld, A.: Multiobjective contaminant sensor network design for water distribution systems. ASCE J. Water Resour. Plan. Manage. **134**(4), 366–377 (2008)
95. Weickgenannt, M., Kapelan, Z., Blokker, M., Savic, D.A.: Risk-based sensor placement for contaminant detection in water distribution systems. ASCE J. Water Resour. Plan. Manage. **136**(6), 629–636 (2010)
96. Aral, M.M., Guan, J., Maslia, M.L.: Optimal design of sensor placement in water distribution networks. ASCE J. Water Resour. Plan. Manage. **136**(1), 5–18 (2010)
97. Dorini, G., Jonkergouw, P., Kapelan, Z., Savic, D.: SLOTS: Effective algorithm for sensor placement in water distribution systems. ASCE J. Water Resour. Plan. Manage. **136**(6), 620–628 (2010)
98. Hart, W.E., Murray, R.: Review of sensor placement strategies for contamination warning systems in drinking water distribution systems. ASCE J. Water Resour. Plan. Manage. **136**(6), 611–619 (2010)
99. Berry, J., Lin, H., Lauer, E., Phillips, C.: Scheduling manual sampling for contamination detection in municipal water networks. In: Proceedings of the ASCE Water Distribution Systems Analysis. Paper 95, vol. 10, pp. 1–16 (2006)
100. Hart, W., Berry, J., Riesen, L., Murray, R., Phillips, C., Watson, J.: SPOT: a sensor placement optimization toolkit for drinking water contaminant warning system design. In: Proceedings of ASCE World Water and Environmental Resources, p. 12 (2007)

Electric Power Systems

Antonello Monti and Ferdinanda Ponci

Abstract The electric power system is one of the largest and most complex infrastructures and it is critical to the operation of society and other infrastructures. The power system is undergoing deep changes which result in new monitoring and control challenges in its own operation, and in unprecedented coupling with other infrastructures, in particular communications and the other energy grids. This Chapter provides an overview of this transformation, starting from the primary causes through the technical challenges, and some perspective solutions.

Keywords Power systems · Smart grid · Renewable energy sources · Monitoring · Control · Distributed resources

1 Introduction

Power systems have been operating for the last about 100 years using the same fundamental principles. Technology has, so far, allowed an improvement of their performance, but it has not revolutionized the basic principles. One fundamental law of physics has been driving the process: because the electrical grid has (nearly) no structural way to store energy, it is necessary that at every instant the amount of power generated to be equal to the power absorbed by the loads. In fact, some energy is naturally stored in the inertia of large generators. This is enough to compensate for small unbalances, which continuously occur and cause small variations of frequency and voltage (remaining within rather restrictive limits). Beyond these, violation of balance leads to voltage perturbation, large frequency variations, and electromechanical oscillations. If corrective measures are not applied in a timely manner, the system may collapse, resulting in widespread blackouts. In this context, automation is the way to determine and actuate these measures via the control of the generators.

A. Monti (✉) · F. Ponci
E.ON ERC, Institute for Automation of Complex Power Systems, RWTH Aachen University, Aachen, Germany
e-mail: AMonti@eonerc.rwth-aachen.de

E. Kyriakides and M. Polycarpou (eds.), *Intelligent Monitoring, Control, and Security of Critical Infrastructure Systems*, Studies in Computational Intelligence 565,
DOI 10.1007/978-3-662-44160-2_2

The loads are predictable only in a statistical sense, hence the automation is designed following a demand-driven approach. Based on the prediction of the load, the bulk of generation is scheduled. At run-time, unexpected deviations are actively compensated, reaching the due balance of generation and demand. Such a principle works very well under some clear assumptions:

- generation is "perfectly" controllable and predictable, so that the correction in the power balance can always be applied to the generation side of the balance
- generation is concentrated as much as possible in large plants, simplifying the problem of scheduling.

For traditional power systems these assumptions hold perfectly well, and have driven the design and construction of large power plants as we know them today.

The growing attention to the environmental impact and the consequent rise of new policies supporting the penetration of renewable energy sources are, vice versa, real game changers. Two points can immediately make the difference clear:

- generation from renewable energy sources is not perfectly predictable, and furthermore only in a limited sense controllable
- generation from renewable energy sources pushes to a more decentralized approach

While it would be incorrect to claim that power systems were not complex systems in the past, the new scenario is definitely raising the bar, and making complexity an even more tangible concern. In fact, the direct consequence of the new scenario is the creation of a stronger link between the electrical energy system and other infrastructures, such as communications or other energy infrastructures (e.g., gas grid. Furthermore, the new scenario is undermining the foundations of the automation principles, calling for a more decentralized approach to system functions, such as monitoring and control. These are the prime features of the smart grids.

In this chapter the consequence of the rise of complexity is analyzed in detail [1]. A review of the basics of power system automation, as implemented today, is provided. This introduces the final part of the chapter, focusing on cutting edge research and emerging methods for future electrical power systems.

2 The Arising of Complexity in Smart Grids

As introduced in the previous paragraph, one of the main reasons of complexity in energy systems is the growing interdependence among infrastructures. We summarize here the types of the interdependent, heterogeneous infrastructures, the points of coupling, and the arising of complexity.

The presence of intermittent, energy sources, such as renewables, which may not be reliably predicted and dispatched, enforces the presence of energy storage (or of other fast dynamic sources) for balancing load and generation [2]. But massive deployment of new dedicated energy storage units is technically and financially challenging, and at some extent may not be even necessary. Instead, for maximum usage of the existing resources, we need to resume to a broader concept of energy storage, involving the non-electrical energy grids, such as gas and heat, and distributed storage resources, such as plug-in electric vehicles. Gas energy is coupled to electrical energy via thermo-electric devices, such as Combined Heat and Power Units (CHP). Heat grids and heat storage in buildings is coupled to the electrical system through CHP and heat pumps. Plug-in electric vehicles are coupled to the electrical power grid through their batteries. Eventually, the electrical system is coupled with the gas, heat, and traffic systems.

Various business models are under consideration for the joint operation of these interconnected systems, including the creation of new markets for reserves and reactive power, and the establishment of Virtual Power Plants, Virtual Storage Systems, Dual Demand Side Management systems. The economic operation of these businesses and related markets represents yet another type of infrastructure, and another source of interdependence between the aforementioned infrastructures.

The coherent operation of all these infrastructures, and the unprecedented interactions between the generation, transmission and distribution sections of the power system, make the communications critical as never before. The communication infrastructure is coupled with all the aforementioned infrastructures and constitutes their glue (one of the main efforts in this direction are the European projects Finseny [3] and FINESCE [4]).

For interdependent grids, the distribution of control and monitoring functions is a necessity. In fact, the central control is infeasible and undesirable, particularly due to the minute granularity of some of the actors in the smart grid.

The resulting energy system is not only a complicated system made of many parts. It is also a complex system, that is, a system whose global behavior may not be inferred from the behavior of the individual components, and where no single entity may control, monitor and manage the system in real time [5]. The effect of the interdependencies is largely unknown and unforeseeable, in the absence of a clear view of the coupling points, of ways to model it, and of models, data and measurements. Besides, no way to predict the behavior means no way to control the behavior. Let alone more practical issues, such as the interdisciplinary harmonization of the standards, shared knowledge, and more. The vision of the smart grid as the complex system outlined above yields practical effects on the enablers: (1) education and training, (2) tools, (3) methods and design.

2.1 Education

Interdisciplinary study (and research, for that matter) has long been a buzz word in academia. Nonetheless the instruments to make it happen are extensively lacking, as the legal and organizational background is not ready for it. The integration of the disciplines of Electrical Engineering alone, and of closely related fields, is in itself a challenge. The creation of truly interdisciplinary areas, with related official degrees, conflicts with a structure that was developed for very different (and sometimes openly opposite) educational needs. In practice, though, besides the implementation issues, there is a more fundamental question to be addressed, and that is: what is the right curriculum for the smart grid engineers? And in particular, what should the ideal new hire of an energy service company know? What are the topics for her/his life-long education program?

Academic education is not the only competence-building environment affected by these challenges. Professional formation and training at all levels are equally affected. The lack of field experience, of the possibility to train in the field, and of trainers themselves, completes the picture.

2.2 Tools

Tools may mean: (a) the numerical tools to carry out the analysis of the complex smart grids, or (b) the technologies for de-risking new devices and algorithms, supporting the transition from numerical to in-field testing. These tools are primarily numerical simulators, and Hardware in the Loop (HIL) and Power Hardware in the Loop (PHIL) testing platforms. An example of application to interdependent heterogeneous systems that are part of a smart grid can be found in [6].

In a nutshell, these tools should support:

- multi-physics, multi-technology environment, to allow for representing all kinds of dynamic interdependencies
- multidisciplinary environment, fit for diverse users, working at different aspects of the same simulation scenario, with universally understood knowledge
- dynamic and reconfigurable model definition, enabling different users to interact with the simulation schematic focusing though on different levels of detail and obtaining results in a reasonable time
- high-level graphic visualization to support system analysis for the different disciplines, hence providing on one hand the favorite individual visualization options (oscilloscope-like, color-code-like,…) and on the other hand able to synthesize a "system-picture"
- uncertainty propagation, from the sources of uncertainty (e.g., renewable generation, loads, prices) through the discipline borders to the entire system.

2.3 Methods and Design

The lack of methods for representing and analyzing the smart grid as a complex system, and the lack of performance metrics for such a complex system, result in lack of methods for designing holistic controls and for designing components fit for this environment. The complex smart grid must be designed to manage uncertainty and inconsistencies, to be resilient and gracefully degrade when necessary. This implies that all the interdependent systems that the smart grid comprises must have these features, and use them in a coherent way.

We have here briefly described the factors that make the smart grid complex and some consequences. These consequences impact very different worlds, from the education and training, to the manufacturing of components, to development of numerical tools, to the philosophy of design. An enormous innovation in technology and tools is required, together with a deep change of mentality, for the transition to the smart grid to occur. The bright side is that a successful undertaking in the energy sector in the aforementioned directions may then be migrated to other sectors, with widespread benefits for all.

3 Challenges of the Future Grid

3.1 Characteristics of the Present Grid

Most of the existing power systems present common characteristics. The main ones, which are expected to be subject to change, are summarized here and pictorially shown in Fig. 1. Here the power system is shown in its main sections: generation in large power plants; transmission grid to transport the energy at high voltage and for long distances to the location of the demand; and medium and low voltage distribution grid that provides energy to the end users, who may be domestic, commercial or industrial.

Generation is highly concentrated in bulk power plants. These plants can be controlled, in a small signal sense, as briefly outlined in the introduction, to compensate for the variations caused by temporary unbalances of energy supply and demand. However, because of the large inertia of the generators in bulk power plants, and the slow dynamics of non-electrical actuators (e.g., fuel injection), the responsiveness to large variations of demand is very slow. So, while this inertia is beneficial to the overall system, as it constitutes a form of energy storage, on the other hand it makes the control of large, fast variations extremely challenging, if not impossible.

The occurrence of such large imbalances of generation and load, by the way, used to be linked primarily to faults or disruptions. Normally, the system could be considered as quasi-static. This characteristic used to be the underlying assumption of control design. Such an assumption relied on statistically foreseeable, slowly changing load profiles, and most of all, fully controllable generation. The latter assumption based on the prompt availability of the primary source of energy, i.e., some kind of fuel, in most cases. But volatile sources, mainly renewable, do not possess this characteristic, and instead they may be themselves responsible for

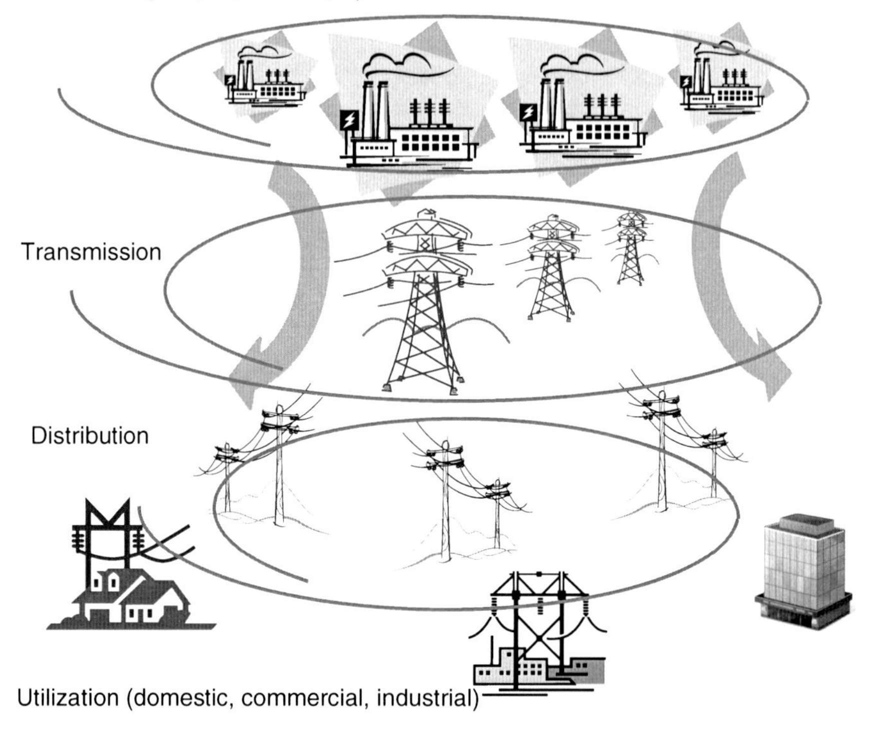

Fig. 1 Scheme of the power system today

large variations, especially in case of high penetration, when they represent a significant fraction of the generation.

From the system viewpoint, the flow of energy in classical systems, is unidirectional, from transmission to distribution and from bulk generators to loads. Generators and loads always preserve their "power injection", "power absorption" nature with respect to the grid. The distribution systems are assumed to be totally passive.

In Europe, the transmission systems are interconnected as shown in Fig. 2, meaning that power can flow between areas managed by different Transmission System Operators (TSO), and across national borders, and in principle virtually traverse the entire continent. The European system supplies about 500 million people. The Continental Europe Synchronous Area comprises most of continental Europe, and comprises systems locked at the same frequency of 50 Hz (with restrictive tolerance to variation, as mentioned in the introduction). These interconnected synchronous systems operate somehow jointly to address unforeseen situations and major disturbances. The occurrence of disturbances is "sensed" far away from the point where they are originated, through frequency deviations and unexpected changes in the power flows at the interconnections. A true joint

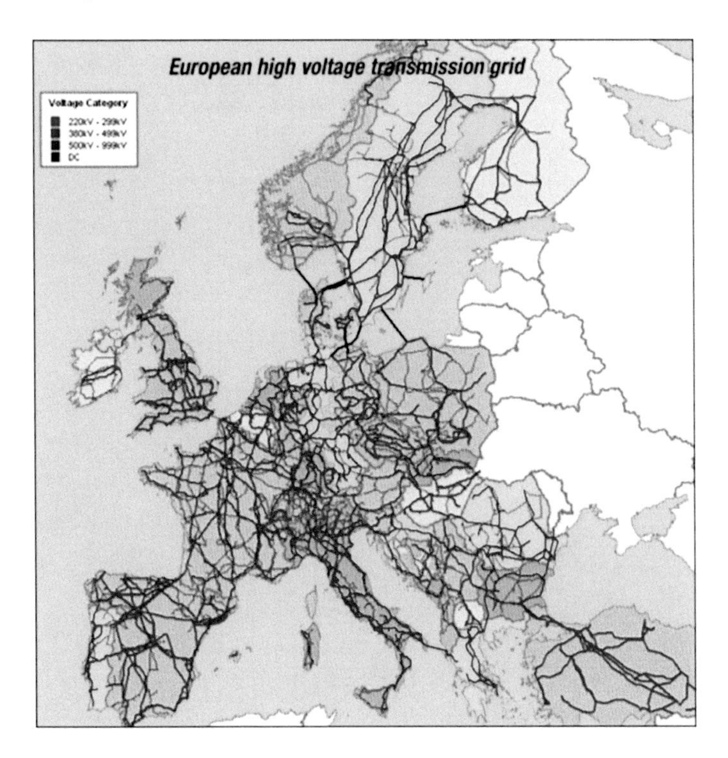

Fig. 2 European high voltage transmission grid from ENTSO-E [10]

operation of such system [7, 8] requires the ability to monitor very wide areas [9], exchange data and information promptly and in a standardized way, to mention just the major needs. More on the planning, regulation and market can be found in [10].

A sample behavior of the power flows within a national transmission grid, the German grid, is visualized in Fig. 3 [10] for the particular case of combined conditions of strong wind, hence large generation from wind farms in the north of the country, and heavy loading. In this scenario, two aspects can be emphasized: the first is the flow from generating areas to loading areas; the second is the consequent "long way" of the power flow, with proportionally large related losses. Furthermore, we point out that in this condition, the status of the grid and of the exchange with the neighboring states depends heavily on a volatile source.

To present the current status of power systems, particularly with reference to penetration of renewables, and provide some coherent quantification, we refer here to the case of Germany, as an exemplary case in Europe.

The German transmission grid comprises the portions at extra-high voltage, 220–380 kV, and at high voltage, 110 kV, with about 350 substations. A high level representation of the German power system is shown in Fig. 4. The distribution grids at medium voltage (MV), 1–50 kV, and low voltage (LV), 0.4 kV, include about 600.000 MV/LV substations. The extra-high voltage transmission system is

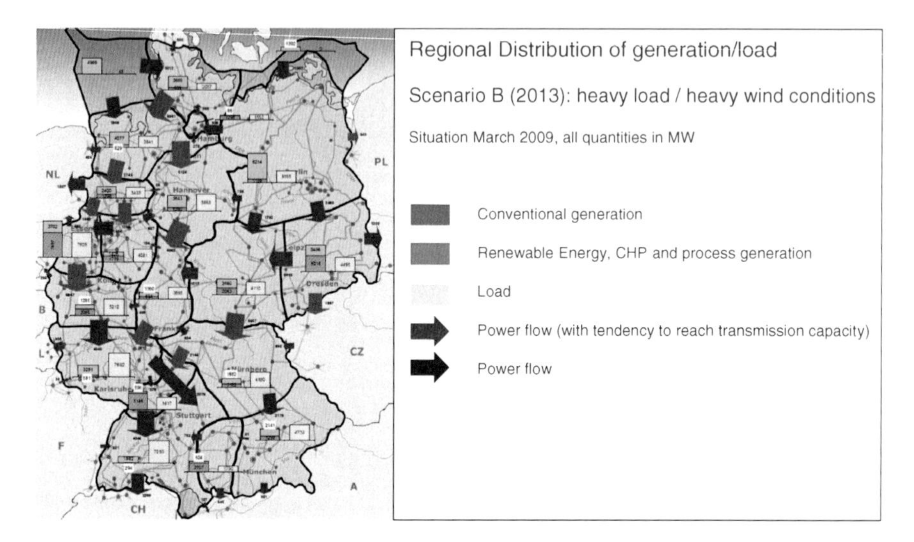

Fig. 3 Scenario of strong-wind combined with heavy load in Germany [10]

part of the European backbone, and it connects the bulk generation. The new generation plants based on renewable sources, together with other minor generation sources (e.g., some industrial plants with generation capacity) are instead connected to the MV. Domestic loads and very small generation are connected to the low voltage distribution systems.

The installed generation capacity amounts to 152.9 GW, as of 2009 (Data from Bundesministerium für Wirtschaft und Technologie), with wind and solar plants representing a fraction of about 27 %. Correspondingly, energy generation in Germany, in 2009, amounted at 593.2 TWh with a consumption of 578.9 TWh. Out of the generated energy, 45.2 TWh came from combined solar and wind sources, thus representing 7.6 % of the total. While, as of now, the penetration of renewables still represents a significant but manageable fluctuating source, things are progressing in the direction of increasing this penetration.

The key to allowing for local fluctuations is the interconnection between systems, such as the case of the European high voltage transmission grid. Thus, in the perspective of increasing the penetration of renewables, the interconnections are to be strengthened further.

A look at the electricity prices provides ground to considerations on the small wiggle room for price variation and for so called grid parity forecasts. The average price in Germany in 2011 is 24.95 c/kWh, of which 45.5 % is taxes.

With reference to the scenario outlined above, the power system of the future is expected to be characterized by: more distributed generation, part of which originating from renewable sources, which are numerous and in part (smaller sizes) not controllable except by the owner, and subject to uncertain fluctuations; active distribution networks; faster dynamics, due to smaller inertia of the sources;

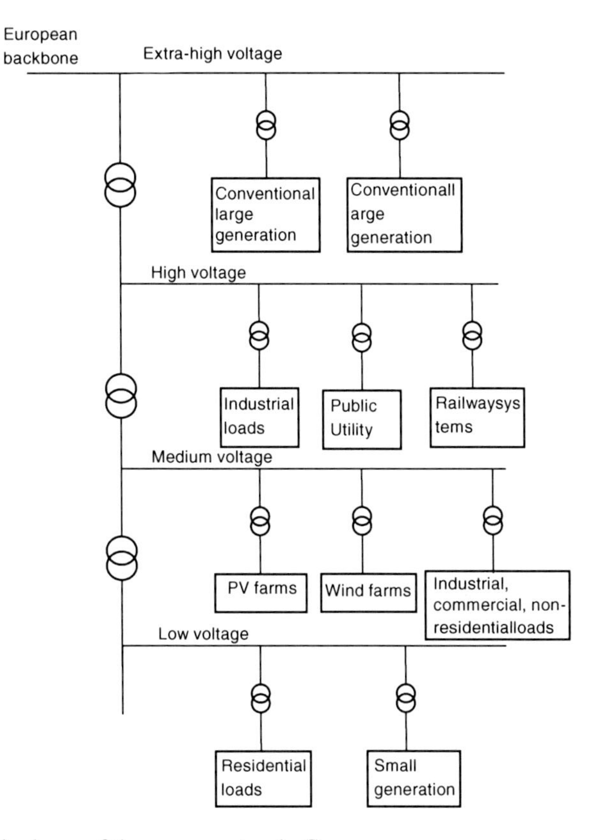

Fig. 4 Notional scheme of the power system in Germany

and potentially sudden source variations (e.g., wind puffs). One additional newly arising characteristic is the interconnectivity between energy systems of different nature, noticeably the connection and coupling between electrical and heat systems, as mentioned with reference to arising complexity. These characteristics give rise to unprecedented challenges.

In the first place, the constant balancing act of the electrical energy system is expected to be more difficult because:

1. if the sources are less predictable the balance is less predictable and
2. if sources are more numerous and decentralized, likely not under the authority of one or very few companies, the balance is more complex

Actually the most developments in the short term are brewing in the distribution system. This is because on one hand that is where the most dramatic changes are occurring, and on the other hand because they may take on more generation and control duties, thus relieving the challenged transmission system. As of now, the distribution systems are in general not ready for automation because of:

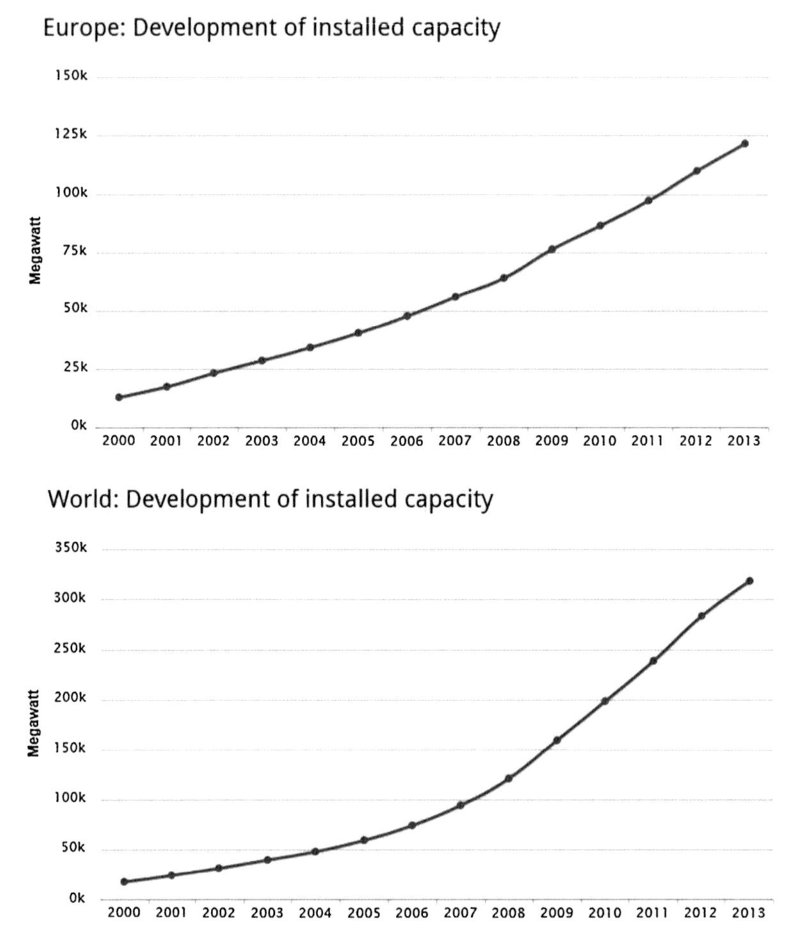

Fig. 5 Installed wind power capacity in Europe and worldwide (© Bundesverband WindEnergie e.V.)

- Very limited monitoring (instrumentation, monitoring functions)
- Majority of resources are not controllable
- Lack of data, information (topology, status, parameters)
- Protection system designed for unidirectional flow of power
- Lack of communication infrastructure
- Scalable methods for monitoring and control in developing phase

The operation of the system is designed under the assumption of slow dynamics and may not accommodate quick variations because of control rooms with human-in-the-loop and insufficient data refresh rates.

To gain an idea of the rate of growth of the phenomena that call for overcoming the limitation of existing power system infrastructure, the following data is provided. The trend in worldwide wind power installation is shown in Fig. 5. Data in Fig. 5 are

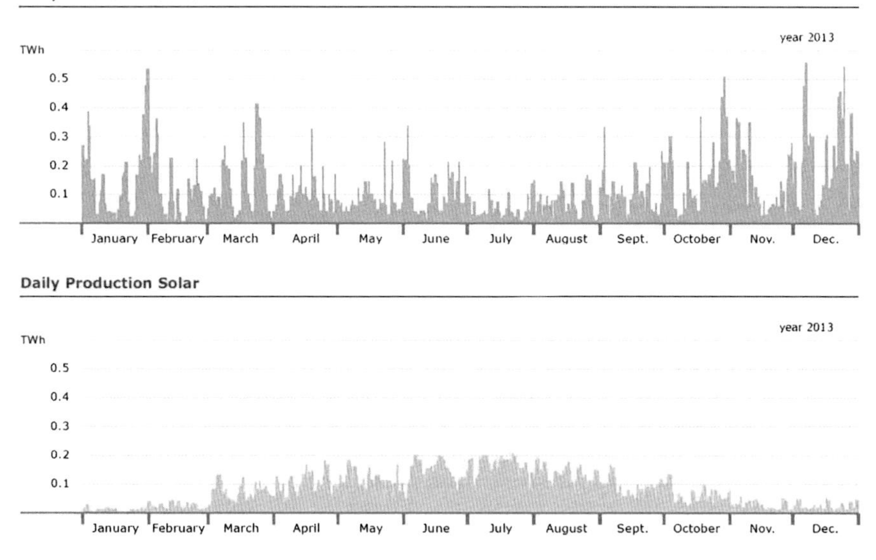

Fig. 6 Wind and solar power daily generation in Germany in 2013 [11]

global, involving power systems in different countries, with very different characteristics and production patterns, so the impact on the operation of the transmission system may not be immediately derived. Instead, to gain a sense of this impact and variations that wind power generation undergoes, a sample case is shown in Fig. 6. Here the 2013 production from wind in Germany is depicted, featuring a maximum peak of 0.56 TWh on 16.12.2013, and a minimum of 0.006 TWh on 17.02.2013, with a power maximum of 26.3 GW on December 5, 2013, at 18:15 (GMT +1:00) and minimum of 0.12 GW on September 4, 2013 at 13:45 (GMT +2:00). Solar production yielded a maximum of 0.20 TWh on 21.07.2013 and a minimum of 0.002 TWh on 18.01.2013. (Data 2013 from Fraunhofer ISE-EXX [11]).

Data in Figs. 5 and 6 point at two main aspects: penetration of renewables is on a steep rise and the volatility is extremely high. Furthermore, volatility affects the other main form of renewable energy, solar energy, as shown in Fig. 7.

The large installed capacity from renewables is generally located where the primary source (e.g., wind or solar irradiation) is maximum, which hardly ever matches the location of maximum load, resulting in a challenge for the transmission system, which was not developed to operate in this way. The volatility adds on the challenge, as it is not a good match for a system that was developed to operate "as planned" rather than "as it comes", the latter requiring real-time monitoring and control.

The challenges of the renewable scenario extend down into the distribution system. In fact, here, many generation units, some of which non-dispatchable, are installed, making the distribution system an active one. The resulting challenges, linked to load-generation balancing and coordination, do not find a natural solution

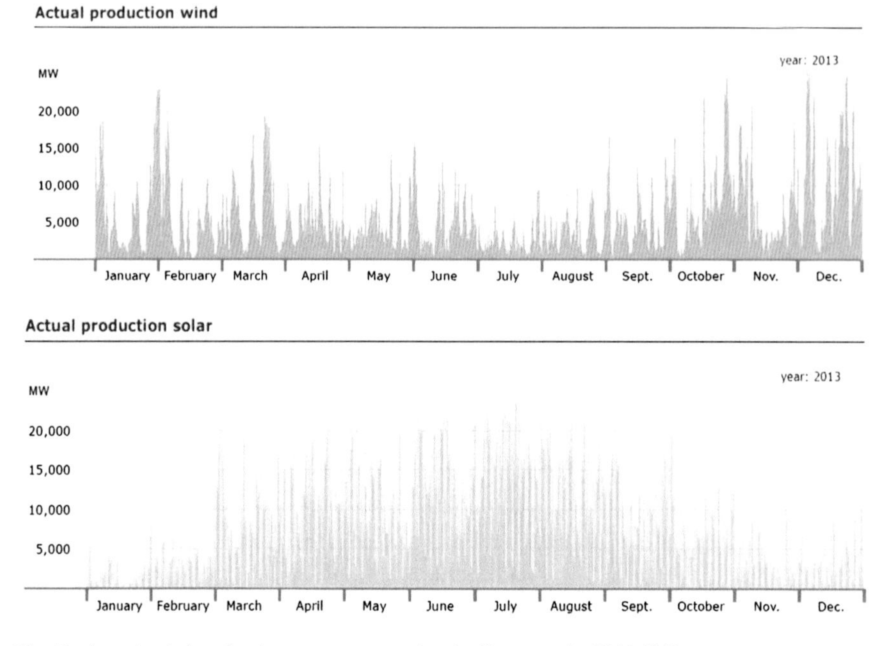

Fig. 7 Actual wind and solar power generation in Germany in 2013 [11]

here, because of the chain of interests of the businesses involved, as they arose after the deregulation, with the Distribution System Operator (DSO), retailer, and possibly other roles, being distinct commercial entities.

In summary, the exemplary case of Germany shows a total installed renewable capacity of 77.36 GW, predicted to rise to 90 GW by 2020, an average load of 60 GW, with peak 70 GW, (2013) [10]. With these numbers, the "100 % renewable mode" is a possibility now, and a likely event after 2020. To guarantee the power quality levels we have today, the most critical factors are going to be energy storage and real-time control, wide-area monitoring and control, and a new architecture for power distribution.

So the new facts influencing the electrical energy scenario and that make it a complex infrastructure interconnected with other complex infrastructures are:

- The storage options (e.g., gas grid, e-vehicles, and thermal capacity of buildings) are heterogeneous; to achieve the amount that is needed, they all must be exploited. As a result, more coupling points would connect the related infrastructures. To quantify the impact of some of these factors, consider for example that 10 million of electrical cars would represent a potential storage in the order of 10–100 GWh.
- The presence of DC technology, in HV and MV, supporting the optimization of energy flows is a realistic option for strengthening the transmission system via

HVDC links. DC is in particular directly applicable in some scenarios, such as an MV DC collector grids for offshore wind farms, so as to easily accommodate connection of renewables.
- Fast, real-time monitoring and control in MV and LV networks.
- Links among energy grids, particularly at district level, where heat, MVDC and LVAC networks may provide a solution that is not invasive at household level but creates the condition of optimal overall use of available energy.

The new interconnections between different infrastructures involve:

- heterogeneous physical forms of energy and related storage capacity (e.g., heat, electrical, gas, environmental)
- heterogeneous forms of each energy type (AC, DC, various voltage levels)
- topological small and large scale (e.g., neighborhood, city, region)
- large and small individual power levels (aggregated small resources, e.g., fleets of e-vehicles, distributed generation, and large wind farms)

In synthesis, the fact that the critical infrastructures are coupled in various combinations of the above list, shows what a challenge is to try to understand the global behavior and model it, test the sensitivity to the occurrence of disruptive events, and design a comprehensive monitoring and control system.

In this framework, what is a Smart Grid? Many definitions are provided, depending on the level of implementation [12–14]. A major common element of these definitions is the presence of two-way flows of energy and information, which enforces new requirements on the measurement/metering infrastructure, and urgently calls for the development of new standards. The pathway to smart grids can be structured according to the natural partition transmission-distribution.

With particular reference to the electrical system, the transmission grid, partly already "smart", may achieve better efficiency via power routing through power electronic devices, such as FACTS (Flexible Alternating Current Transmission Systems). At the distribution level, the full deployment of automation would enable the involvement of the customers in peak shaving, generation (via the Virtual Power Plant VPP), storage, and possibly in grid supporting services.

Several standards address areas that are relevant for Smart Grids, the main ones being

- Automation within the Substations: IEEE 61850 completed and active
- IEC 61970-301 with extension IEC 61968 (in particular part 11)—Common Information Model (CIM)
- Series of IEEE 1547 for Distributed Resources

Gaps in the standards have been identified and reference documents to address them have been published, leveraging on the most recent research work. The USA and EU roadmaps can be found in NIST Special Publication 1108—NIST Framework and Roadmap for Smart Grid Interoperability Standards [15], and the

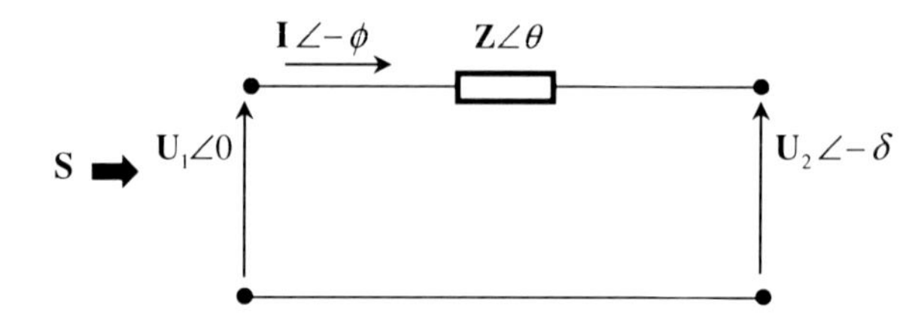

Fig. 8 Model of a line and complex through power

Standardization Mandate (M/490) to European Standardization Organisations (ESOs) to support European Smart Grid deployment [16], respectively. In particular, the Standardization Mandate (M/490) of the European Commission caters to European Standardisation Organisations (ESOs) to support European Smart Grid deployment. These organizations have eventually produced the Final Report of the CEN/CENELEC/ETSI Joint Working Group on Standards for Smart Grids [17].

In the power system infrastructure, the distribution system is expected to undergo the deepest transformation. The presence of power injections at all voltage levels and the presence of controllable resources calls for methods for state estimation, distributed control, and requires new infrastructures (for measurement, computation, communications). Such methods and infrastructures have been, up to now, typical of transmission systems. And the migration of the related transmission system technologies to distribution systems is a challenge [18] in terms of size of the system and characteristics. In fact the distribution system has orders of magnitude more nodes than the transmission system; at present time very limited instrumentation and actuators, especially in the LV section, where it is also poorly known and modeled (in terms of topology and parameters of existing systems) and features different characteristics of the lines.

Consider, for example, the impact of the assumption, commonly accepted for transmission lines, that lines are primarily "reactive", meaning that $R \ll X$, where R is the resistance and X is the reactance of the line. This assumption influences the model of the dependence between active and reactive power flow, and the terminal voltages. Such model is then used in monitoring and control, and so if this assumption does not hold, the related analytical derivations are not applicable, and the same can be said of the derived monitoring and control methods. Hence, for distribution systems, where this assumption usually does not hold, such existing methods developed for transmission are not in general directly applicable.

With reference to Fig. 8, the active and reactive power through a line with impedance Z are given by:

$$\begin{aligned} P &= \frac{U_1^2}{Z}\cos\theta - \frac{U_1 U_2}{Z}\cos(\theta+\delta) \\ Q &= \frac{U_1^2}{Z}\sin\theta - \frac{U_1 U_2}{Z}\sin(\theta+\delta) \end{aligned} \tag{1}$$

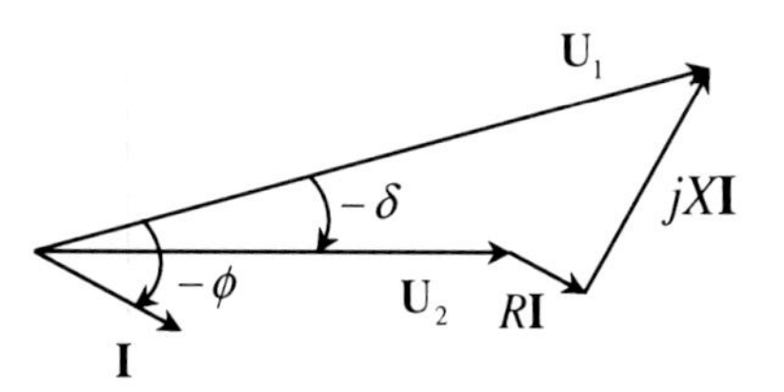

Fig. 9 Terminal voltage and current angles for line in Fig. 8

where

$$Ze^{j\theta} = R + jX$$

The phasor diagram of voltages and currents in this example is shown in Fig. 9. The active and reactive power can also be expressed as:

$$\begin{aligned} P &= \frac{U_1}{R^2 + X^2}[R(U_1 - U_2\cos\delta) + XU_2\sin\delta] \\ Q &= \frac{U_1}{R^2 + X^2}[-RU_2\sin\delta + X(U_1 - U_2\cos\delta)] \end{aligned} \tag{2}$$

The assumption mentioned above, that R ≪ X, allows for the approximation δ ≈ sin(δ), and thus:

$$\begin{aligned} P &\cong \frac{U_1 U_2}{X}\delta \quad (\text{hence } P \propto \delta) \\ Q &\cong \frac{U_1^2}{X} - \frac{U_1 U_2}{X} \quad (\text{hence } Q \propto U_1 - U_2) \end{aligned} \tag{3}$$

These simplified relations are the basis for control methods and state estimation decoupling in transmission systems. As the underlying assumption does not hold in distribution systems, where instead R ≈ X, then such control and monitoring methods can not be directly adopted.

4 Overview of Power Systems Control

In a broader perspective the management of the power system can be summarized in market, business management and energy management, as in Fig. 10.

The focus here is the Energy Management System (EMS), which pursues two main goals:

- Production cost minimization
- Loss minimization

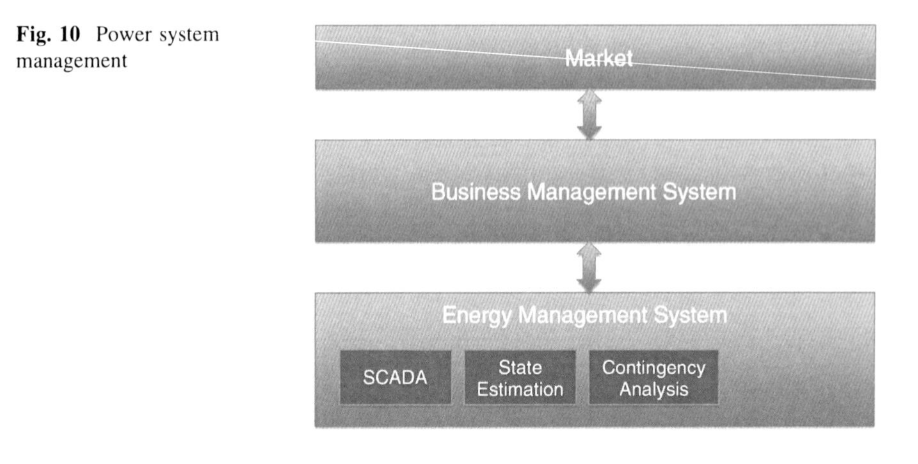

Fig. 10 Power system management

To achieve these goals, the following elements of action are used:

- Generator output control
- Phase shift control (to control voltage phase angle and hence power)

The EMS manages the power system and the power flows. This is done through the central control station and the distributed control centers with delegated control (collecting data and acting on a portion of the system). The management process relies on the Supervisory Control and Data Acquisition system (SCADA) and on the communication links from the control center to primary devices (generators, circuit breakers and tap changers). To realize this link, the devices are equipped with IED (Intelligent Electronic Device) interface actuators, to enable the remote control. Remote Terminal Units (RTU) are computer based interfaces to IEDs and other equipment, capable of local processing and control functions and of maintaining the communication with the SCADA.

The SCADA is a platform, with the basic functionality to classify and handle events, alarm processing, monitoring, and some system level control actions. It comprises the control room system, communication and remote terminal units, and its functions can classified in process database, Man Machine Interface (MMI), and application software. In particular:

- Data acquisition (from the system equipment) and processing
- State estimation (on the basis of the data collected at the substation level) and related ancillary functions
- Control of the transmission system equipment and alarm notifications to operators
- Event and data logging
- MMI
- Voltage control

Considering the layers utility-network-substation-distribution-consumer, the SCADA may be deployed to control different layers in an integrated manner as one system, or as separate systems, passing on data and information to the layer that is hierarchically superior.

The EMS operates through SCADA to achieve the following:

- Generation reserve monitoring and allocation
- Quality and security assessment, and state estimation
- Load management
- Alarm processing
- Automatic generation control (comprising load frequency control (LFC), economic dispatch (ED), and control for limiting the tie line flows to contractual values)
- Inter-control center communications

Reserve monitoring and allocation is needed to guarantee stability by addressing imbalances, as later explained in detail, and requires awareness of resource availability, location and timings. State estimation yields the current state of the system and it is the basis for contingency analysis ("what if" scenarios, to preempt and tackle large disturbances caused by loss of generation or lines) and assessment of the operating conditions, in terms of quality and security. Load management deals with the emergency procedure of balancing load and generation by dropping some of the loads, to rescue the system from potential instability. Alarm processing deals with alarms originating from tripping of protection breakers and violation of thresholds. Automatic generation control dispatches generators according to day ahead set points, but also for real-time regulation in presence of frequency deviations or deviations of tie line flows between areas when exceeding contractual limits. Finally, communication to other control centers guarantees a certain level of coordination in the above actions, that affect the operation of the system over a broad area.

While the functions above are typical of transmission system control centers, the same concept of "automation" applies to the distribution system, covering the entire range of functions, from protection to SCADA operation. The Distribution Automation System is a set of technologies that enables an electric utility to remotely monitor, coordinate and operate distribution system components in a real-time mode from remote locations.

The Distribution Management System (DMS) in the control room coordinates the real-time function with the non-real-time information needed to manage the network. Two basic concepts of distribution automation can be defined:

- Distribution Management System enables the Control Room operation and provides the operator with the best possible view of the network "as operated"
- Distribution Automation System fits hierarchically below the DMS, and it includes all the remote-controlled devices and the communication infrastructure.

The DMS comprises:

- SCADA of substations and feeders
- Substation automation
- Feeder automation

- Distribution system analysis
- Application based on Geographic Information System (GIS)
- Trouble-call analysis management
- Automatic Meter Reading

The SCADA of substations and feeders coupled with RTUs serve as supporting hardware to the DMS in monitoring, providing data collection and event logging, and generating reports on system stations.

The functions of the distribution automation can be summarized as follows, given that not all of them may be active in all distribution control centers:

- Demand-side management
- Voltage/VAr regulation
- Real-time pricing
- Dispersed generation and storage dispatch
- Fault diagnosis and location
- Power quality
- Reconfiguration
- Restoration

5 Control Methods

Existing Primary-Secondary-Tertiary control schemes and state estimation methods are reviewed in this section. Their applicability to distribution systems is under investigation; in fact, active distribution systems are expected to require new controls to limit perturbations. These include propagation of power fluctuations to the transmission system, and internally active and continuous regulation of voltage, regulation of frequency, load and generation balance. Some of these control modes of the distribution systems are rather futuristic, but may help contain the adverse effect of some of the issues caused by the distribution grid transformation mentioned in previous sections.

In the context of frequency control, as reported in Fig. 11, the control levels are so defined and characterized [19]:

- Primary Control (reaction time <30 s) represents the autonomous reaction of generators. Generators exhibit an individual speed-droop characteristic, and their reaction, accordingly, is individual too (independent for each turbine). This control method results in a steady-state frequency deviation different from zero.
- Secondary Control (reaction time 15 s–15 min) consists of enforcing a change or shift of the speed-droop characteristic of the generators. This centralized direction to the turbines enables the return to the nominal frequency of the system, under different loading conditions, preventing undesired power flows within different areas. Eventually, this control eliminates deviations through the Area Control Error.

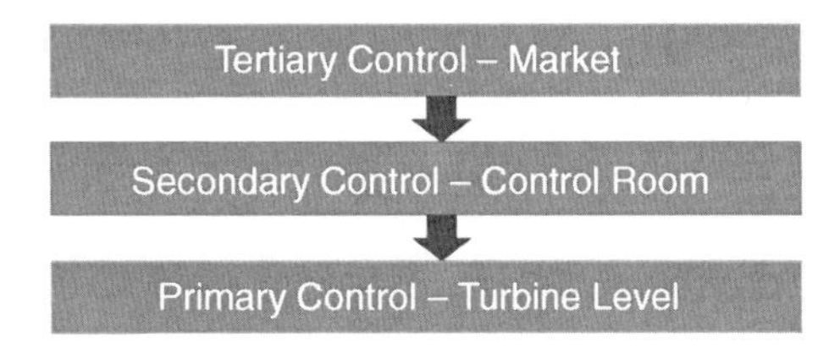

Fig. 11 Power system control scheme

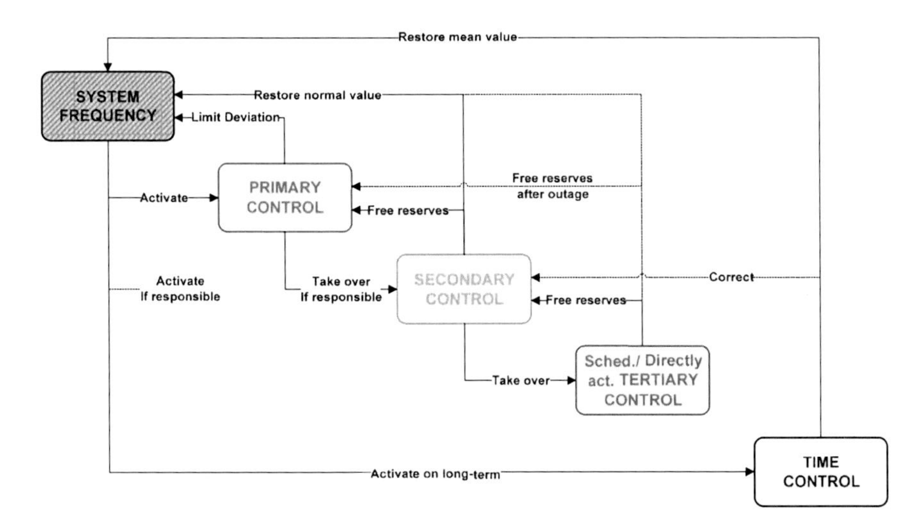

Fig. 12 General scheme of load-frequency control, in UCTE Handbook ENTSO-E [20]

- Tertiary Control (reaction time >15 min) optimizes the overall system operating point (set-points). This is achieved by setting the reference generation for individual generation units so that it is sufficient for secondary control. Reserve is distributed among the available generators in a way that is geographically distributed, and accounts for the various economic factors of each generation unit.

The relation between Primary, Secondary and Tertiary control in response to an outage is shown in Fig. 12.

The generation and load characteristics, resulting in the determination of the operating point are exemplified in Fig. 13 through Fig. 16. Figure 13 shows the characteristic frequency versus active power of the controlled turbine-generator set, which can be written as follows (for the linear portion):

$$\frac{\Delta P_G}{P_{GN}} = -k_G \frac{\Delta f}{f_N} \tag{4}$$

with P_{GN} being the nominal power of the generator, ΔP_G the deviation from nominal of the actual active power of the generator, f_N the nominal system

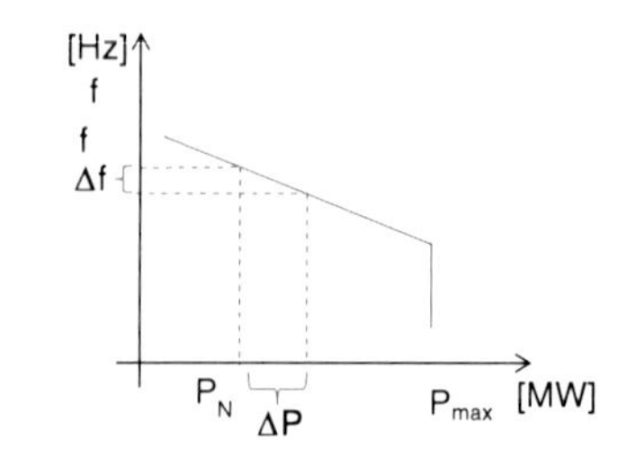

Fig. 13 Generation characteristic of one individual generator

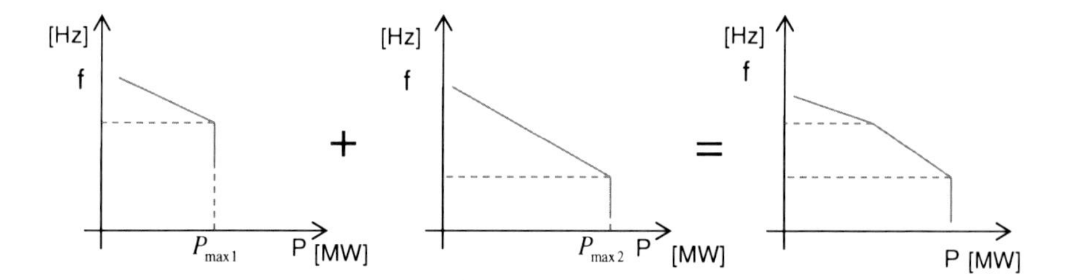

Fig. 14 Combination of two generation characteristics

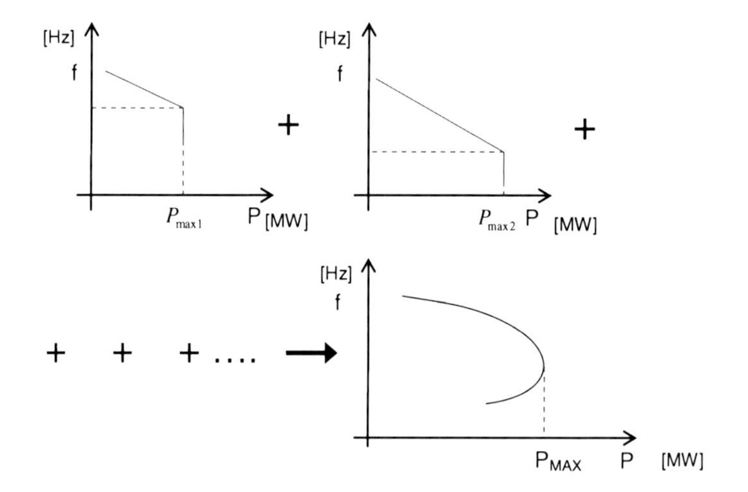

Fig. 15 Combination of generation characteristics

frequency in Hz, Δf the frequency deviation and k_G the characteristic coefficient. Notice that a reasonable assumption is that $k_G \approx 20$, which means that a 5 % frequency deviation corresponds to 100 % deviation power output.

The simplified combination of two such generation characteristics within an interconnected system is graphically shown in Fig. 14. In a system with multiple

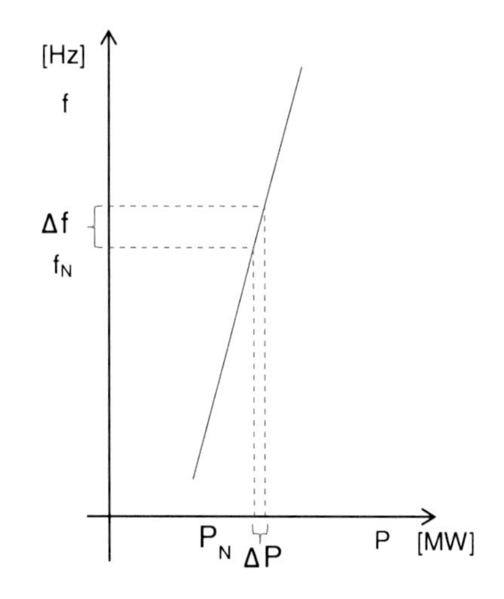

Fig. 16 Load characteristic

generation units, the combination of the characteristics results in the so called droop nose curve, as shown in Fig. 15. This curve shows the capability limits of the primary control, which has a stability limit at power P_{TMAX}.

The active characteristic frequency vs active power of the load is shown in Fig. 16, where ΔP_L is the variation from the nominal load P_{LN}, and k_L is the load coefficient. Notice that the load characteristic shows a weaker dependence than the generation characteristic, in fact, the load coefficient k_L is usually within the range $0.5 \div 3$. This characteristic can be expressed as follows:

$$\frac{\Delta P_L}{P_L} = k_L \frac{\Delta f}{f_N} \tag{5}$$

Given the generation and load characteristics, the operating point is determined by the intersection point, ideally at nominal frequency. Figure 16 shows the effect of the Primary Control, in case of a load increment. The initial operating point is (P_1, f_1) on the characteristic P_L^{old}.

From here, we assume that the load increases by ΔP_{Demand}. In other words, the load characteristic changes to P_L^{new}. The new desired operating point, at frequency f_1, lies on the characteristic P_L^{new}, but cannot be reached with the given generation characteristic. Instead, given the generation characteristic, the new operating point is (P_2, f_2), where the generation provides extra ΔP_G, the load comes short of ΔP_L and the frequency deviates to f_2 from the initial frequency f_1. This is the result of the Primary Control of the turbine. If the new operating point is sustainable, and the frequency within the acceptable limits, the control action has been sufficient and successful. Otherwise, if the new operating point is not sustainable, the secondary control must kick in.

Notice that only generation units that are not already fully loaded can contribute with the behavior exhibited in Fig. 17. The margin ΔP_G was in this case available, implying that at the initial operating point (P_1, f_1) the machine was not fully

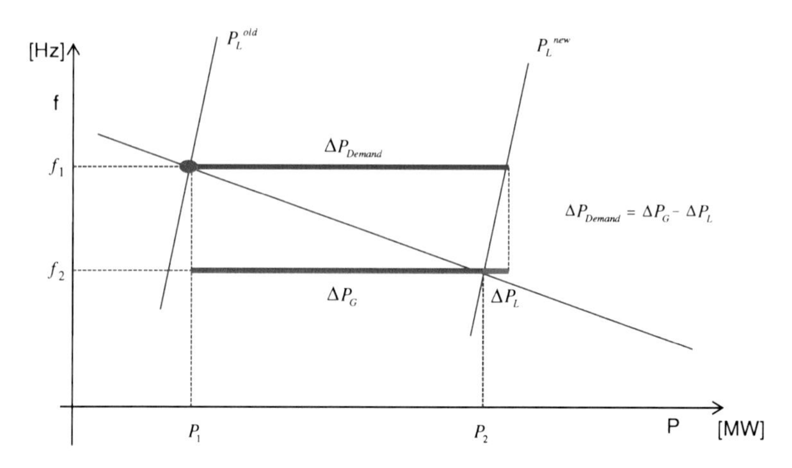

Fig. 17 Generation, load and frequency deviations in response to primary control

loaded, i.e., it had “spinning reserve”. Hence it can be seen that the role of the spinning reserve is critical. Furthermore, the spinning reserve must be properly sized, to address most of the normal load variations, and well distributed, to avoid overloading of the lines.

The intervention of the Secondary Control consists of the modification of the characteristic of the individual generators, which results in the modification of the global generation characteristic. The new references are originated centrally with the objective of taking the frequency back to the original value, and are distributed to the generation units.

The effect of this control is shown in Fig. 18. To sustain the new load demand while maintaining frequency f_1, the new operating point should be (P_2', f_1). Hence, the generation characteristic must be different from P_G^{old}, in particular, it must be P_G^{new}. The new operating point eventually is (P_2', f_1), belonging to the characteristics P_G^{new} and P_L^{new}.

The power through the tie lines represents a constraint centrally enforced here. The scheme of a Secondary Control with proportional-integral (PI) structure is shown in Fig. 19, where the Area Control Error (ACE) is obtained in terms of power shortage, combination of power demand inferred from the frequency deviation and deviation from reference power flow through the border of the area. The ACE is then the input to the PI controller which produces the total power deviation to be compensated for by the controlled generators. This quantity is then split among available generators according to indices α_i.

In conclusion, initially, the operating point of each individual power generation unit is set according to the Optimal Power Flow, to minimize global generation cost while satisfying network constraints. Corrections to instantaneous deviations between generation and load are in first place addressed by the Primary Control, hence adequate spinning reserve must be available for units participating in Primary Control. The resulting frequency offset, is corrected by the Secondary

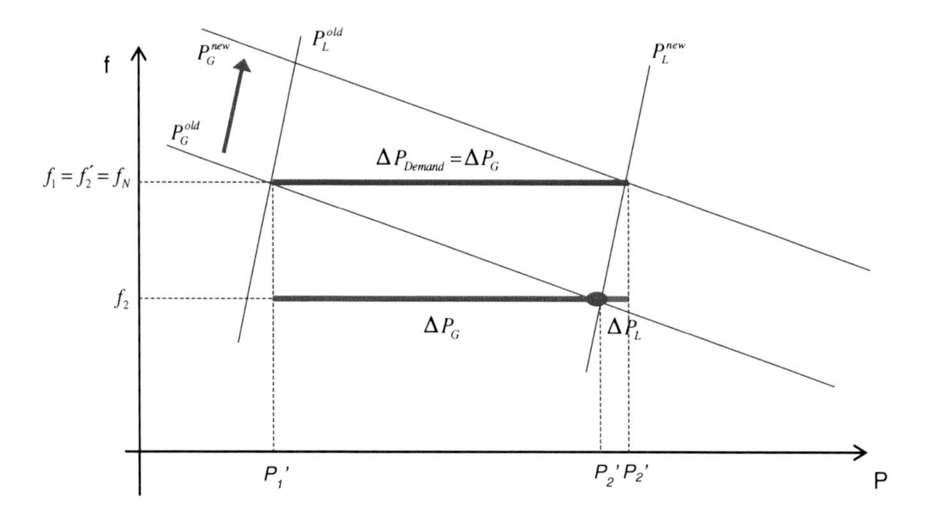

Fig. 18 Generation, load and frequency deviations in response to primary and secondary control

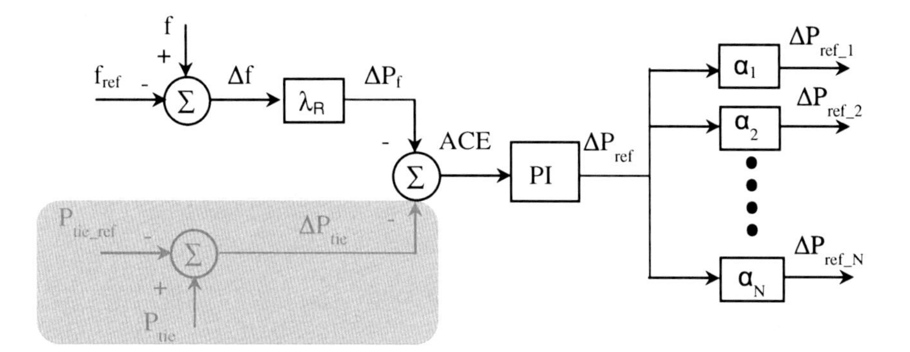

Fig. 19 Secondary PI control

Control that is expected to realize the optimal dispatch of units participating in Secondary Control.

Given the increasing number of generating entities, extending the control concept presented above may be a challenge, because of the large amount of data to be handled in a centralized manner. To increase the survivability of the system, single points of failure, such as the centralized secondary control, should be avoided. Furthermore, future control approaches should support opening the market to new players, for example aggregators of virtual power plants, and should support the Plug and Play concept, especially when extending these control approaches down to the distribution level. Distributed control technologies are under investigation for this purpose.

6 Monitoring of the Future Grid

The most comprehensive monitoring function is the state estimation, yielding the state of the system in terms of voltage phasors at all nodes. This knowledge is used to determine security conditions and as a starting point for contingency analysis. While state estimation is a well established practice in transmission systems, its application to distribution systems is in the developing phase.

The state estimation process is based on the following steps: definition of the current topology of the system, based on breaker and switch status; observability analysis and determination of observable islands; solution of the state estimation problem as determination of the most likely state of the system given the measurements, assumed to be affected by random errors; bad data detection and correction; determination of errors of initially adopted parameters and topology.

In general (although methods to overcome these limits are in full development), some assumptions are made that can be difficult to sustain in certain cases, particularly when applied to distribution networks. These assumptions are:

- The system is balanced, hence it can be represented in terms of direct sequence, and the system evolves slowly.
- Measurements are collected during one scan of the SCADA system, hence a certain time skew is to be expected and tolerated. Usually no dynamic model is included.
- Measurements may originate from the field, but may also consist of statistical information, such as load profiles. These so called pseudo-measurements, are usually affected by large uncertainty, particularly when they refer to grids with pervasive presence of renewable sources, for which historical record over a significant period of time is not yet available.

In distribution systems in particular, measurements from the field are often limited to primary substations, and possibly in the near future secondary substations, and in principle (experimentally used) the measurements from smart meters at the customer premises.

The classic formulation of the state estimation problem, formalized here in terms of measurement equations (where measurements may be active and reactive power flows and injections, rms voltages and currents) with measurements z_i, states x_i, (non-linear) measurement functions h_i and measurement errors e_i. The state of the system consists of voltage rms and phase at all nodes (with respect to a reference bus whose phase is not estimated, but assumed or measured).

$$\begin{bmatrix} z_1 \\ z_2 \\ \\ z_m \end{bmatrix} = \begin{bmatrix} h_1(x_1, x_{2,} \ldots, x_n) \\ h_2(x_1, x_{2,} \ldots, x_n) \\ \\ h_m(x_1, x_{2,} \ldots, x_n) \end{bmatrix} + \begin{bmatrix} e_1 \\ e_2 \\ \\ e_m \end{bmatrix}$$

Some common assumptions are that the measurement error has an expected value equal to zero, with a normal distribution and a known covariance matrix, often assumed to be diagonal. This over-determined, non-linear problem is then solved recursively according to the maximum likelihood criterion, with a Weighted Least Squares approach. The setup of the numerical problem is to be preceded by an update of the topology of the model of the network, observability analysis, and followed by bad data detection and identification, and verification of topology and parameters, as mentioned above. For details and other possible formulations refer to [21].

In this framework, the synchrophasor measurements, via Phasor Measurement Units (PMUs) [22, 23] are one of the most interesting developments, for transmission grids and, although at some extent futuristic, for the distribution grids. These measurements consist of synchronized rms voltages and currents and phases with a common reference. These measurements can support state estimation by augmenting the classical set of measurements, and find application in advanced monitoring and control applications [24], particularly over wide areas, where the synchronization feature can be well exploited.

A further development in state estimation consists of the implementation of dynamic state estimation. This is increasingly needed in systems that feature rapidly changing conditions, with much smaller inertia than in the past, due to the presence of many power converter interfaced sources, and less large rotating masses. Various approaches to dynamic state estimation and dynamic estimation of the error, have been proposed, many through Kalman filters [25].

While a well-established process for transmission grids, substantial challenges remain in the application to distribution grids. Here, current estimation as opposed to voltage estimation and the use of three-phase estimators to accommodate the presence of imbalances are the two major differences from established transmission state estimators. The primary issue in distribution remains the lack of accurate knowledge of topology and parameters and the lack of measurements. In fact, the "full" state estimation is affected by a lack of actual measurements, which forces an extensive use of pseudo-measurements with poor accuracy. Furthermore, classical state estimation yields phase angles, whose use in applications in distribution systems is at the moment rather limited. And finally, if applied at all nodes in the system, the classical estimation procedure may be cumbersome and at the present time not necessary to the Distribution System Operators.

The potential and performance of the state estimator is heavily dependent on the measurements, in terms of location, type and quality. And such measurements must fulfill other needs besides state estimation, for example metering for billing, as the customer smart meter, or protection. Hence, the decisions on where and how to deploy new measurement equipment, besides satisfying regulatory requirements, is to be seen as a multi-objective optimization. This is an ongoing research topic that accounts for technical issues (of the power and communication system jointly) together with economic issues (accommodating incremental deployment through near optimal "temporary" stages).

Finally, the distribution of the state estimation function may become necessary due to the size of the problem and the amount of measurements to be communicated, for networks with a large number of nodes, for dynamic estimation and in applications where the state estimate is used in time critical functions. A framework to assess state estimation methods, needs, and requirements, particularly in terms of accuracy and dynamics, and particularly for distribution systems is not yet shared knowledge.

7 Advances in Controls for Future Grids

The advances in controls outlined in this section address the increasing complexity of the power system and its dependence on the coordination of distributed resources, thus highlighting the role of communication. The quality of service and stability in generation, transmission and distribution are now linked to one another, while each group is actively controlled for local and global objectives. In general, we can say that dependencies across levels, which once were hierarchically and unidirectionally controlled, are now reconsidered to improve the operation of the system and accommodate new types of sources. However, these same dependencies also create new and easier paths for the propagation of disturbances. This is the scenario that new controls must be designed for.

7.1 Integration of Distributed Energy Resources: The Sample Case of Photovoltaic Systems

The integration and management of distributed energy resources (DERs) is one of the most significant challenges of future grids. In this context, the microgrid concept may facilitate the integration by coordinating the automation of a sub-section of the grid. At present, most of the DERs are not involved in the automation process, hence they do not increase local efficiency and they are not involved in the reactive power management, so they do not support network stability.

The case of photovoltaic (PV) systems in the low voltage (LV) grid is here taken as an example to demonstrate the feature and potential of DERs. From the regulatory standpoint, codes are defined in EN 50438:2007, now superseded by FprEN 50438:2013 in the approval stage. Voltage constraints are set by EN 50160 and result in low voltage being regulated.

In the future, PV sources may likely be asked to provide reactive power for voltage control along the feeder, through one of the following methods:

- $\cos\phi$(P) characteristic
- fixed $\cos\phi$ method
- fixed Q method
- Q(U) droop function

During normal grid operation, the active power flows from sources to loads (grid node S > L). The magnitude and direction of reactive power flow can be influenced by reactive power injection via PV inverters. Due to this reactive power control the voltage magnitude at the Point of Common Coupling (PCC) "L" can be actively influenced.

A voltage violation larger than the acceptable limit can be compensated for by setting the PV inverter to an "inductive angle" equal to $\varphi_{PV} = arctan(Q_{PV}/P_{PV})$, which implies shifting the operating point along the characteristic. This control method requires the ability of the PV inverter to control reactive power injection, which is not currently a common feature. Furthermore, the presence of multiple PV systems connected to the same PCC requires coordination, which may be better realized in a distributed manner, for example as in [26].

7.2 Microgrids

A microgrid is a section of a power grid connected through a main switch (MS) to the main power line as in Fig. 20, with local distributed generation and loads, and characterized by the following properties:

- It is equipped with local generation and it is then able to operate independently from the main grid
- It is equipped with a hierarchical control that can manage two main operating states:
  - Parallel mode
  - Islanding mode
- It is equipped with a central controller, in charge of detecting conditions that require to switch from parallel to islanding mode

In parallel mode the microgrid controller coordinates power dispatch of local sources (economic optimization) so the microgrid operates in parallel to the grid implementing synchronization action. In islanding mode, the local sources operate typically according to a droop control logic for P and Q, while the central controller may implement functionalities similar to secondary control or/and change droop coefficients.

The control structure of a microgrid may be summarized as in Fig. 21, where the central controller (CC) functions are Energy Management and Protection Coordination, while the local controller of the DER realizes the local implementation of the control commands from the central controller.

The control structure is schematically shown in Fig. 22, where the feedback variables have the following meaning with reference to Fig. 23:

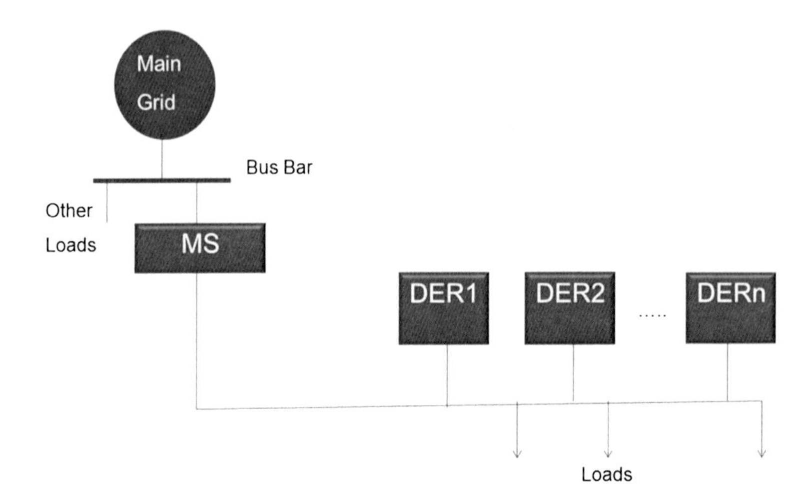

Fig. 20 Generation and load in a microgrid

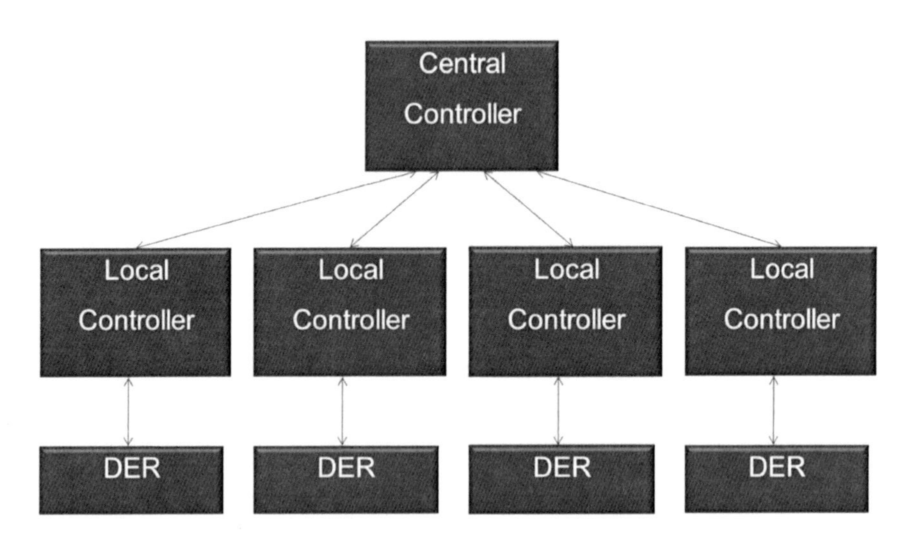

Fig. 21 Microgrid control structure

- i_l: current through L_f
- v_c: voltage across C_f
- i_o: current through L_o
- p, q: active and reactive power reference in the Park domain for the inverter

The more modern state space-based approach can be adopted instead of the nested loop control. In this case Optimal Control and Linear Quadratic Regulator are the typical choices. These control methods are presented next in the more general framework of distributed generation management.

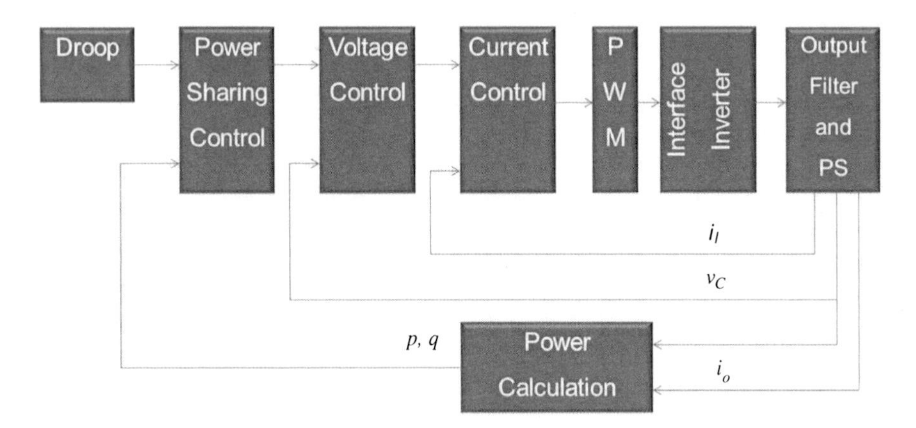

Fig. 22 Control structure of a DER in a microgrid

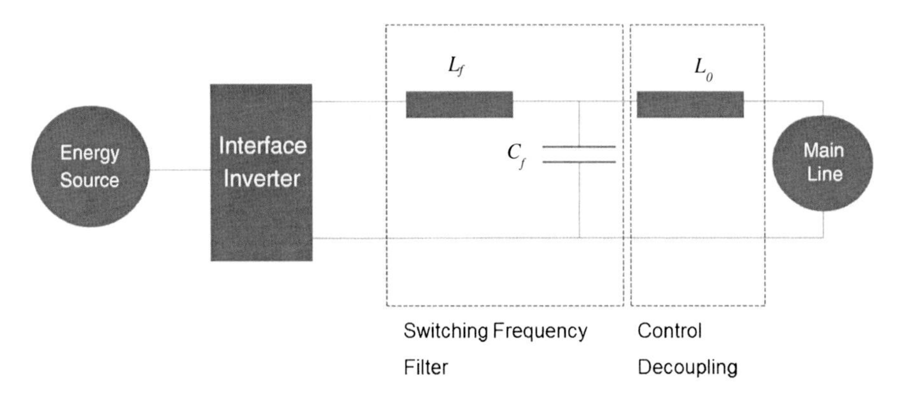

Fig. 23 Interface between each distributed resource and the main line

7.3 Control of Distributed Generation

Distribution of monitoring and control functions in general [27], and control of distributed power generation (DPG) in particular, has been investigated widely regarding stability, grid connection behavior, voltage control and power flow, as for example in [28, 29]. In particular, distributed control methods for parallel power inverters have been investigated in [30–32] where the conventional PI controller is used. Advanced control methodologies, in particular decentralized optimal state feedback [33] and distributed model predictive control (MPC) in [34, 35] have also been proposed. The following section discusses in particular the linear quadratic Gaussian control, based on the theory of optimal linear quadratic regulator (LQR), in its decentralized version.

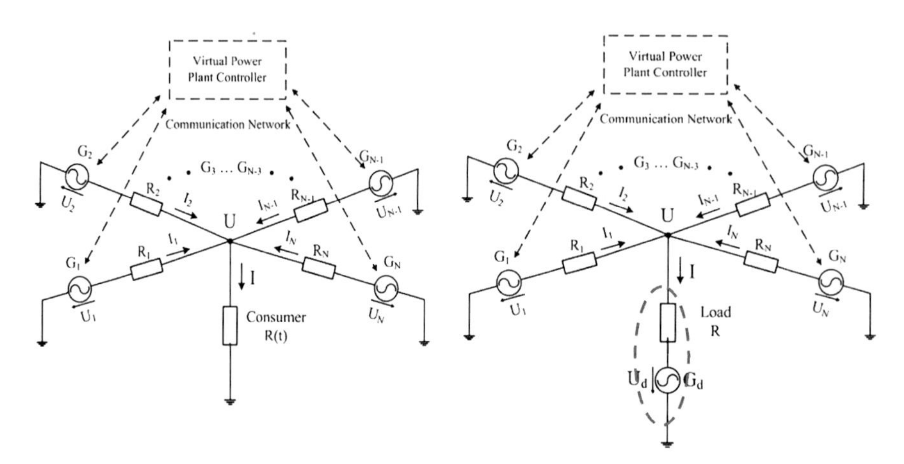

Fig. 24 Simplified model of a DPG network [38]

7.4 *Distributed Optimal Control*

7.4.1 Decentralized Control Based on Incomplete System Knowledge: Local LQG + Local on-Line Learning of $U_d(t)$ + Local Set-Point Adaptation

Although at present decentralized/distributed controllers are assumed to operate independently, their design is based on a global model of the overall system. In [36] a decentralized linear quadratic Gaussian (LQG) control approach is investigated for droop control of parallel converters through emulation of a Finite-Output Impedance Voltage Source. The design of the decentralized controllers presented in such work is based on local models of the inverters, without communication, and response to disturbance, such as load variation, is not considered. Similarly, in [37] a decentralized LQG control approach is designed, based on local models of a simplified DPG network with incomplete system knowledge. In this approach the local controllers operate individually without knowing and communicating with each other. The control objective is to reject the load variation, modeled as a step disturbance arising in the network. In line with this, each local controller adopts online learning of the disturbance and adapts its own set-point individually. Rejection of the disturbance is achieved after a short transient period, which is also corrected for by the controllers automatically.

The model of a DPG network [38] is shown in Fig. 24 with N linear electrical DC power generators $G_1,\ldots,G_N$. The constant load R can be interpreted as the average of many consumers. The variation of the load is modeled by the generator G_d that produces a time varying voltage U_d.

The power lines are modeled as resistances $R_1, \ldots, R_N$. Due to Kirchhoff's laws, the voltages and currents of all generators are coupled. As opposed to conventional decentralized control synthesis, in which the network is fully known,

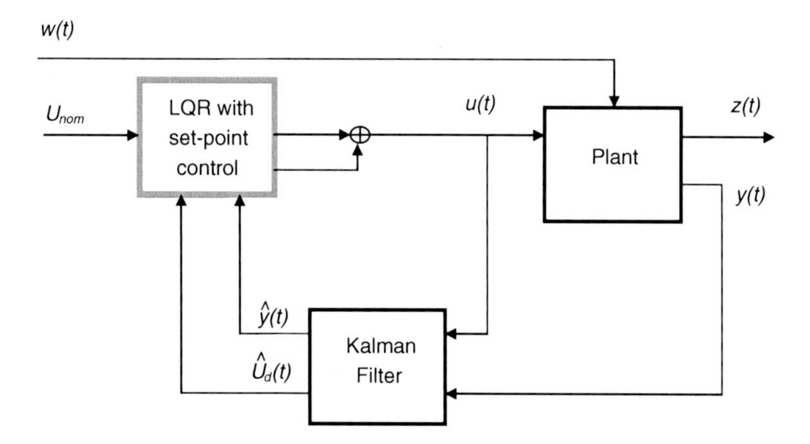

Fig. 25 Structure of the local LQG controller of an individual generator

it is assumed here that each generator is unaware of the presence of any other generator in the network, and it operates as the only generator providing power. In this case, the behavior of each generator is regulated by a local controller based on a local model with incomplete knowledge of the network. The local model from the viewpoint of generator G_i only consists of the elements that are depicted and connected by solid lines, while the rest of the real power grid is invisible to G_i.

The local controller of each generator implements the linear quadratic Gaussian control method, designed with the control objective to reject the influence of the disturbance voltage U_d on the system voltage U, to be kept equal to the constant nominal reference voltage U_{nom}. Each local LQG controller measures the current value of U instantaneously and pursues the objective individually. Thus, each local LQG controller aims at minimizing the squared output error voltage (U-Unom) over the infinite time horizon in the presence of disturbance and measurement noise. Since we model the disturbance *Ud* as a step signal representing on-off switching of loads, the set-points of the LQG controllers have to be adapted. For the set-point adaptation, each local controller estimates the disturbance by regarding it as an additional state in the system dynamics. The structure of the local LQG controller of an individual generator is shown in Fig. 25. Notice that the local model differs from the full model of the power network, which also implies an inherent modeling error of the local model. Hence, the estimation of the disturbance does not lead to the estimation of the actual disturbance U_d, but of a fictitious time-varying disturbance resulting from U_d and from the unknown dynamics of the other generators and their controllers. Nevertheless, it can be shown that this process is fast and it leads to the stable behavior of the system.

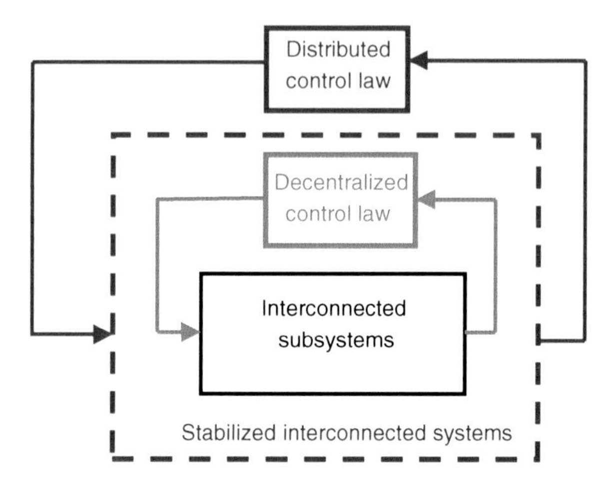

Fig. 26 Two-layer architecture for small signal stabilization

8 Wide Area Monitoring and Control

The small-signal stability of the power system is an important requirement for power system operation. Low-frequency oscillations may possibly cause blackouts or limitations in the power flows. According to the classical control approach, local generator controllers are expected to intervene through the Automatic Voltage Regulator (AVR) and the Power System Stabilizer (PSS). The concept of the Wide Area Measurement and Control System (WAMCS) is to collect accurate information about power system dynamics to support efficient control strategy and enhance the small-signal stability. These information, for example from synchronized phasor measurements or remote signal and information sources are enabled by a wide area communication network.

This points at how the communication network represents an important design parameter, and how it affects the control performance. In fact, adding links may degrade performance [39].

A joint design of controller-communication topology for distributed damping control may be performed and realized via a two-layer architecture for robustness against communication failures, comprising decentralized and distributed control, as shown in Fig. 26. In a first step, the local feedback control law is designed for the i-th generator in order to stabilize the overall system only with a local controller and to provide the minimal damping performance. In a second step, the distributed control law that operates on top of the decentralized control for the i-th generator, is designed to enhance the damping performance in terms of increasing the decay rate of the system. In this step, communication is accounted for as a cost of available communication links and a cost of new communication links, yielding to a mixed integer optimization problem.

In conclusion, this control architecture achieves:

- Stability, guaranteed by decentralized control
- Damping performance, improved by distributed control
- Robustness to permanent link loss

9 Conclusions

This Chapter presents the evolution of electrical power systems towards a complex system, deeply coupled with other critical infrastructures. New monitoring and control challenges, the consequence of the penetration of renewable sources, and the active behavior of most power system actors are reviewed, together with the basics of current control strategies for power systems.

Some examples of the technologies capable of addressing these challenges are described. Among the underlying commonalities, we identified the distribution of functions and the need to coordinate distributed resources. These features emphasize the dependence on the communication network. Furthermore, the utilization of volatile energy sources require the exploitation of all available flexibility, and the active participation of distributed resources starting from the renewable sources themselves.

References

1. Amin, M.: Automation, control, and complexity: An integrated approach. In: Samad & Weyrauch (eds.) Wiley, pp. 263–286 (2000)
2. Monti, A., Ponci, F.: The Complexity of Smart Grid. IEEE Smart Grid e-Newsletter, May 2012
3. Future Internet for Smart Energy (Finseny): Future Internet PPP, FP7. http://www.fi-ppp-finseny.eu/
4. Future Internet Smart Utility Services (Finesce): Future Internet PPP, FP7. http://www.finesce.eu/
5. Ericsen, T.: The second electronic revolution (It́s all about control). IEEE Trans. Ind. Appl. **46**(5), 1778–1786 (2010)
6. Molitor, C., Benigni, A., Helmedag, A., Chen, K., Cali, D., Jahangiri, P., Muller, D., Monti, A.: Multi-physics test bed for renewable energy systems in smart homes. IEEE Trans. Ind. Electron. **60**(3), 1235–1248 (2013)
7. Karlsson, D., Hemmingsson, M., Lindahl, S.: Wide area system monitoring and control—terminology, phenomena, and solution implementation strategies. IEEE Power Energy Mag. **2**(5), 68–76 (2004)
8. Atanackovic, D., Clapauch, J.H., Dwernychuk, G., Gurney, J., Lee, H.: First steps to wide area control. IEEE Power Energy Mag. **6**(1), 61–68 (2008)
9. Chakrabarti, S., Kyriakides, E., Bi, T., Cai, D., Terzija, V.: Measurements get together. IEEE Power Energy Mag. **7**(1), 41–49 (2009)
10. Data provided by ENTSO-E. http://www.entsoe.eu

11. Fraunhofer Institute for solar energy systems ISE – Electricity production from solar and wind in Germany. http://www.ise.fraunhofer.de/de/downloads/pdf-files/aktuelles/stromproduktion-aus-solar-und-windenergie-2012.pdf (2013)
12. http://www.iec.ch/smartgrid/background/explained.htm
13. http://energy.gov/oe/downloads/smart-grid-introduction-0
14. http://ec.europa.eu/energy/gas_electricity/smartgrids/doc/expert_group1.pdf
15. NIST Special Publication 1108 NIST Framework and Roadmap for Smart Grid Interoperability Standards, Release 1.0. http://www.nist.gov/public_affairs/releases/upload/smartgrid_interoperability_final.pdf. Accessed Jan 2010
16. http://ec.europa.eu/energy/gas_electricity/smartgrids/doc/2011_03_01_mandate_m490_en.pdf
17. ftp://ftp.cencenelec.eu/CENELEC/Smartgrid/SmartGridFinalReport.pdf
18. Della Giustina, D., Pau, M., Pegoraro, P.A., Ponci, F., Sulis, S.: Distribution system state estimation: Measurement issues and challenges. IEEE Instrum. Meas. Mag. (2014)
19. Machowski, J., Bialek, J.W., Bumby J.R.: Power System Dynamics: Stability and Control. Wiley, New York (2008)
20. Continental Europe Operation Handbook, © ENTSO-E (2014)
21. Abur, A., Exposito, A.G.: Power System State Estimation, Theory and Implementation. Marcel Dekker, Inc., New York (2004)
22. IEEE Standard for Synchrophasor Measurements for Power Systems (IEEE Std. 37.118.1-2011) and IEEE Standard for Synchrophasor Data Transfer for Power Systems (IEEE Std. 37.118.2-2011)
23. IEEE Guide for Synchronization, Calibration, Testing, and Installation of Phasor Measurement Units (PMUs) for Power System Protection and Control (IEEE Std. C37.242-2013)
24. Lira, R., Mycock, C., Wilson D., Kang, H.: PMU performance requirements and validation for closed loop applications. In 2nd IEEE PES International Conference and Exhibition on Innovative Smart Grid Technologies (ISGT Europe), pp. 1–7, Manchester, UK (2011)
25. Junqi, L., Benigni, A., Obradovic, D., Hirche, S., Monti, A.: State estimation and branch current learning using independent local Kalman filter with virtual disturbance model. IEEE Trans. Instrum. Meas. **60**(9), 3026–3034 (2011)
26. Baran, M.E., El-Markabi, I.M.: A multi-agent based dispatching scheme for distributed generators for voltage support of distribution feeders. IEEE Trans. Power Syst. **22**(1), 52–59 (2007)
27. Monti, A., Ponci, F., Benigni, A., Liu, J.: Distributed intelligence for smart grid control. In: 2010 International School on Nonsinusoidal Currents and Compensation (ISNCC), pp. 46–58. Lagow, Poland (2010)
28. Marwali, M.N., Keyhani, A.: Control of distributed generation systems—part I: Voltages and currents control. IEEE Trans. Ind. Electron. **19**(6), 1541–1550 (2004)
29. Blaabjerg, F., Teodorescu, R., Liserre, M., Timbus, A.V.: Overview of control and grid synchronization for distributed power generation systems. IEEE Trans. Ind. Electron. **53**(5), 1398–1409 (2006)
30. Macken, K.J.P., Vanthournout, K., Van den Keybus, J., Deconinck, G., Belmans, R.J.M.: Distributed control of renewable generation units with integrated active filter. IEEE Trans. Power Electron. **19**(5), 1353–1360 (2004)
31. Tuladhar, A., Hua, J., Unger, T., Mauch, K.: Control of parallel inverters in distributed AC power systems with consideration of line impedance effect. IEEE Trans. Ind. Appl. **36**(1), 131–138 (2000)
32. Karlsson, P., Svensson, J.: DC bus voltage control for a distributed power system. IEEE Trans. Power Electron. **18**(6), 1405–1412 (2003)
33. Xie, S., Xie, L., Wang, Y., Guo, G.: Decentralized control of multimachine power systems with guaranteed performance. IEE Proc. Control Theory Appl. **147**(3), 355–365 (2000)
34. Venkat, A.N., Hiskens, I.A., Rawlings, J.B., Wright, S.J.: Distributed MPC strategies with application to power system automatic generation control. IEEE Trans. Control Syst. Technol. **16**(6), 1192–1206 (2008)

35. Negenborn, R.R.: Multi-agent model predictive control with applications to Power networks. Doctoral Dissertation, TU Delft (2007)
36. De Brabandere, K., Bolsens, B., Van den Keybus, J., Woyte, A., Driesen, J., Belmans, R.: A voltage and frequency droop control method for parallel inverters. IEEE Trans. Power Electron. **22**(4), 1107–1115 (2007)
37. Liu, J., Obadovic, D., Monti, A.: Decentralized LQG control with online set-point adaptation for parallel power converter systems. In: IEEE Energy Conversion Congress and Exposition (ECCE 2010), pp. 3174–3179. Atlanta, GA, USA (2010)
38. Liu, J.: Cooperative control of distributed power grids using a multi-agent systems approach. Master thesis, Technische Universität München (2009)
39. Gusrialdi, A., Hirche, S.: Performance-oriented communication topology design for large-scale interconnected systems. In: 49th IEEE Conference on Decision and Control (CDC), Atlanta, GA, USA, pp. 5707–5713 (2010)

Telecommunication Networks

Rasmus L. Olsen, Kartheepan Balachandran, Sara Hald, Jose Gutierrez Lopez, Jens Myrup Pedersen and Matija Stevanovic

Abstract In this chapter, we look into the role of telecommunication networks and their capability of supporting critical infrastructure systems and applications. The focus is on smart grids as the key driving example, bearing in mind that other such systems do exist, e.g., water management, traffic control, etc. First, the role of basic communication is examined with a focus on critical infrastructures. We look at heterogenic networks and standards for smart grids, to give some insight into what has been done to ensure inter-operability in this direction. We then go to the physical network, and look at the deployment of the physical layout of the communication network and the related costs. This is an important aspect as one option to use existing networks is to deploy dedicated networks. Following this, we look at some generic models that describe reliability for accessing dynamic information. This part illustrates how protocols can be reconfigured to fulfil reliability requirements, as an important part of providing reliable data access to the critical applications running over the network. Thereafter, we take a look at the security of the network, by looking at a framework that describes the digital threats to the critical infrastructure. Finally, before our conclusions and outlook, we give a brief overview of some key activities in the field and what research directions are currently investigated.

Keywords Communication networks · Smart grid · Inter-operability · Dynamic information access · Reliability · Availability · Cyber security

---

R.L. Olsen (✉) · K. Balachandran · S. Hald · J.G. Lopez · J.M. Pedersen · M. Stevanovic
The Faculty of Engineering and Science, Department of Electronic Systems,
Aalborg University, Fredrik Bajers Vej 7, Room A4-212, 9220 Aalborg, Denmark
e-mail: rlo@es.aau.dk

E. Kyriakides and M. Polycarpou (eds.), *Intelligent Monitoring, Control, and Security of Critical Infrastructure Systems*, Studies in Computational Intelligence 565,
DOI 10.1007/978-3-662-44160-2_3

1 Introduction

Communication networks have become an essential part of our everyday lives and currently, our society is highly dependent on the proper functioning of communication networks. In connection with our private lives, the number of services that are being delivered over these networks is progressively increasing, and probably in the future, new services will appear, all converging over the same infrastructure [1].

Regarding professional aspects, communication networks play a key role in efficiently developing economical activities in a fast, secure, and reliable way. In addition, it is possible to find very powerful companies in the world having Internet services and applications as their main activity, such as Google or Facebook, and they are continuously expanding. In relation to critical applications, the question is whether the existing communication infrastructure can provide the necessary reliability, or whether new networks that allow the harsh requirements of such applications will be needed?

2 The Role of Telecommunication

In the early days of telecommunication, communication between two entities was ensured through a physical communication channel between the two entities by so-called circuit switched networks [2]. This type of network ensures a guaranteed bandwidth between the entities, since once setup there are no interferences, and is kept available during the communication. However, setup time is required to ensure the connection is established. Besides the setup time, it should also be fairly easy to imagine the great limitations of these types of networks when considering millions and millions of connected end devices that need communication.

The introduction of packet switched networks allowed the sharing of the limited physical medium among several communicating entities [2], and allowed for a much more flexible communication. The road from the first initial baby steps of a few machines connected over some copper wires done in the late 1960s as a response to the cold war, up to today's full scale hyper complex networks of networks is a study worth in itself [3]. Today, end users have grown accustomed to have access to communication networks more or less everywhere and whenever needed, leaving them with a perception that the Internet is something that just is. This impression has not only been driven by the fact that wired communication offers very high speeds to the individual, but also mobile communication today offers a high data rate due to the technological development [4]. Now, the next step has come, that we want to use the existing communication infrastructure to mission critical applications. One thing is to say "my Internet works nicely at home, and at work it also works nice", but another thing is to put so much trust into the network that we allow critical systems such as water, electricity management, or eHealth applications to run on top of these networks of networks.

2.1 Basic Communication Between Two Entities

Critical infrastructures are to a large extent also distributed by nature. In such setting, communication is often required to be able to have elements in the system interact in a synchronized and organized matter. Subsystem A may need to react upon events that happen in subsystem B or vice versa. At the same time, critical infrastructures are often characterized by deadlines due to their connection to the physical world and the properties of the physical world which a critical infrastructure interacts with. Figure 1 shows the example of two communicating physical devices located at different geographical and network locations. These two entities could for example be a water management system that has to interact with remote control units located at strategic points along water pipes.

The data that is required to be passed from A to B and vice versa has to undergo a long way through several routers due to the packet switched approach we have today. At each passing point (router), the received data needs inspection to decide which router the packet should go to next. This may not always be the same for all packets even though their sources and destinations are the same as clever traffic load balance algorithms may be applied to avoid congestions among routers. A data packet route example is illustrated in Fig. 1 as data going up and down in the different levels of the OSI model [2]. In some parts, packets need to be addressed at a network level, while in others only a link level is required. But it is clear that each hop takes some time for the routers to process packets. This leads to an end-to-end delay even though we experience the communication as a direct communication between A and B. The end-to-end delay depends on many factors, such as the quality of each link between routers, network traffic conditions (there may be bottlenecks between some routers), the route through which the data packets are

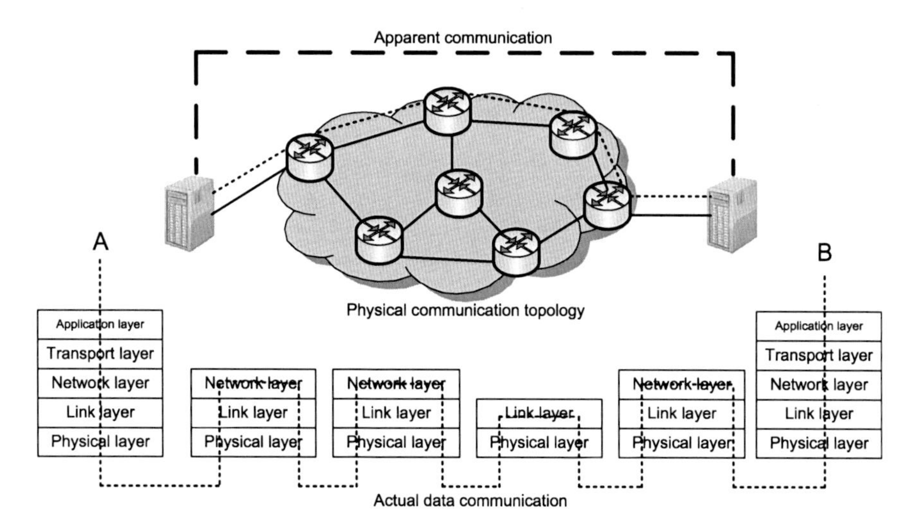

Fig. 1 Communication between two entities [5]

being sent (this is not controlled by the application, but is impacted by the decisions made by the network). These factors also mean that the end-to-end delay usually shows a highly stochastic behaviour, rather than a desirable deterministic behaviour, and even changes over time, e.g., in some parts of the network there is more traffic during work hours than in night time or weekends.

This complexity of intercommunication needed to transport data from A to B illustrates some key problems that critical applications have when dealing with communication over networks today. There is no or very little control of the data streams going between A and B. When routers receive a significant amount of traffic, they may start to drop incoming packets. Some transport protocols, as TCP aim at providing reliable transport by ensuring retransmission of missing packets, but at the cost of end-to-end delay because it first needs to detect missing packets, and then ask for retransmission. Others such as UDP offer to send data with crossed fingers that it appears at the receiving side. In that case, the application must be able to tolerate packet losses.

Therefore, some of the key challenges that communication systems face, and even considering only two entities communicating, to support critical applications are not necessarily only classical data throughput, but definitely also latency and reliable communication. These key requirements are often assumed, because they are to a large extent hidden to the everyday user, but as communication developers we need to take these issues seriously if critical infrastructures shall be supported by (existing) communication technologies.

2.2 *Communication over Heterogeneous Networks*

As a further complication, networks are heterogeneous and full of new and legacy systems. Figure 2 shows a conceptual example of how different devices and applications may be connected via a large set of networks of networks. The heterogeneity covers some challenges that may limit some use cases as the following example illustrates:

In order to communicate between two entities an addressing scheme is required (a basic requirement for any type of communication). The most prominent addressing scheme today is IP addressing. In principle, all addresses should be uniquely defined, in order for packets to be sent to the right destination at all times. When the Internet Protocol version 4 was designed and implemented, the space allocated for the address in the protocol allowed for 'only' approximately 4.29 billion unique devices. At the time of development, that was enough, but with the current development this amount has shown to be too small, since all sorts of sensors, mobile devices, multiple interface devices, etc., has ultimately led to an address starvation. This has not been acceptable, so several solutions have been invented to overcome this address starvation, such as Network Address Translation, use of private network addresses, tighter control of Internet registries, network renumbering of existing networks, etc. Even a new IP version 6 has been developed

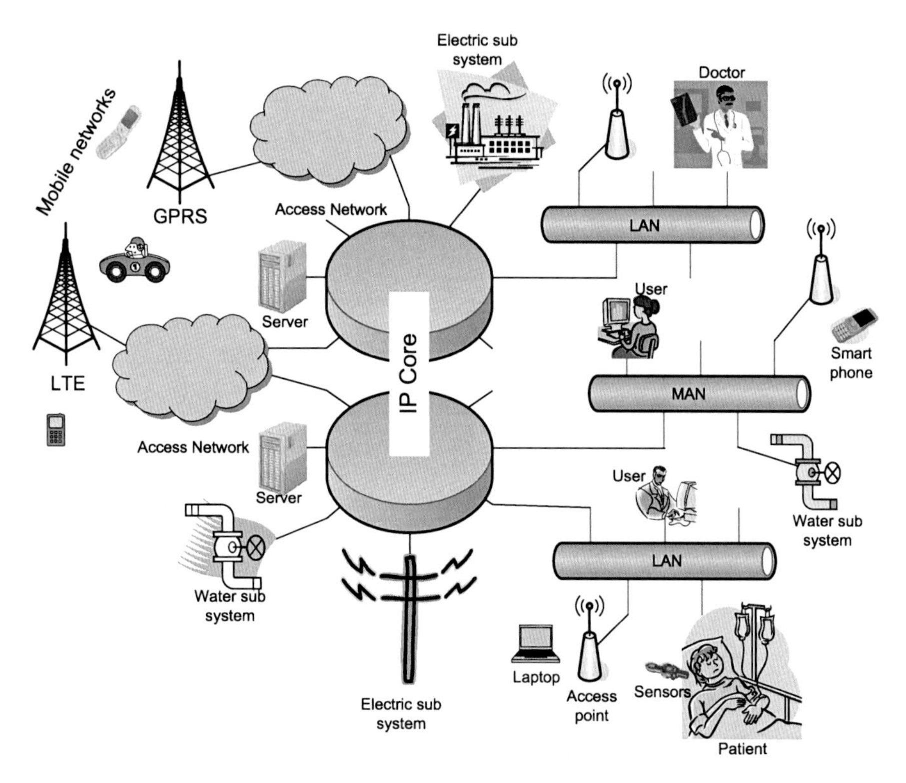

Fig. 2 Simplified example of networks and connected applications

with a much larger address capacity, but still the earlier IP is being used, since so many devices depend on the stability of this protocol's address space, that in fact no one knows exactly what happens if suddenly software started to use IPv6 [6].

The IP address issue is just one of many examples of the complexity of networks of networks shown in Fig. 2 and how networks even today are challenged by more or less invisible problems. Therefore, once again, one could ask the following question: do we trust these networks to serve as communication media for mission critical applications? Do we dare to let this patchwork of networks, software patches, numerous of standards, protocols and configurations be the bearing part of critical, life depending elements such as electricity, water and health in our everyday life?

Referring to Fig. 2, the vision of several critical applications which interact with the network is shown:

- Doctors who have online meetings with patients, eHealth [7]
- Water detection and control, [8]
- Electric power control, smart grids [9].

On top of this, there will surely be other applications using the network, such as web browsing, video streams, emails, online games, etc. The question is, why should these critical applications use the same network? For example, cars and air planes have several dedicated internal communication networks so as not to mix the application traffic with real time critical data traffic. However, for simplification and cost reduction reasons, research is ongoing on how merge the different data traffic [10].

The time, effort and cost of deploying the Internet as it is today is immense. Deploying a separate network for each of the applications that is envisioned does not appear to be a very attractive solution if we somehow are able to ensure the existing infrastructure can support the requirements of the critical applications. A major key to the solution is the flexibility of the network as it is today. As a corner stone design idea of the early Internet (from ARPANET [3]), the idea of robustness to link failures has been eminent, which is a key feature for critical infrastructure as well as not having to deploy net networks from scratch. However the complexity and heterogenic nature of communication due to legacy systems requires interoperability and standards. In the following we take a closer look at how this is being addressed in the smart grid as an example.

2.3 Communication Standards for Smart Grids

In smart grids, data from consumers and potentially also from the power grid infrastructure is required to be collected for control purposes. This not only means communication over heterogeneous networks, but also with a wide range of communicating entities, smart meters for example, produced by different vendors. Today, the core purpose of the meters already developed is to monitor power usage for e.g., electricity bills [11]. However, the data also have value for the utility company.

In smart grid terminology, being able to remotely read the meter is called automatic meter reading (AMR) and is a part of the automatic metering infrastructure (AMI), i.e., the network between the smart meter and the utility company [12]. With an AMI network the monitored data from a household can be sent to the utility company, which can use it to optimise their production of power and thereby avoiding over/underproduction.

AMI has been addressed in various projects like Power Matching City [13] and others [14, 15] and it has been shown, that using the data from monitoring systems can lead to optimisation of the demand and response (DR) [16, 17]. However, it is not clear what are the requirements for the communication network. For example, the authors in [13] have made a living lab to demonstrate a smart grid. In the chosen setup, dedicated ADSL communication lines are used for AMI in order to avoid human interference. The purpose of this is to make sure that they have enough bandwidth for the communication. Nevertheless, the requirements for the networks are not mentioned. Implementing dedicated communication lines only for the smart grid communication is very expensive and not a feasible solution for all cases of AMI.

Furthermore, the data can be used by the utility company to act as an energy consultant, to advice the customers how to save money based on their power usage data and additionally offer free or cheaper electricity during overproduction periods to encourage their customers to use power consuming devices, for example heat pumps, or allow the utility company to perform Direct Load Control (DLC) by controlling the customers devices and Distributed Energy Resources (DER) [18]. Being able to control e.g., heat pumps, electric cars and other DER's, makes it possible for the utility company to control and level the peak periods in the power grid by remotely turning on/off specific units in order to take off load or supply more power to the grid appropriately [16].

In Denmark, the Danish Ministry of Climate, Energy and Building has published a report in October (2011) about Smart Grid in Denmark [70]. In the report they encourage further research and development in smart grid and also mention communication as a vital part of the smart grid.

2.4 *Standards for Smart Grids*

Mapping requirements for critical infrastructures is a challenging task, because the smart grid is a large complex infrastructure with different layers of networks, which gives a diversified communication performance expectation [18]. IEEE and IEC have already proposed a number of standards regarding the communication in smart grid in different layers. One of the most commonly used standard is IEC 61850, which focuses on the substation automated control and is used in the Danish smart grid project in Bornholm called ECOGRID [71] and for the AMI there is IEC 62056 [19, 20]. In the following some communication standards are mentioned which are proposed for different parts of the smart grid.

IEC 61968-9 and 61970: Defines the common information model for data exchange between devices and networks in the power distribution domain and the power transmission domain respectively. They are used in: Energy management system [21].

IEC 60870-6: Defines the data exchange model between the control center and power pools. It is used in: Inter-control center communication [21].

IEC 61850: Defines the communication between devices in transmission, distribution and substation automation system. It is used in: Substation automation [21].

IEEE P2030: Defines the inter-operability of energy devices and IT operation with electric power systems. It is used in: Customer side application [21].

IEEE 1646: Defines the communication delivery times to substation. It discusses the requirements of the system to deliver real-time support, message priority, data criticality and system interfaces. It is used in: Substation automation [21].

Challenges for the smart grid network: Other challenges in the communication network in smart grids, relates to delay, availability and security [12, 15, 18].

If the utility company is expected to control the distributed energy resources (DER), delay becomes a critical metric in the network performance. The delay in the network will have a high impact on the grid if the DERs are not activated or shutdown on time; thus, some sort of message priority scheme in the communication protocol will be required [18, 22]. Availability and reliability of the network are of importance to ensure the grid operation and also plays a vital role in the demand response [23, 24]. Security is crucial to the smart grid. When extracting information from the user, the privacy of the user has to be secured. The grid can also be vulnerable to terrorist attacks if the control messages to control various electric devices are hijacked [25]. Adding security can have an impact on the delay, as the messages have to be encrypted. Another way to add security is by not letting the hacker know where these control messages come from or go to. For this, an anonymous packet routing with minimum latency has been proposed [26].

There are a number of challenges in the communication network for smart grid. The standards proposed still have to be implemented and tested in many scenarios, in order to examine the network performance. There is still a need for research in protocols that can deliver messages safely according to the time constraints specified by the different standards.

3 Design of Critical Optical Transport Infrastructure

Communication networks have become an essential part of our everyday lives and our society is highly dependent on the proper functioning of communication networks. In connection with our private lives, the number of services that are being delivered over the data networks is progressively increasing, and probably in the future new services will appear, all converging over the same infrastructure.

Regarding professional aspects, networks play a key role in efficiently developing economical activities in a fast, secure, and reliable way. In addition, it is possible to find very powerful companies in the world having Internet services and applications as their main activity, such as Google or Facebook, and they are continuously expanding.

Telecommunication systems have been evolving for the past 10 years, towards the unification of services and applications over the same infrastructure [27]. Currently, it is possible to identify how this initiative is partially followed by operators providing voice, data and video transmissions over the same access connection, known as Triple Play. However, the distribution networks of these services are not unified physically. For example, the TV and telephony infrastructures are traditionally separated. The unification of these infrastructures implies ambitious requirements to be supported by the network, for example due to the heterogeneity of the traffic flowing through, or the significant profit losses, or the number of affected users when loss of connectivity occurs [28]. Examples of recent cable cuts in optical networks are:

- In the beginning of 2008, the cable connecting Europe and Middle East suffered four single cuts, affecting millions of users [29].
- In April 2009 AT&T suffered cable cuts, perhaps due to vandalism, in the area of San Jose and Santa Clara, California, leaving many of Silicon Valley businesses and customers without phone and data services [30].
- In July 2011 35,000 broadband customers were affected for several hours by a cable cut caused by a truck in Washington State, USA [31].

Looking back it is possible to realize how fast the world of communications has evolved. For example, in 2009 the 40th anniversary of the first ever data transmission over ARPANET was celebrated. This fast evolution of communication technologies and services is causing a gradual increment of the bandwidth and reliability requirements to be fulfilled by the network infrastructure [32].

In addition, the traffic supported by the Internet has significantly grown over the last few years, shaping up the requirements that future networks need to handle. In order to be able to support all the traffic and quality of service demands, there is a need for high performance transport systems, and focusing on the specific field of this work, **a highly reliable optical backbone infrastructure** [33].

The interest is especially increasing regarding optical transport networks, where huge amounts of traffic are continuously traversing their links. Inefficiencies or disruptions on delivering the information become critical at this level, due to the number of simultaneously affected users. All these huge flows of information must be efficiently distributed using reliable high capacity transport networks. Bandwidth requirements clearly indicate that the optical network technology based on wavelength division multiplexing (WDM) will play a key role regarding this issue [34].

4 Planning the Physical Infrastructure

The deployment of optical networks is a long and expensive process, due to the trenching tasks involved, especially for large geographical areas such as national or continental territories. It can take 10–15 years to deploy such networks under the cost of millions of euros [35]. Moreover, the lifetime of the physical infrastructure is rather long, between 30–50 years [36, 37]. These features make such a network deployment a long term investment project, which should be carefully planned. In addition, when this high investment is combined with a **reliable**, **preventive** and **green** planning, a better outcome can be achieved [35].

Optical network systems are very complex; each of the network layers can have great impact on the overall performance, going from physical to application layers. In relation to backbone networks, the infrastructure can be limiting the global performance of a network just by the fact that the physical interconnection scheme has not been carefully planned [28]. In fact, when the network requires some kind of physical upgrade, the solution's costs in economic terms might be much more significant, due to trenching and deployment tasks, than at higher layers where software updates might be enough to solve the problem [38].

Hence, networks should be planned and designed to provide high performance and high availability in transmissions. As the economic relevance of these networks is increasing, it is also feasible to increase the investment for their deployment. This capital increment opens up a whole new space of possibilities when designing the network's interconnection. Planning is crucial and even small improvements may have high economic impact. However, no well documented methods exist for the whole interconnection planning process leaving room for the development of this work.

The main overall challenge can be described as the physical interconnection decision problem of a set of nodes. This interconnection can be configured in many different ways, and several of these might be "optimal", depending on the objective function and constraints of the optimization. In this case, two of the most relevant objective functions are: *Deployment cost minimization* and *average connection availability maximization.*

The main problem is how to cover these two aspects in the same optimization process, as they are contradictory. Usually, the minimization of deployment costs would imply a negative effect on the availability of the designed networks. The number of deployed links to interconnect the nodes is compromised, in order to achieve the minimization goal. This can be avoided by conveniently selecting the proper constraints in the search process. The feasible solutions spectrum can be reduced to graphs that a priori will provide good solutions in terms of availability and not be extremely costly. Three-connected graphs are a good possibility to implement such transport networks. This type of graphs is discussed below, but first the models regarding deployment costs and availability should be introduced.

Definition 1 *A graph is k-connected when any* $k-1$ *elements, nodes or links, can be removed from the network and still maintain a connected graph.*

4.1 Deployment Cost

There are three main contributing elements regarding the economic costs for deploying optical transport networks using WDM technology: trenching, nodes, and fiber spans. These are considered to define the following deployment cost model used in this work but new elements may be included at a later stage. Some of these concepts can be found in [39] and a complete review on the architecture of optical networks can be found in [34].

Let $I_{NT} = I_{links} + I_{nodes}$ be the total cost of deploying a network. I_{links} is the cost of deploying the links and I_{nodes} is the cost of deploying the nodes. Each existing link is characterized by its length, Lm_{ij}, and the traffic traversing it, Ld_{ij}. Not all pairs $i-j$ have an existing link. Basically, three cost parameters can be defined to calculate I_{links}: I_{trench}, I_{fix}, and I_{span}.

I_{trench} corresponds to the price for the trenching tasks per km and its value can significantly vary, depending on the region or landscape. I_{fix} corresponds to the fiber terminating equipment, and I_{span} is the cost related to each fiber span where ducts, fiber and amplifier costs are included. The length of each span L_{span} can vary from 50 to 100 km [36].

Therefore, CW being the wavelength (λ) capacity and W being the number of $\lambda's$ per fiber, the number of fibers nf_{ij} for the link between nodes i and j is defined in Eq. (1). The number of amplifiers, nla_{ij} in a link is defined in Eq. (2). Consequently, the economic costs for deploying the links of a network Top, I_{links}, is formally defined in Eq. (3).

$$nf_{ij} = \left\lceil \frac{Ld_{ij}}{CW \cdot W} \right\rceil \tag{1}$$

$$nla_{ij} = \left\lfloor \frac{Lm_{ij}}{L_{span}} \right\rfloor \tag{2}$$

$$I_{links} = \sum_{\forall\, i,j \in S_N} \left(I_{trench} \cdot Lm_{ij} + 2 \cdot nf_{ij} \cdot I_{fix} + \left(\left\lfloor \frac{TRF_{ij}}{L_{span}} \right\rfloor + 1 \right) \cdot I_{span} \right) \tag{3}$$

Regarding the nodes, I_{nodes} can be divided in two parts; the facility cost, I_{fal}, and the switching cost, I_{swch}, related to each switch size. The switch size is given by the number of incoming and outgoing fibers to each node and the number of wavelengths per fiber. Usually standard switch sizes are given by $2^m x 2^m$ for $0 < m \leq 5$, and I_{swch} is not linearly proportional to m [40]. Concluding, I_{nodes} is formally defined as Eq. (4).

$$I_{nodes} = N \cdot I_{fal} \sum_{\forall\, i \in S_N} I_{swch}(i) \cdot W \tag{4}$$

4.2 Availability

Availability is a convenient parameter for measuring the efficiency of the network infrastructure regarding failure support. Availability is defined in [36] as follows: "*Availability is the probability of the system being found in the operating state at some time t in the future, given that the system started in the operating state at time t = 0. Failures and down states occur, but maintenance or repair actions always return the system to an operating state*".

Availability in optical networks has been widely studied. For example, [41] presents an interesting availability review of Wavelength-Division Multiplexing (WDM) network components and systems. In terms of connection availability analysis, [42] compares the availability results between simulation and analytical environments for a shared protection scenario. Also, in [43] an interesting approach is followed to evaluate the availability in optical transport networks. It is measured in expected traffic losses when failures occur.

Availability is mathematically defined as the *working time/total time* ratio. Let *MTTF* be the Mean Time To Fail of any element or system and *MFT* be the Mean Failure Time. Availability Av is presented in Eq. (5), resulting in a numerical value of $0 \leq Av \leq 1$.

$$Av = \frac{MTTF}{MTTF + MFT} \tag{5}$$

The availability in relation to an $s - d$ pair connection is given by the availability calculation of sets of elements in series and in parallel. For each provided disjoint path to be available, all of its elements must be available. For a connection to be available, at least one path must be available.

Let $Av_{pj(s,d)}$ be the availability of each j of the provided k disjoint paths between s and d. m_j is the number of elements of path $p_j(s,d)$, each of these characterized by an availability Av_i. Equation (6) presents the calculation of the availability for each path. The connection availability for an $s - d$ pair, $Av_{C(s,d)}$, is presented in Eq. (7).

$$Av_{pj(s,d)} = \prod_{i=1}^{m_j} Av_i \tag{6}$$

$$Av_{C(s,d)} = 1 - \prod_{j=1}^{k} (1 - Av_{pj}) \tag{7}$$

4.3 The Graphs

The decision of how to deploy optical links to interconnect network nodes is a complex problem. Its combinatorial nature makes it impractical to make the interconnection decision using exhaustive search methods. Heuristics may provide a good approximation to optimal results while Integer Linear Programming would be the ultimate approach.

In relation to the distribution of the links, currently it is widely accepted that these ring interconnections are reliable enough for the demands of the users. However, if the current evolution of telecommunication demands keeps heading towards a more IT dependent society, higher degree physical networks (>2) can significantly contribute to the improvement of failure support, congestion control, or delay propagation aspects due to multipath options. For example, in the long term, it might be cheaper to build more reliable networks than less reliable networks that require higher maintenance investment to keep similar availability levels. Therefore, 3-connected graphs could be the natural evolution for this type of networks [44].

Consequently, networks formed by 3-connected graphs are capable of supporting two simultaneous failures, links or nodes, and still maintain connectivity between any pair of nodes.

4.4 Illustrative Example

The **main goal** of this example is to identify the deployment cost versus availability consequences of using 3-regular, 2- and 3-connected graphs to interconnect several sets of nodes. The number of links in all options is the same but their different distribution of the links implies different performance characteristics. In order to obtain concrete numerical results, three scenarios are presented. These consist of sets of 16 nodes in Europe, US, and Asia to be interconnected.

This experiment consists of designing the interconnection for these sets of nodes following these topologies: Single Ring, *SR*; 3-regular 2-connected, $D3_{2C}$; and 3-regular 3-connected, $D3_{3C}$. The *SR* is the shortest 2-connected topology and it is used as a lower bound reference.

Downtime and capacity allocation are determined considering two disjoint paths (1:1 protection) in the *SR* case. In the $D3_{2C}$, three disjoint paths (1:1:1 protection) are provided between pairs of nodes, if possible; for the rest, two disjoint paths are used. In the $D3_{3C}$ case, three disjoint paths are provided between each pair of nodes. The interconnections are optimized in terms of deployment costs, and the cost and availability models presented above are used. Details about this experiment, and Figs. 3 and 4 can be found in [45].

Figure 3 presents the comparison of deployment cost vs. downtime in the three scenarios. The pattern followed by the results in the three cases is similar, and it can be noticed how the improvement of moving from the $D3_{2C}$ to the $D3_{3C}$ is more significant than moving from the *SR* to the $D3_{2C}$. The slope of the lines between points can be interpreted as the availability benefit of increasing the deployment costs, the steeper the better. Figure 4 illustrates the resulting $D3_{3C}$ graphs for the three scenarios.

In summary, to deploy $D3_{3C}$ graphs is slightly more costly (between 2 and 11 % higher) than the $D3_{2C}$ option, but the improvement on availability pays off the extra investment by reducing the yearly downtime between 230–400 times.

5 Reliable Access to Dynamic Information in Critical Infrastructures

As a part of dependability, reliability of the system and information being accessed is critical. In many situations in critical systems, events are happening and require reporting to a server that will take action upon the event. In this matter, it is critical

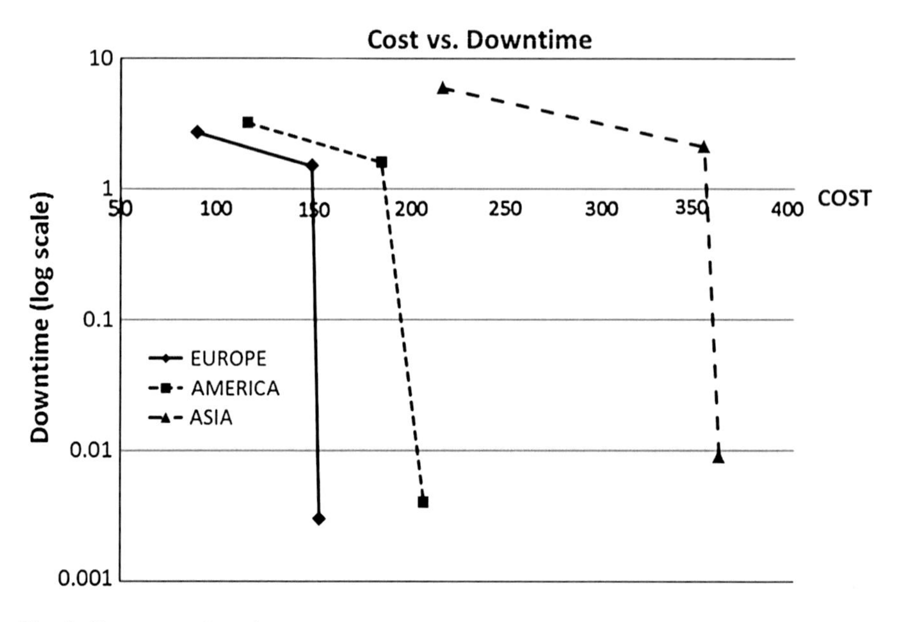

Fig. 3 Cost versus downtime

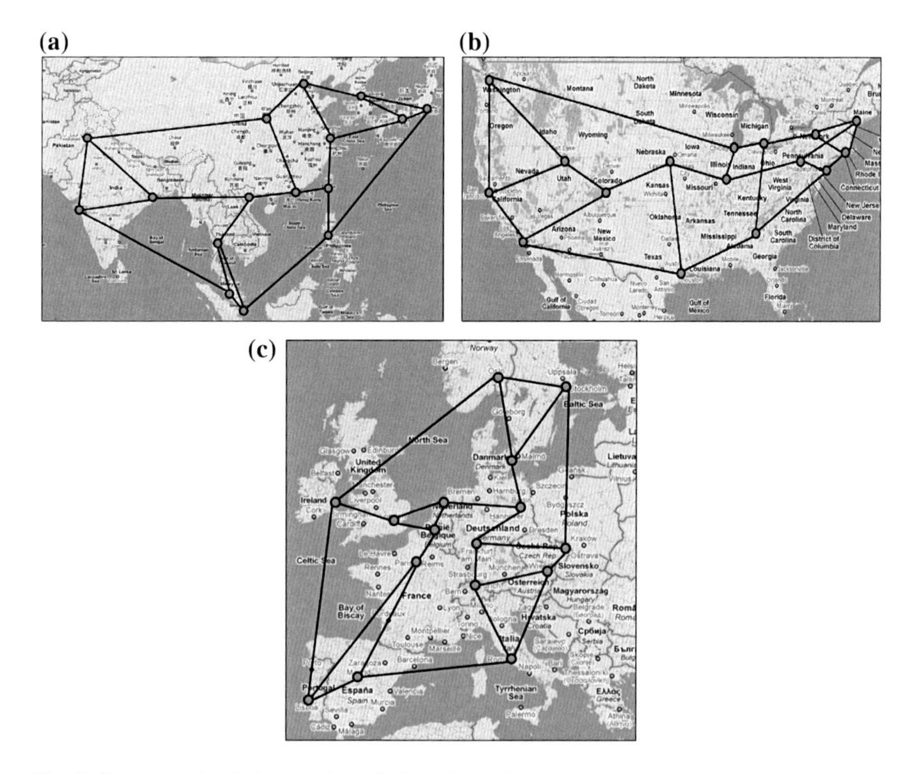

Fig. 4 3-connected solutions. **a** Asia, **b** America, **c** Europe

that the server reacts upon the situation as it is, and not how it was moments ago. In the following we consider the communication between two entities A and B, at different geographical and network locations, with A having a need to obtain information from B which happens to be dynamic, and this access needs to happen over a (complex) network with a stochastic end-to-end delay (see e.g., Fig. 1).

In [46] we evaluated the impact on different access strategies to dynamic information elements over a network. The access scheme models are generic meaning that they do not model any particular protocol, but rather the behaviour of one. The delays involved are described statistically such that any stochastic models based on MAC, IP or Application layer can be incorporated. The models consist of the following three basic schemes: reactive access, proactive event driven and proactive periodic.

Then, for the following we consider the stochastic processes

- An Event process $\mathcal{E}$, if Poisson with rate λ_e. $E = \{E_i, i \in \mathbb{Z}\}$, where i the ith event number.
- An upstream (A–B) and downstream (B–A) delay $\mathcal{D}$, if Poisson with rate ν, and indices u and d for upstream and downstream, respectively. $D = \{D_j, j \in \mathbb{Z}\}$, where j the jth delay.
- An Access Request process $\mathcal{R}$, if Poisson with rate μ_r. $R = \{(R_k, k \in \mathbb{Z})\}$, where k indicates the kth request.

The definition of an event may not be unique, but by event we here mean a *significant* change in the value of some attribute or information element of interest. Significant can for example be if a signal exceeds some threshold or simply takes another enumerated value.

5.1 Reactive Access

The reactive access is characterized by A sending a request for information to B. Once B receives the request, it sends back the response to A containing the information and A will use this information for some purpose, for example as input for a smart grid control. Then we can calculate the probability of A using outdated and mismatching information as [46],

$$mmPr_{rea} = 1 - \int \overline{B}_E(t) F_D(dt) \tag{8}$$

where $\bar{B}_E(t)$ is the CDF of the backwards recurrence time (see [46] for details) of the event process, and $F_D(t)$ the CDF of the delay (with the bar indicating the reliability function, i.e. $\bar{F}_X = 1 - F_X$). Here traffic is only generated whenever needed. The average waiting time is entirely defined by the sum of the upstream and downstream delay. The result obtained from B may also be cached at A, on which for some time period any requests are fetched from the local cache rather than sending requests to B. The mismatch probability model for that case is rather complex (see [47]) for details. Applying a cache means also that network traffic is in average reduced, as well as the average waiting time.

5.2 Proactive: Event Driven Update

Another option is that B, which collects the data, sends an update to A if it detects an event has occurred. In this way, network traffic is only generated whenever events occur, however, if the event process is rather fast this can become a rather large amount of traffic. The mismatch probability for this approach can be calculated by considering two types of updates: (1) if each update contains full information, i.e., each update completely overwrites existing value at A, or (2) if each update only provides the incremental value since the previous update. For case (1), the mismatch probability becomes exactly the same as for the reactive strategy, but for (2), the mmPr can be modelled by considering the probability of finding a $G/G/\infty$ queue being busy, with queue elements modelling updates in transit. If just one is in transit, then there will be a mismatch (see [46] for further details).

$$\text{mmPr}_{pro,evn}^{(inc)} = \mathbb{P}(E/D/\infty \quad queue \quad is \quad busy) \tag{9}$$

5.3 Proactive: Periodic Update

This approach relies on B sending the current value or state of the information to A with a time period specified in one way or another. The traffic generated by this approach is entirely determined by the update rate. The mismatch probability for this approach is based on a thinned Poisson update process. This also means that the update process is stochastic, while it normally is deterministic. However, due to clock and scheduling drift in operating systems, and the fact that a deterministic update process (via simulation studies) shows to provide better reliability, this assumption serves as a worst case scenario. Thus, the model becomes

$$\text{mmPr}_{pro,per} = \int_0^\infty \exp\left(-\int_0^t \tau F_D(s)ds\right) A_E(dt), \tag{10}$$

with τ the update rate, F_D the delay distribution and A_E the backward recurrence time, [46].

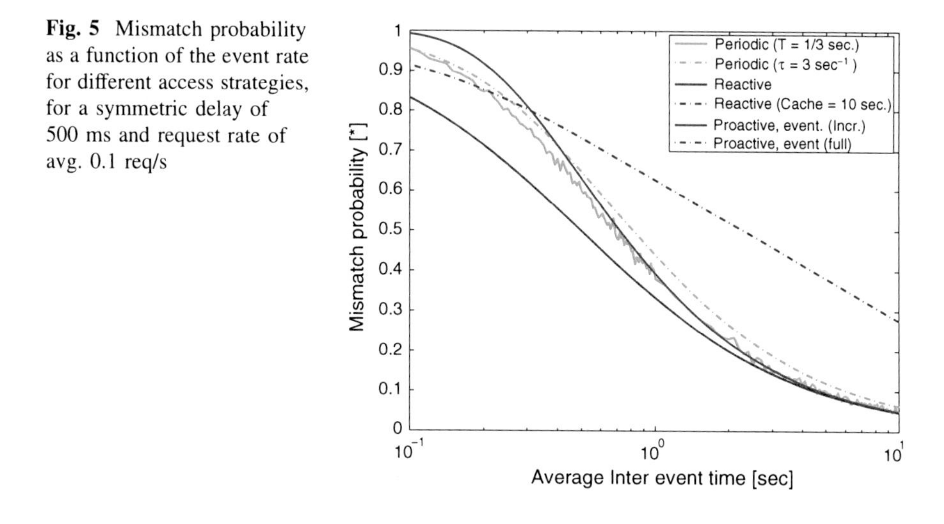

Fig. 5 Mismatch probability as a function of the event rate for different access strategies, for a symmetric delay of 500 ms and request rate of avg. 0.1 req/s

5.4 Impact of Access Strategy on Reliability

Figure 5 shows the mismatch probabilities for the different access strategies with varying event process rate, under the assumption that the involved processes are all exponentially distributed (for simplicity). The effect is clear for this situation: there is not only a significant impact, but also a very different impact on the different strategies, which makes it less clear which approach is actually the best.

Although it appears as if the pure reactive and event driven (using full update) appears to perform best, one also has to consider that the traffic generated is different (reactive in this case is 0.2 messages/s, event driven in the range of 0.1–10 messages/s compared with the periodic one which is 1/3 message/s). Thus, in case of scalable systems such as smart grids where several thousands of customers provide data regularly, the solution of which strategy to take is not necessarily clear.

5.5 Impact of Event and Delay Processes

It is also not just the event rate that has an impact on the reliability. If there are restrictions to which values an information source can attain, e.g., a lamp is either on or off, and that is the information needed for e.g., a smart grid solution, then this additional information on restriction has an additional effect on the reliability. Figure 6 shows an example of a case where information can only be in one of two states (ON or OFF), with the ratio of time spent in one or the other state ($\zeta = \lambda/\mu$, with λ and μ as the rates at which the information changes states in a two state Markov Chain).

The figure shows that there is a maximum mismatch probability at the ratio of 1, i.e., when there is equal probability of finding the information in one of the two states. Any other ratio gives a lower mismatch probability due to the biased time

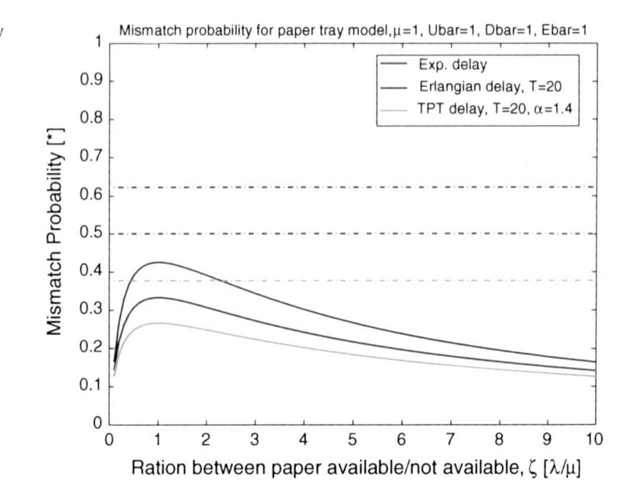

Fig. 6 Mismatch probability for an ON/OFF type of information model

spent in one or the other state. In comparison, the same process modelled as a Markov jump process instead, shows not only the obvious that the ratio does not have an impact, but that the resulting mismatch probability is higher than the two state model at any point.

Another important aspect which is shown in Fig. 6 is that the delay process is important too (in fact the following observation is equally valid for the event process too—that is as long it is not a recurrent process). The figure shows three results of the same plot, coming from three different distribution types: (1) an exponential distribution, (2) an Erlang distribution with 20 phases, and (3) a Truncated Power Tail (TPT) with 20 phases, all with the same mean value. The results, which are consistent with other results shown in [46], show that the deterministic process (modelled by the Erlang distribution) is far the worst case, i.e., resulting in the highest mismatch probability. The best case is the highly stochastic type of information modelled by the TPT distribution.

This is indeed good news for example for smart grid solutions, where the most stochastic elements are those which will need to be observed, while the deterministic elements are often those which will need or can be controlled, and thus do not need much other observation than perhaps a feedback whether the element has been activated or not. Take the example of a lamp, TV, or other similar device such as a household device which has a level of stochastic behaviour which may be modelled e.g., as an ON/OFF model as shown in Fig. 6.

5.6 Network Adaptive Access Strategies

One example of how the models can be useful is the selection of the update rate of the periodic strategy. Figure 7 shows how an appropriate choice of update rate (τ) in the periodic scheme can lead to a consistent reliability, set here to 0.3. However,

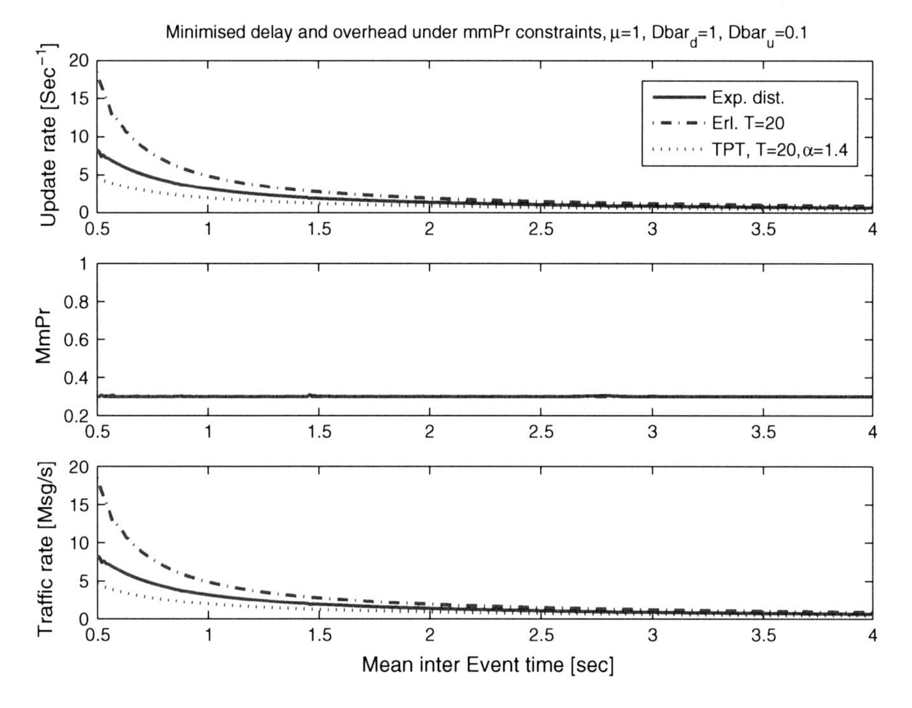

Fig. 7 Mismatch probability for an ON/OFF type of information model

this comes at a cost of an increased network overhead which in theory can rise to an infinitely fast update rate (this is of course unrealistic).

Similarly, an appropriate caching period can be selected to keep reliability (see [47]). Therefore, these models offer a large range of possibilities for reconfiguration of protocols which have the above mentioned type of interaction for dynamic information. However, as already discussed, networks are not static, and end-to-end delays are not necessarily similar over the days, weekdays or months. Therefore, these processes should be monitored and estimated, which is not trivial and requires software components, [67–69]. But, if done properly, algorithms via observations of information access request rates, data sizes of requests, and knowledge of the delays and event processes can be provided for proper selection and configuration of access to dynamic information for reliable operation. Then, the selection of metrics becomes a matter of how much traffic a system is allowed to create and how high requirements to the reliability of the information one has. A proposal for such an algorithm that utilizes the mentioned aspects has been described and evaluated in [48].

5.7 Summary

From this section we learned that the information access, the dynamics of the information, and the delay are closely related to reliability. Mismatch probability is a probabilistic notion on how certain dynamic information accessed remotely is, when being used for whichever purpose. The more likely it is that the information is correct, the more likely a correct system behaviour will occur. Unless dedicated networks are set up where data can be reliably scheduled, future critical infrastructure will have to face such challenges thus only probabilities for success can be given, and the cost of deploying networks is not to be underestimated as we will look into later in this chapter.

6 Security and Threats to Critical Infrastructure

Networks are controlled by software, and as such they are vulnerable to flawed designs in protocols, implementation bugs, exploits and so forth, and every opened communication channel generates principally for a possible attack opportunity. In today's world where for example terrorism is in focus, securing networks when used in critical infrastructures is more important than ever. Securing networks is a continuous battle against an invisible and sometimes unknown enemy, and in war, one of the most fundamental elements for a successful defense is to know and understand the enemy. Therefore, in this section we propose a framework for analyzing digital hacker threats to critical infrastructures [49, 50].

As the digital aspect of society and our lives in general becomes ever more important, the defence hereof becomes an increasingly higher priority. Billions of dollars are spent each year to protect the systems that form the basis for our everyday lives against malicious attackers who would steal our data or sabotage our critical infrastructures. It is an uneven battle—attackers need to be successful only once, while the defenders must be successful every time. Therefore it is imperative that we know as much as possible about the potential attackers and their methods to be able to prioritize the defence efforts. If we know the potential attackers, we are able to predict, detect and manage possible attacks.

To do this, it is necessary to make a threat assessment for the systems in question [51]. Threats can be divided into three categories: natural disasters, accidents, and attacks. Traditionally, attackers are considered to be the same attackers who would attack the installation/system in the physical world (e.g., criminals and terrorists), with an added category of "hackers" [52, 53]. However, "hackers" is such a loose term that by grouping all attackers into this one category, it is very difficult to say anything specific about important threat properties.

For an assessment of the hacker threat to critical infrastructures to be useful, it should have a greater granularity with respect to the attacker identification. By dividing the "hacker" category into a number of subcategories, for which we can determine specific threat properties, it is possible to give a much more precise picture of the threat. We divide the hackers into eight categories:

- Script kiddies
- Insiders
- Gray hats
- Hacktivists
- Cyber punks
- Petty thieves
- Professional criminals
- Nation states.

The motivations behind each of the categories of hackers can be represented by a circumplex, as illustrated in Fig. 8. Each of these categories will be evaluated separately, whether or not they represent a sabotage threat against critical infrastructures. In this context, the sabotage is considered as reducing or removing the availability of the service provided by the critical infrastructure. In terms of cyberattacks, this can be achieved for instance, by a malicious use of controls, (Distributed) Denial of Service (DDOS) attacks, or worms [54]. These are the vectors that will be considered. The threat assessment is performed by comparing hacker motives (intent) and capabilities (competencies and available resources) [51], with the profile of the targeted critical infrastructure. This profile is described widely in the literature, e.g., in [55] and [56], and will be summarized in the following section. Obviously, the first step is to generally protect the infrastructure, by making sure that it is not vulnerable to common security threats. Therefore, in

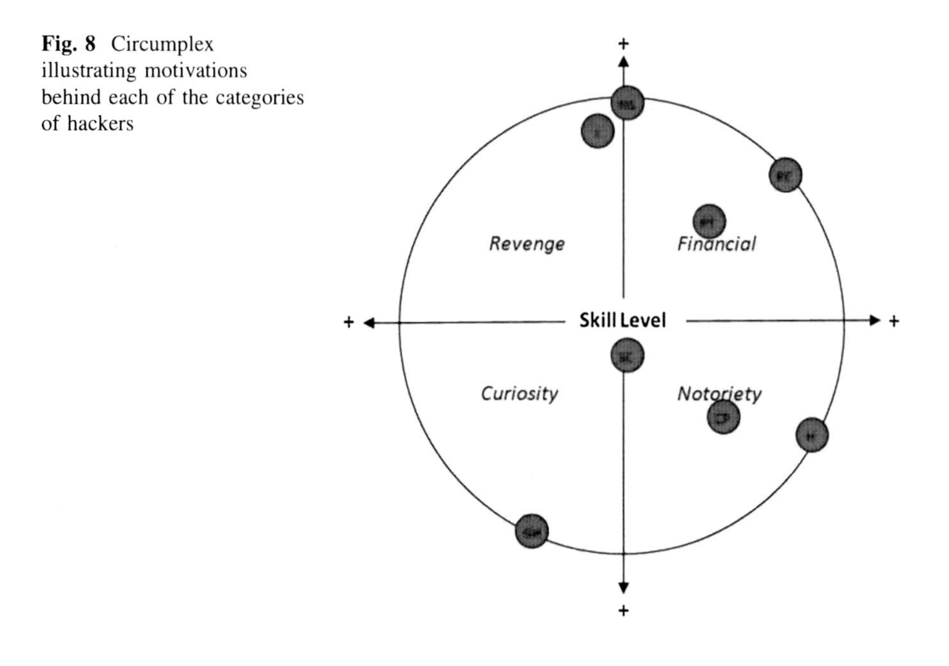

Fig. 8 Circumplex illustrating motivations behind each of the categories of hackers

the rest of the section we will assume that appropriate action has been taken to ensure this, and we will focus on more advanced/severe kind of attacks.

Intent: Critical infrastructure is a high-profile target that provides critical services to the area it is located in, indicating that attackers with Notoriety or Revenge motives would find critical infrastructure tempting targets. The return on investment on sabotaging critical infrastructure for money is very low because of the high security and the lack of monetary gain, so Financial motives are not likely to drive a hacker to this type of attack. And while Curiosity might cause an attacker to try and break into critical infrastructure, it is unlikely that it would spur the hacker to perform sabotage [57].

Triggers: Critical infrastructures, being high-profile targets, do not need any special triggers to be attacked. However, since critical infrastructures are often state-run and a necessity for a state to run, attackers with revenge motives could be triggered to attack at a perceived threat or insult from the government of the state in question.

Capabilities—skills: Critical infrastructures should be and usually are assumed to be protected against common vulnerabilities; the systems are updated and protected by security solutions. Therefore it takes either a significant amount of very specialized skills to break the defences and/or considerable resources [58].

Capabilities—resources: For sabotage of critical infrastructures to be really effective, several facilities need to be hit for prolonged periods. This requires either huge amounts of resources to be able to hit several systems in parallel and sustain attacks, or very specialized skills and insider knowledge to create cascading effects between networked systems.

Methods: While critical infrastructures are generally well protected, no system is completely secure. Critical infrastructures are vulnerable to all considered attack vectors (malicious use of controls, DDoS attacks, and worms) if the attackers utilize them with enough resources and/or skills.

Trends: The increased focus on cyber security in the society in general also has a positive effect on the security awareness of critical infrastructures. One example is that industrial control systems are generally better protected after the Stuxnet incident made everyone aware that they are a potential target [59].

In the following we will use these threat properties to discuss to which extent the different hacker categories have the will and the capabilities to execute successful attacks specifically against critical infrastructures.

6.1 Script Kiddies

Attackers of the Script kiddies category are novices with low hacking skills and limited understanding of technical consequences, who use tools or scripts downloaded from the internet.

Intent: Primary motivations for Script kiddies are notoriety and curiosity, and as such critical infrastructure being a high-profile target would be attractive, but the tools used by Script kiddies often choose targets at random (for example, The Honeynet Project (2004)).

Triggers: The only trigger Script kiddies need to attack is the opportunity. If they find a vulnerability in their random, automated search, they will exploit it.

Capabilities—skills: Low technical competencies makes it highly improbable that a Script kiddie could execute a successful attack against a target with a minimum of up-to-date defences. This is expected to exclude critical infrastructures from their targets.

Capabilities—resources: Having very little resources available and operating solo further decreases the probability of a successful attack.

Threat assessment: Have the will but lack the capabilities.

6.2 Cyber Punks

Members of the Cyber punks category are medium-skilled but mostly solitary hackers and virus writers.

Intent: As primary motivations of Cyber punks are notoriety and curiosity, they might target critical infrastructures.

Triggers: Cyber punks need no trigger to attack.

Capabilities—skills: Cyber punks have some technical skills and understanding, and they will be able to use and even improve tools available on the Internet. A Cyber punk might write their own malware exploiting well-known vulnerabilities, but they are not likely to develop their own 0-days or perform very advanced attacks. However, some virus writers also fall under this category, and a novel virus might compromise one or more critical infrastructure systems [60]. It is not likely that systems could be compromised in such a manner as to cause cascading failures without being specifically designed to do so, but if a worm spreads aggressively enough to infect a large number of critical infrastructure systems, then it could cause widespread denial of service.

Capabilities—resources: Cyber punks have limited resources since they are mostly solitary, or in small groups. They are likely to able to perpetrate simple, isolated attacks, although nothing sustained or large-scale.

Threat assessment: Have the will and the capabilities.

6.3 Insiders

Insiders are malicious but trusted people with privileged access and knowledge of the systems in question.

Intent: Insiders are motivated by revenge and to some degree notoriety, and the former part makes them likely to try to conduct sabotage.

Triggers: A malicious Insider mostly needs to be triggered to attack, however since the trigger can be any perceived insult or slight at the workplace, it can be very hard to determine whether or not a potential attacker has been triggered—especially considering that the potential attacker is typically a trusted employee.

Capabilities—skills: An Insider can have extensive knowledge of the systems, including vulnerabilities, as well as privileged access to controls. An Insider might even have the skills to perpetrate an effective cascading attack.

Capabilities—resources: While insiders very often work alone, they usually have the resources needed to make an effective attack, namely privileged access to controls, and physical access to the systems.

Threat assessment: Have the will and the capabilities.

6.4 Petty Thieves

Members of the Petty thieves category commit low-level fraud and theft, usually by using existing tools and scripts.

Intent: Petty thieves are primarily motivated by financial gain, and as such, critical infrastructures do not constitute an attractive target to members of this category.

Triggers: This group needs no other trigger than opportunity and a viable business case to attack.

Capabilities—skills: Petty thieves use a standard portfolio of tools and techniques primarily focused around phishing, scamming, and credit card fraud. They are not likely to possess the skill set nor the tools needed to attack critical infrastructures.

Capabilities—resources: Members of this group work alone or in small groups, and considering their focus on low-level crime, it is not likely they will have the resources needed to commit a successful attack.

Threat assessment: Lack the will and the capabilities.

6.5 Grey Hats

Grey hats are often skilful hackers with limited criminal intent but a lack of respect for limitation on information flow and a large curiosity.

Intent: This category of attackers is primarily motivated by curiosity, and they are very unlikely to perpetrate any form of sabotage.

Triggers: Rumors of secret information or "impenetrable" defences might increase the risk of attack.

Capabilities—skills: Grey hat hackers have very specialized technical skill sets and an extensive exchange of information, and as such it is likely they would be able to execute an attack successfully.

Capabilities—resources: While there is a high degree of knowledge exchange in this group, most work alone and as such do not have the manpower to make widespread and persistent attacks. They do have the skills and equipment however, to gain insider access to the systems in question, which could enable them to execute a cascading attack.

Threat assessment: Lack the will but have the capabilities.

6.6 Professional Criminals

Professional criminals are organised groups of hackers with a strict business approach to attacks.

Intent: Professional criminals are purely financially motivated. There is currently no business model that makes the reward of an infrastructure attack worth the risk, since most governments do no negotiate with terrorists, so they are unlikely to attack.

Triggers: Like Petty thieves, the Professional criminals need no specific trigger apart from a viable business case.

Capabilities—skills: Members of this group of attackers possess a wide variety of technical skills and knowledge, and they are willing to recruit or hire people with the necessary competencies to complete an operation.

Capabilities—resources: This group has many resources in the form of money, equipment, and manpower—enough to perpetrate a successful attack against critical infrastructures.

Threat assessment: Lack the will but have the capabilities.

6.7 Hacktivists

Hacktivists are groups of ideologically motivated hackers with varying technical skills, but many and geographically distributed members.

Intent: Since Hacktivists are motivated by ideological agendas and notoriety and known for a lack of regard for consequences, they are likely to target critical infrastructures. The US Department of Defense warns that it believes one of the biggest hacktivist groups, Anonymous, have both the will and the capability to perform such an attack [61]. However, Anonymous have publicly declared [62], that they are not interested in attacking the power grid because they realize the adverse effect it would have on the general population, and as such the intent in this regard is not completely clear. There are many groups, though, and not all of them have the same moral scruples as Anonymous claims to have. On the other hand, they do not alone have the necessary resources available.

Triggers: Hacktivists are triggered by perceived threats or insults to their ideology.

Capabilities—skills: While the levels of technical skills are diverse within the Hacktivist groupings, they usually have members with high technical competencies, although they might not have the highly specialized skills needed for a cascading attack.

Capabilities—resources: Hacktivists have a large geographical spread and sometimes vast amounts of manpower. They may also have the attack resources in the form of botnets available to execute a widespread and sustained attack. However, only Anonymous is big enough at this point in time to conduct a sustained attack, and they have declared a lack of interest in doing so. This is subject to change though.

Threat assessment: Have the will or the capabilities, but not both.

6.8 Nation States

Nation states or representatives hereof have been known to perpetrate everything from industrial espionage over acts of terrorism to devastating nation-wide attacks in the cyber arena.

Intent: In case of a conflict, any disruption of the enemy's infrastructure is desirable, and as such, critical infrastructures represent an extremely attractive target to a hostile Nation state.

Triggers: Nation states are almost exclusively triggered by disputes in the physical arena, most often geopolitical in nature.

Capabilities—skills: Many nation states have substantial presence in cyberspace and commands many highly skilled experts in the critical infrastructure field.

Capabilities—resources: They have vast resources in the form of money, manpower, specialized knowledge and intelligence, and equipment. They are very likely to be able to conduct a successful attack.

Threat assessment: Have the will and the capabilities.

6.9 Threat Picture

Three of the hacker categories—Insiders, Hacktivists, and Nation states, see Table 1—can be considered a substantial sabotage threat to critical infrastructures at present time, but this is subject to change. Currently, only these three categories have both the will and the capabilities to execute a successful attack on critical infrastructures.

Table 1 Threat matrix indicating will and capabilities of attackers

Categories	Will	Capabilities	Threat
Script kiddies	Yes	No	No
Cyper punks	Yes	Yes	Yes
Insiders	Yes	Yes	Yes
Petty thieves	No	No	No
Gray hats	No	Yes	No
Professional criminals	No	Yes	No
Hacktivists	Yes/No	Yes/No	No
Nation states	Yes	Yes	Yes

6.10 Defence Priorities

Working to defend critical infrastructure in a resource-constrained environment means that limited budget funds must be applied to achieve the best effect. While the thorough "basic" security is assumed in place, there are many areas in which security officers could focus their attention.

Cyber punks methods: While most of the attacks performed by this category of attackers are limited in scope and sophistication, the Cyber punks do have one weapon that is a legitimate sabotage threat to critical infrastructures, namely viruses. A worm exploiting an unanticipated attack vector might plausibly infect several of critical infrastructure systems and causing widespread denial of service. Viruses such as Melissa (1999), ILOVEYOU (2000), Nimda (2001), Slammer (2003), and Conficker (2008) (most contributable to Cyber punks) show that it is certainly possible to reach a critical amount of infection in a very short while. To defend against such a worm may prove difficult; see Wiley, Brandon (circa 2002). An Intrusion Detection and Prevention System (IDPS) that uses statistical anomaly-based detection and/or stateful protocol analysis detection would have a good chance of catching such a threat, and there are different ways of hardening the network, depending on how fast the worm propagates [63, 64].

Insider methods: Attacks by Insiders will likely take the form of malicious use of controls. Based on the [65], 43 % of Insider attacks were executed while the attacker still had legitimate access to the systems. The majority of Insider attacks were, however, executed while the attacker should no longer have the access. The defence mechanisms that would help defend against such attacks should be based on the principle of least privilege in order to limit the amount of damage that can be done with malicious use of legitimate access. Also, having tight management of personnel access would limit the amount of damage previous employees could cause, since they would lose access as soon as they were no longer employed. More detailed advice on how to mitigate the threat of Insider attacks can be found in [66].

Nation state methods: The methods employed by Nation states vary, but there seems to be a prevalence towards spear-phishing and zero-day exploits combined with worms (e.g., Night Dragon and Stuxnet attacks) as well as devastating DDoS attacks (e.g., Georgia, 66, and 10 Days of Rain). Defence priorities include educating staff and increasing awareness to avoid anyone falling victim to spear-fishing or similar social engineering attacks. It is in the nature of things quite difficult to detect the use of zero-day exploits, but good intrusion detection systems might be able to pick up the change in network behaviour. DDoS prevention should also be a priority, and the volume of the attacks can be expected to be severe, so cooperation with for instance large ISPs might be an option to consider.

6.11 Summary

Based on the threat picture described in this section, defence efforts against digital sabotage of critical infrastructures should focus on dealing with malicious Insiders as well as Cyber punks, and hostile Nation states. This could be done by prioritizing access management, including implementing the principle of least privilege, as well as installing an Intrusion Detection and Prevention System, educating staff to be wary of spear-phishing and similar social engineering attacks, and protecting systems from DDoS attacks. Defence efforts should be particularly focused in times of geopolitical conflict, where the risk of a Nation state attack is high. Furthermore, an eye should be kept on the development in the Hacktivist category, since there is a risk that this group will develop into a threat in the near future. The analysis presented can also be used when designing and deploying new infrastructures such as Smart Grid, Intelligent Transportation Systems, and infrastructures supporting Tele-Health in larger scales. Knowing which groups are the most probable attackers, and knowing their capabilities in terms of skills and resources, can help identify which kinds of attacks are important to prepare for. There is a big difference between being attacked by Insiders, Nation states, or Script kiddies. For researchers in the domain of cyber security this knowledge can be used to indicate where further research and development is needed, e.g., when developing technologies for intrusion detection systems for general or specific critical infrastructures.

7 Previous and Ongoing Research Activities

The purpose of this section is not to provide an elaborate list of projects, but give a short overview of some of the activities that have been carried out, are running or will be running in the near future, at European level in the area of communication role in Critical infrastructures and related topics. For more details, the reader should visit the IST website: http://www.ist-world.org/ or http://www.cordis.europa.eu/ist/.

7.1 Brief Overview of Selected Projects and Activities

CRUTIAL This project, which ran from 2006–2008, has focused on key issues related to trust establishment, access control and fault diagnosis for critical infrastructures. Inspired by Intrusion tolerant system architecture, CRUTIAL is based on two facts; Critical Information Infrastructure features a lot of existing legacy systems, and two existing security solutions may jeopardize operational and functional requirements to critical infrastructure systems. Key components in their architecture are (1) configuration aspects, (2) middleware for automatic fault tolerance inclusive intrusion, (3) trustworthy monitoring mechanism, (4) security and access policy management and enforcement.

MAFTIA was a project running between 2000–2003, focusing on fault and intrusion tolerance network support for critical infrastructure systems. The project aimed to develop an architecture based on hybrid failure assumptions, recursive use of fault prevention and fault tolerance techniques, and the notion of trusting components to the extent of their trustworthiness. MAFTIA distinguishes between *attacks*, *vulnerabilities*, and *intrusions* as three types of interrelated faults. Selected system components were implemented and validated through a set of text scenarios with success.

HIDENETS The aim of HIDENETS, running from 2006 to 2008, was to develop and analyse end-to-end resilience solutions for distributed applications and mobility-aware services in ubiquitous communication scenarios. The idea has been to make use of off-the-shelf components and wireless communication links to dramatically decrease the costs of market entry and enable ubiquitous scenarios of ad hoc car-to-car communication with infrastructure service support commercially feasible. Dependability has been a keyword in this project, and many of the lessons learned here will be beneficial to critical infrastructure communication systems.

GRIDCOMP GridComp main goal was to design and implement a component based framework suitable to support the development of efficient grid applications. Such frameworks are important since the automation and complexity of critical infrastructures makes it necessary to have a structural way of developing software components and middleware on top of networks. The key feature of the developed framework was its level of abstraction perceived by programmers by hierarchically composable components and advanced, interactive/integrated development environments. The framework was supposed to allow a faster and more effective grid application development process by considering advanced component (self-) management features.

NESSI-GRID The objective of this project was to contribute to the activities of the Networked European Software and Service Initiative (NESSI) with special focus on next generation Grid technologies; Services not necessarily only for user interaction, but also for critical subsystems. The aspect regarding services in the communication infrastructure and inter operability is not trivial, and it was expected that output of this activity would be aligned with that of Service Oriented Knowledge Utilities (SOKU) as defined by the Next Generation Grid expert group.

INTEGRIS A project which has run since 2010 and ended in 2012, addressing a novel and flexible ICT infrastructure based on a hybrid power line communication/wireless integrated communication system for smart electricity grids. The project covers monitoring, operation, customer integration, demand side management, voltage control, quality of service control and control of distributed energy resources. On the communication side, in particular interoperability of the power line communication, wireless sensor networks and radio frequency identification are in focus.

REALSMART This project, running from 2010 to 2014, aims to look at smart grid solutions. From a communication point of view, the solutions sought in this project relate to measurement collection procedures for phasor measurements to allow an improved observation of the transmission grid. Among other questions that need to be addressed, is the handling of the massive amounts of real time data that are to be collected.

EDGE EDGE is a Danish funded project running from 2012 to 2017, which aims to develop complex control algorithms for smart grid systems based on power flexibility from consumers. The role of the communication in this project is in particular on the interaction between these advanced control algorithms and strategies and the network dynamics that exist in heterogenous networks.

SmartC2Net This project is a new smart grid project starting in the last part of 2012 until 2015. This project aims to provide an ICT platform that supports flexible interaction possibilities between control algorithms and strategies, and the physical entities distributed in the power grid. This entails information reliability models, flexible reconfiguration of the network, control strategy based QoS control, network monitoring and security threat analysis.

7.2 Summary

A common denominator of all these projects is the focus of ICT and its role in the various aspects of critical applications that run over the network. As investigated in this part, this is necessary as networks are complex, and far from perfect. In most cases, and in particular for large scaled, general purpose networks (e.g., the Internet) only probabilistic guarantees can be given, which may or may not be sufficient for the critical applications. Projects such as CRUTIAL, MAFTIA and HIDENETS focus to a large extent on the dependability of services in the network infrastructure, which all concerns both reliability and availability in a distributed setting elements. Availability and security is strongly linked. These are key elements when supporting critical infrastructures, and surely lessons learned here will in some way find the way in future networks.

Projects such as GRIDCOMP and NESSI-GRID focus also on service interaction and interoperability as well as on how programmers can integrate software solutions. This is absolutely a key feature for reliable operation of networks, since as also seen in many of the other projects, middleware and software components in the network will play a key role in future critical networks.

Finally, projects such as INTEGRIS, EDGE and SmartC2Net aim specifically to address the interaction between control algorithms for smart grids and the network. For the network, the running application behaviour is critical, as this sets the requirements to the network that provides the end-to-end communication. Therefore, understanding the interaction between application and network is critical, and is manifested to some degree in the need for such projects.

These examples as well as the other non-European efforts, national projects and initiatives, provides valuable insight in the challenges and solutions of using the existing network for critical applications. At the end this benefits to save costs for redeployment of dedicated networks to each critical application that exists. This makes it necessary to explore and spend funding on research to achieve a reliable communication infrastructure.

8 Conclusions and Outlook

Communication networks are complex and pose a challenge to distributed systems, and although these networks offers great advantages of cheap exchange of information for various purposes, it is critical not to under estimate the role communication plays in distributed applications. In particular not for critical infrastructures which require communication due to their distributed nature. The complexity and dynamics of the networks are challenged at all levels. Through this chapter, we looked at basic communication, standards for smart grids, network deployment, reliability of protocols, security and threats, and also gave a brief overview of what type of research is or has been ongoing at a European level to address the issues that need to be tackled for networks to support critical infrastructures.

References

1. Future Internet 2020: Visions of an Industry Expert Group, DG Information Society and Media Directorate for Converged Networks and Service—"The Internet People", May 2009, European Commission, Information Society and Media. ISBN: 978-92-79-11320-8, doi:10.2759/4425
2. Tannenbaum, A.S.: Computer Networks, 4th edn. Prentice Hall, Upper Saddle River, Internation Edition, ISBN: 0-13-038488-7
3. http://www.netvalley.com/history_of_internet.html
4. Prasad, R., Mihovska, A.: New Horizons in Mobile and Wireless Communications: Reconfigurability, ISBN: 978-1-60783-971-2, New Horizons in Mobile and Wireless Communications series, Artech House (2009)
5. Wang, W., Xu, Y., Khanna, M.: A survey on the communication architectures in smart grid. Comput. Netw. **55**(15), 3604–3629
6. http://www.ipv6vsipv4.com/

7. Murray, et al.: Why is it difficult to implement ehealth initiatives? A qualitative study. Implementation Sci. **6**, 6 (2011)
8. Strobla, R.O., Robillardb, P.D.: Network design for water quality monitoring of surface freshwaters: a review. J. Environ. Manag. **87**(4), 639–648 2008. http://dx.doi.org/10.1016/j.jenvman.2007.03.001
9. Mattern, F., Staake, T., Weiss, M.: ICT for green—how computers can help us to conserve energy. In: Proceedings of the 1st International Conference on Energy-Efficient Computing and Networking (e-Energy 2010), ACM, pp. 1–10. Passau (2010)
10. Lim, H.-T., Volker, L., Herrscher, D.: Challenges in a future IP/ethernet-based in-car network for real-time applications. In: Design Automation Conference (DAC), 2011 48th ACM/EDAC/IEEE, pp. 7–12, 5–9 June 2011
11. Karjalainen, S.: Consumer preferences for feedback on household electricity consumption. Energy Build. **43**(23), 458–467 (2011). ISSN 0378-7788, doi:10.1016/j.enbuild.2010.10.010
12. Ye, Y.; Yi, Q., Sharif, H.: A secure and reliable in-network collaborative communication scheme for advanced metering infrastructure in smart grid. In: Wireless Communications and Networking Conference (WCNC), 2011 IEEE, pp. 909–914, 28–31 March 2011
13. Bliek, F., van den Noort, A., Roossien, B., Kamphuis, R., de Wit, J., van der Velde, J., Eijgelaar, M.: PowerMatching City, a living lab smart grid demonstration. In: Innovative Smart Grid Technologies Conference Europe (ISGT Europe), 2010 IEEE PES, pp. 1–8, 11–13 Oct 2010
14. Benzi, F., Anglani, N., Bassi, E., Frosini, L.: Electricity smart meters interfacing the households. IEEE Trans. Ind. Electron. **58**(10), 4487–4494 (2011)
15. Depuru, S.S.S.R., Wang, L., Devabhaktuni, V., Gudi, N.: Smart meters for power grid—challenges, issues, advantages and status. In: Power Systems Conference and Exposition (PSCE), 2011 IEEE/PES, pp. 1–7, 20–23 Mar 2011
16. Byun, J., Hong, I., Kang, B., Park, S.: A smart energy distribution and management system for renewable energy distribution and context-aware services based on user patterns and load forecasting. IEEE Trans. Consum. Electron. **57**(2), 436–444 (2011)
17. LeMay, M., Nelli, R., Gross, G., Gunter, C.A.: An integrated architecture for demand response communications and control. In: Proceedings of the 41st Annual Hawaii International Conference on System Sciences, pp. 174, 7–10 Jan 2008
18. Wang, W., Xu, Y., Khanna, M.: A survey on the communication architectures in smart grid. J. Comput. Netw. **55**(15), 3604–3629 (2011)
19. Sidhu, T.S., Yin, Y.: Modelling and simulation for performance evaluation of IEC61850-based substation communication systems. IEEE Trans. Power Delivery **22**(3), 1482–1489 (2007)
20. Kanabar, M.G., Sidhu, T.S.: Reliability and availability analysis of IEC 61850 based substation communication architectures. In: Power & Energy Society General Meeting, 2009. PES '09. IEEE, pp. 1–8, 26–30 July 2009
21. Gungor, V.C., Sahin, D., Kocak, T., Ergut, S., Buccella, C., Cecati, C., Hancke, G.P.: Smart grid technologies: communication technologies and standards. IEEE Trans. Ind. Inf. **7**(4), 529–539 (2011)
22. Zaballos, A., Vallejo, A., Selga, J.M.: Heterogeneous communication architecture for the smart grid. IEEE Netw. **25**(5), 30–37 (2011)
23. Zhang, R., Zhao, Z., Chen, X.: An overall reliability and security assessment architecture for electric power communication network in smart grid. In: 2010 International Conference on Power System Technology (POWERCON), pp. 1–6, 24–28 Oct 2010
24. Moslehi, K., Kumar, R.: A reliability perspective of the smart grid. IEEE Trans. Smart Grid **1**(1), 57–64 (2010)
25. McDaniel, P., McLaughlin, S.: Security and privacy challenges in the smart grid. IEEE Secur. Priv. **7**(3), 75–77 (2009)
26. Venkitasubramaniam, P., Tong, L.: Anonymous networking with minimum latency in multihop networks. IEEE Symposium on Security and Privacy, 2008. SP 2008, pp. 18–32, 18–22 May 2008

27. Doshi, B., Harshavardhana, P.: Broadband network infrastructure of the future: roles of network design tools in technology deployment strategies. IEEE Commun. Mag. **36**, 60–71 (1998)
28. To, M., Neusy, P.: Unavailability analysis of long-haul networks. IEEE J. Sel. Areas Commun. **12**, 100–109 (1994)
29. Singel, R.: Fiber optic cable cuts isolate millions from internet, future cuts likely wired. http://www.wired.com/threatlevel/2008/01/fiber-optic-cab/ (2008). Accessed January 2008
30. Hachman, M.: Sabotage suspected in silicon valley cable cut PCMag. http://www.pcmag.com/article2/0,2817,2344762,00.asp (2009) . Accessed 9 April 2009
31. Farley, J.: Bremerton fiber optic cable cut knocks out service for wave broadband customers. http://www.kitsapsun.com/news/2011/jul/06/bremerton-fiber-optic-cable-cut-knocks-out-for/#axzz36lWFNBmE(2011).
32. Zhang-shen, R., Mckeown, N.: Designing a predictable internet backbone with valiant load-balancing. IWQoS **2005**, 178–192 (2005)
33. Raza, K., Turner, M.: CCIE Professional Development Large-Scale IP Network Solutions. Cisco Press, Indianapolis (1999)
34. Iniewski, K., McCrosky, C., Minoli, D.: Network Infrastructure and Architecture: Designing High-Availability Networks. Wiley, New York (2008)
35. Riaz, T.: SQoS based planning for network infrastructures. Ph.D. thesis (2008)
36. Grover, W.D.: Mesh-Based Survivable Networks, Options and Strategies for Optical, MPLS, SONET and ATM Network, vol. 1. Prentice Hall PTR, Upper Saddle River (2003)
37. Ecobilan: FTTH solutions for a sustainable development (2008)
38. Madsen, O.B., Knudsen, T.P., Pedersen, J.M.: SQOS as the base for next generation global infrastructure. In: Proceedings of IT&T 2003, Information Technology and Telecommunications Annual Conference 2003, pp. 127–136 (2003)
39. Caenegem, B.V., Parys, W.V., Turck, F.D., Demeester, P.: Dimensioning of survivable wdm networks. IEEE J. Sel. Areas in Commun. **16**, 1146–1157 (1998)
40. Gutierrez, J.M., Katrinis, K., Georgakilas, K., Tzanakaki, A., Madsen, O.B.: Increasing the cost-constrained availability of WDM networks with degree-3 structured topologies. In: 12th International Conference on Transparent Optical Networks (ICTON), 2010, pp. 1–4 (2010)
41. Rados, I.: Availability analysis and comparison of different wdm systems. J. Telecommun. Inf. Technol. **1**, 114–119 (2007)
42. Zhou, L., Held, M., Sennhauser, U.: Connection availability analysis of shared backup path-protected mesh networks. J. Lightwave Technol. **25**, 1111–1119 (2007)
43. Booker, G., Sprintson, A., Zechman, E., Singh, C., Guikema, S.: Efficient traffic loss evaluation for transport backbone networks. Comput. Netw. **54**, 1683–1691 (2010)
44. He, W., Somani, A.K.: Path-based protection for surviving double-link failures in mesh-restorable optical networks. In: Proceedings of IEEE Globecom 2003 (2003)
45. Gutierrez, J.M., Riaz, T., Pedersen, J.M.: Cost and availability analysis of 2- and 3-connected WDM networks physical interconnection. In: Proceedings in ICNC 2012 (2012)
46. Hansen, M.B., Olsen, R.L., Schwefel, H.-P.: Probabilistic models for access strategies to dynamic information elements. Perform. Eval. **67**(1), 43 (2010)
47. Schwefel, H.-P., Hansen, M.B., Olsen, R.L.: Adaptive Caching strategies for Context Management systems, PIMRC07, Athens, Sept 2007
48. Shawky, A., Olsen, R., Pedersen, J., Schwefel, H.: Network Aware Dynamic Context Subscription Management, Computer Networks, vol. 58, pp. 239–253. 15 January 2014, ISSN 1389-1286. http://dx.doi.org/10.1016/j.comnet.2013.10.006.
49. Hald, S.L.N., Pedersen, J.M.: The Threat of Digital Hacker Sabotage to Critical Infrastructure. Submitted for GIIS 2012 (2012)
50. Hald, S.L.N., Pedersen, J.M.: An updated taxonomy for characterizing hackers according to their threat properties. In: 14th International Conference on Advanced Communication Technology (ICACT) 2012, IEEE (2011). ISBN 978-8955191639

51. Moteff, J.: Risk Management and Critical Infra-structure Protection: Assessing, Integrating, and Managing Threats, Vulnerabilities and Consequences. Congressional Research Service, Washington D.C. (2005)
52. Devost, M.G.: Current and emerging threats to information technology systems and critical infra-structures. Glob. Bus. Brief. (2000)
53. The White House: The National Strategy to Secure Cyberspace, p. 5. The White House, Washington D.C. (2003)
54. Vatis, M.A.: Cyber Attacks During the War on Terrorism: A Predictive Analysis. Institute for Security, Dartmouth College, Hanover (2001)
55. Shea, Dana A.: Critical Infrastructure: Control Systems and the Terrorist Threat. Congressional Research Service, Washington D.C. (2004).http://fas.org/irp/crs/RL31534.pdf
56. Lewis, James A.: Cybersecurity and Critical Infrastructure Protection. Center for Strategic and International Studies, Washington D.C. (2006)
57. Rogers, M.: A two-dimensional circumplex approach to the development of a hacker taxonomy. Digit. Investig. **3**(97–102), 2006 (2006)
58. Rollins, J., Wilson, C.: Terrorist Capabilities for Cyberattack: Overview and Policy Issues. Congressional Research Service, Washington D.C. (2007)
59. Hunt, J.: Stuxnet, Security, and Taking Charge, Industrial Ethernet Book Issue 62/53, IEB Media GbR, Germany (2011). ISSN 1470-5745
60. Eronen, J., Karjalainen, K., et al.: Software vulnerability vs. critical infrastructure—a case study of antivirus software. Int. J. Adv. Secur. **2**(1) (2009). ISSN 1942-2636 (International Academy, Research, and Industry Association)
61. Department of Homeland Security: National Cybersecurity and Communications Integration Center Bulletin: Assessment of Anonymous Threat to Control Systems. Department of Homeland Security, Washington D.C. (2011)
62. Anonymous, youranonnews: Available at https://twitter.com/youranonnews/status/171941104860672000 (2012)
63. Antonatos, S., Akriditis, P., et al.: Defending Against Hitlist Worms Using Network Address Space Randomization, WORM '05, ACM 1-59593-229-1/05/0011, USA (2005)
64. Lai, S.-C., Kuo, W.-C., et al.: Defending against Internet worm-like infestations. In: Proceedings of the 18th International Conference on Advanced Information Networking and Application (AINA'04), ISSN 0-7695-2051-0/04, IEEE (2004)
65. Keeney, M., Cappelli, D., et al.: Insider Threat Study: Computer System Sabotage in Critical Infrastructure Sectors. United States Secret Service and Carnegie Mellon Software Engineering Institute, Washington D.C. (2005)
66. Capelli, D., Moore, A., et al.: Common Sense Guide to Prevention and Detection of Insider Threats, 3rd edn. Version 3.1, Software Engineering Institute, Carnegie Mellon University (2009)
67. Hernandez, J.A., Phillips, I.W.: Weibull mixture model to characterise end-to-end Internet delay at coarse time-scales. IEE Proc. Commun. **153**(2), 295–304 (2006). doi:10.1049/ip-com:20050335
68. Bolot, J.-C.: Characterizing end-to-end packet delay and loss in the Internet. J. High Speed Netw. IOS Press. ISSN 0926-6801 (Print), 1875-8940 (Online), Comput. Sci. Netw. Secur. **2**(3), 305–323 (1993)
69. Bovy, C.J., Mertodimedjo, H.T., Hooghiemstra, G., Uijterwaal, H., Van Mieghem, P.: Analysis of end-to-end delay measurements in Internet. In: Proceedings of the Passive and Active Measurement Workshop-PAM 2002 (2002)
70. Klima-, Energi- og Bygningsministeriet, HOVEDRAPPORT for Smart Grid Netværkets arbejde, available online at http://www.kebmin.dk/en
71. ECOGRID Bornholm: Official websitehttp://ecogridbornholm.dk/

Water Distribution Networks

Avi Ostfeld

Abstract A water distribution system is a complex assembly of hydraulic control elements connected together to convey quantities of water from sources to consumers. The typical high number of constraints and decision variables, the non-linearity, and the non-smoothness of the head—flow—water quality governing equations are inherent to water supply systems planning and management problems. Traditional methods for solving water distribution systems management problems, such as the least cost design and operation problem, utilized linear/nonlinear optimization schemes which were limited by the system size, the number of constraints, and the number of loading conditions. More recent methodologies employ heuristic optimization techniques, such as genetic algorithms or ant colony optimization as stand alone or hybrid data driven—heuristic schemes. This book chapter reviews some of the more traditional water distribution systems problem algorithms and solution methodologies. It is comprised of sub sections on least cost and multi-objective optimal design of water networks, reliability incorporation in water supply systems design, optimal operation of water networks, water quality analysis inclusion in distribution systems, water networks security related topics, and a look into the future.

Keywords Water distribution systems · Operation · Least cost design · Water quality · Optimization · Multi-objective

A. Ostfeld (✉)
Faculty of Civil and Environmental Engineering, Technion—Israel Institute of Technology, 32000 Haifa, Israel
e-mail: ostfeld@tx.technion.ac.il
URL: http://www.technion.ac.il/~avi/avi.htm

E. Kyriakides and M. Polycarpou (eds.), *Intelligent Monitoring, Control, and Security of Critical Infrastructure Systems*, Studies in Computational Intelligence 565,
DOI 10.1007/978-3-662-44160-2_4

1 Introduction

A water distribution network is an interconnected collection of sources, pipes and hydraulic control elements (e.g., pumps, valves, regulators, tanks) delivering to consumers prescribed water quantities at desired pressures and water qualities. Such systems are often described as a graph, with the links representing the pipes, and the nodes defining connections between pipes, hydraulic control elements, consumers, and sources. The behavior of a water distribution network is governed by: (1) the physical laws which describe the flow relationships in the pipes and the hydraulic control elements, (2) the consumer demands, and (3) the system's layout.

Management problems associated with water supply systems can be classified into: (1) layout (system connectivity/topology); (2) design (system sizing given a layout); and (3) operation (system operation given a design).

On top of those, problems related to aggregation, maintenance, reliability, unsteady flow and security can be identified for gravity, and/or pumping, and/or storage branched/looped water distribution systems. Flow and head, or flow, head, and water quality can be considered for one or multiple loading scenarios, taking into consideration inputs/outputs as deterministic or stochastic variables. Figure 1 is a schematic description of the above.

The typical high number of constraints and decision variables, the nonlinearity, and the non-smoothness of the head—flow—water quality governing equations are inherent to water supply systems planning and management problems. An example of this is the least cost design problem of a water supply system defined as finding the water distribution system's component characteristics (e.g., pipe diameters, pump heads and maximum power, reservoir storage volumes, etc.), which minimize the system capital and operational costs, such that the system hydraulic laws are maintained (i.e., Kirchoff's Laws no. 1 and 2 for continuity of flow and energy, respectively), and constraints on quantities and pressures at the consumer nodes are fulfilled.

Traditional methods for solving water distribution system management problems used linear/nonlinear optimization schemes which were limited by the system size, the number of constraints, and the number of loading conditions. More recent methodologies employ heuristic optimization techniques, such as genetic algorithms or ant colony optimization as stand alone or hybrid data driven—heuristic schemes.

This book chapter reviews part of the topics presented in Fig. 1. It consists of sub sections on least cost and multi-objective optimal design of water networks, reliability incorporation in water supply systems design, optimal operation of water networks, water quality analysis inclusion in distribution systems, water networks security related topics, and a look into the future.

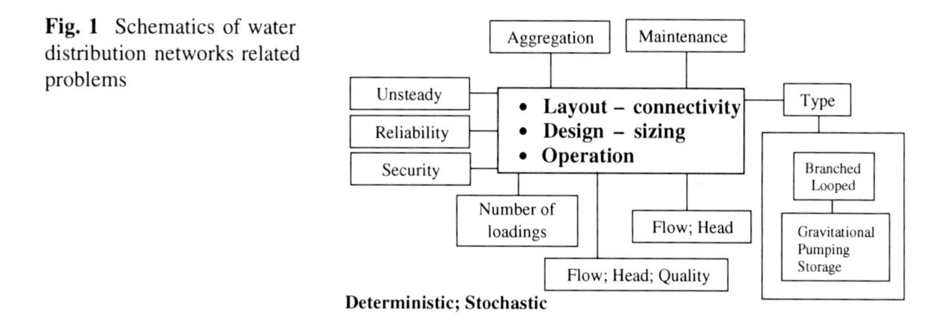

Fig. 1 Schematics of water distribution networks related problems

2 Least Cost and Multi-Objective Optimal Design of Water Networks

2.1 Least Cost Design of Water Networks

The optimal design problem of a water distribution system is commonly defined as a single objective optimization problem of finding the water distribution system component characteristics (e.g., pipe diameters, pump heads and maximum power, reservoir storage volumes, etc.), which minimize the system capital and operational costs, such that the system hydraulic laws are maintained (i.e., Kirchoff's Laws no. 1 and 2 for continuity of flow and energy, respectively), and constraints on quantities and pressures at the consumer nodes are fulfilled.

Numerous models for least cost design of water distribution systems have been published in the research literature during the last four decades. A possible classification for those might be: (1) *decomposition:* methods based on decomposing the problem into an "inner" linear programming problem which is solved for a fixed set of flows (heads), while the flows (heads) are altered at an "outer" problem using a gradient or a sub-gradient optimization technique [1–6]; (2) *linking simulation with nonlinear programming:* methods based on linking a network simulation program with a general nonlinear optimization code [7–9]; (3) *nonlinear programming:* methods utilizing a straightforward nonlinear programming formulation [10, 11]; (4) *methods employing evolutionary/meta-heuristic techniques:* genetic algorithms [12–16], simulated annealing [17], the shuffled frog leaping algorithm [18], ant colony optimization [19]; and (5) *other methods*: dynamic programming [20], integer programming [21].

Decomposition methods [1–6] are limited in the number of loading conditions that can be considered, to converging to local optimal solutions, and to fixed flow directions in the pipes as of the non-smoothness properties of the "outer" problem (excluding [2, 4] who used a sub-gradient scheme to minimize the "outer" problem), but can account for split pipe diameter solutions. Methods based on linking a network simulation program with a general nonlinear optimization code [7–9] divide the overall problem into two levels. In the lower level the system is analyzed for flows, pressures, and cost using a network simulation program, while in the upper level the system design variables: pipe diameters, pump heads, and reservoir volumes are modified according to the information provided by successive runs of the simulation program. The upper level is a general purpose optimization package, such as MINOS [22] or GRG2 [23]. The optimization algorithm uses values of the objective function generated in successive runs of the simulation program, and information on constraint violations to determine the next solution to be tested. Methods based on nonlinear programming simultaneously solve the optimal heads and flows, using general optimization schemes: [11] solved the least cost design of a water distribution system with pipes and pumps under one loading condition, with the design problem transformed into an unconstrained optimization problem using an exterior penalty method; [10] developed a methodology for the least cost design/operation of a water distribution system under multiple loading conditions based on the general reduced gradient (GRG) [24]. Methods based on using a straightforward nonlinear code are limited with respect to the water distribution system size that can be handled, the user intervention, the number of loading conditions, and most likely their convergence to local optimal solutions.

The capabilities of solving water distribution systems optimization problems have improved dramatically since the employment of genetic algorithms [25]. Genetic algorithms are domain heuristic independent global search techniques that imitate the mechanics of natural selection and natural genetics of Darwin's evolution principle. The primary idea is to simulate the natural evolution mechanisms of chromosomes, represented by string structures, involving selection, crossover, and mutation. Strings may have binary, integer, or real values. Simpson et al. [14] were the first to use genetic algorithms for water distribution systems least cost design. They applied and compared a genetic algorithm solution to the network of [26], to enumeration and to nonlinear optimization. Savic and Walters [13] used genetic algorithms to solve and compare optimal results of the one-loading gravity systems of the Two Loop Network [1], the Hanoi network [27], and the New York Tunnels system [28]. Salomons [12] used a genetic algorithm for solving the least cost design problem incorporating extended period loading conditions, tanks, and pumping stations. Vairavamoorthy and Ali [15] presented a genetic algorithm framework for the least cost design problem of a pipe network which excludes regions of the search space where impractical or infeasible solutions are likely to exist, and thus improves the genetic algorithm search efficiency. Wu and Walski [16] introduced a self-adaptive penalty approach to handle the transformation from a constrained into a non-constrained framework of the least cost design and rehabilitation problems of a water distribution system, as applied in a genetic algorithm scheme. Loganathan et al. [17] used the decomposition idea proposed by [1] but with minimizing the "outer" problem through a simulated annealing scheme, showing substantial improvements over previous decomposition methods which used a gradient type procedure to minimize the "outer" problem. Eusuff and Lansey [18] developed a swarm based meta-heuristic algorithm, entitled the Shuffled Frog Leaping Algorithm (SFLA). The SFLA was applied and compared to the same problems as in [13]. Maier et al. [19] applied an ant colony algorithm based on [29, 30] to the gravitational network of [26] and to the New York Tunnels system [28]. Singh and Mahar [20] used dynamic programming to solve the optimal design problem of a multi-diameter, multi-outlet pipeline satisfying pressure outlet constraints. Samani and Mottaghi [21] employed branch and bound integer linear programming to solve the least cost design problem of one loading water distribution systems.

2.2 Multi-Objective Optimal Design of Water Networks

In reality the design problem (as almost any engineering problem) of a water distribution system involves competing objectives, such as minimizing cost, maximizing reliability, minimizing risks, and minimizing deviations from specific targets of quantity, pressure, and quality. The design problem is thus inherently of a multi-objective nature. In a multi-objective optimization problem there is not a single optimal solution but a set of compromised solutions, which form a Pareto optimal solution set. Thus, incorporating multiple objectives in the optimal design of water distribution systems provides a substantial improvement compared to using a single design objective, as a broader range of alternatives is explored, thus making the design outcome much more realistic.

Halhal et al. [31] were the first to introduce a multi-objective procedure to solve a water distribution systems management problem. Minimizing network cost versus maximizing the hydraulic benefit served as the two conflicting objectives, with the total hydraulic benefit evaluated as a weighted sum of pressures, maintenance cost, flexibility, and a measure of water quality benefits. A structured messy genetic algorithm was implemented to solve the optimization problem. Kapelan et al. [32] used a multi-objective genetic algorithm to find sampling locations for optimal calibration. The problem was formulated as a two-multi-objective optimization problem with the objectives been the maximization of the calibrated model accuracy versus the minimization of the total sampling design cost. The problem was solved using a Pareto ranking, niching, and a restricted mating multi-objective genetic algorithm. Karmeli et al. [33] applied a hybrid multi-objective evolutionary algorithm to the optimal design problem of a water distribution system. The hybrid approach employed a non-dominated sorting genetic algorithm coupled with a neighborhood search technique. Two objectives were considered: minimum cost versus minimum head shortage at the consumer nodes. Prasad and Park [34] applied a non-dominated sorting genetic algorithm for minimizing the network cost versus maximizing a reliability index. The reliability index used combined surplus consumer nodes pressure heads with loops having a minimum pipe diameter constraint. Prasad and Park [34] presented a multi-objective genetic algorithm approach to the optimal design of a water distribution network with minimizing the network cost versus maximizing the network resilience, where the network resilience is defined as a reliability surrogate measure taking into consideration excess pressure heads at the network nodes and loops with practicable pipe diameters. Farmani et al. [35]

compared three evolutionary multi-objective optimization algorithms for water distribution system design through visualizing the resulted non-dominated fronts of each of the methods and by using two performance indicators. Vamvakeridou-Lyroudia et al. [36] employed a genetic algorithm multi-objective scheme to tradeoff the least cost to maximum benefits of a water distribution system design problem, with the benefits evaluated using fuzzy logic reasoning. Babayan et al. [37] used a multi-objective genetic algorithm to solve the design problem of a water distribution system under uncertainty. Two objectives were considered: minimum cost versus the probability of the network failure due to uncertainty in input variables. The first objective was evaluated by minimizing the total system cost, while the second by maximizing the nodal pressures above a minimum value. The stochastic problem which simulated the uncertainty of the system inputs was replaced with a deterministic numerical approach which quantified the uncertainties.

3 Reliability Incorporation in Water Supply Systems Design

Reliability considerations are an integral part of all decisions regarding the planning, design, and operation phases of water distribution systems. Quantitatively, the reliability of a water distribution system can be defined as the complement of the probability that the system will fail, where a failure is defined as the system's inability to supply its consumers' demands.

A major problem, however, in reliability analysis of water distribution systems is to define reliability measures which are meaningful and appropriate, while still computationally feasible. While the question, "Is the system reliable?", is usually understood and easy to follow, the question, "Is it reliable enough?", does not have a straightforward response, as it requires both the quantification and evaluation of reliability measures. Much effort has already been invested in reliability analysis of water supplies. These examinations, however, still commonly follow heuristic guidelines like ensuring two alternative paths to each demand node from at least one source, or having all pipe diameters greater than a minimum prescribed value. By using these guidelines, it is implicitly assumed that reliability is assured, but the level of reliability provided is not quantified or measured.

Reliability of water distribution systems gained considerable research attention over the last three decades. Research has concentrated on methodologies for reliability assessment and for reliability inclusion in least cost design and operation of water supply systems. A summary of these two major efforts is provided below.

3.1 Reliability Evaluation Models

Shamir and Howard [38] were the first to propose analytical methods for water supply system reliability. Their methodology took into consideration flow capacity, water main breaks, and maintenance for quantifying the probabilities of annual shortages in water delivery volumes.

Vogel [39] suggested the average return period of a reservoir system failure as a reliability index for water supply. A Markov failure model was utilized to compute the index, which defined failure as a year in which the yield could not be delivered. Wagner et al. [40] proposed analytical methods for computing the reachability (i.e., the case in which a given demand node is connected to at least one source) and connectivity (i.e., the case in which every demand node is connected to at least one source) as topological measures for water distribution systems reliability. Wagner et al. [41] complemented [40] through stochastic simulation in which the system was modeled as a network whose components were subject to failure with given probability distributions.

Reliability measures such as the probability of shortfall (i.e., total unmet demand), the probability of the number of failure events in a simulation period, and the probability of inter-failure times and repair durations were used as reliability criteria. Bao and Mays [42] suggested stochastic simulation by imposing uncertainty in future water demands for computing the probability that the water distribution system will meet these needs at minimum pressures. Duan and Mays [43] used a continuous-time Markov process for reliability assessment of water supply pumping stations. They took into consideration both mechanical and hydraulic failure (i.e., capacity shortages) scenarios, all cast in a conditional probability frequency and duration analysis framework. Jacobs and Goulter [44] used historical pipe failure data to derive the probabilities that a particular number of simultaneous pipe failures will cause the entire system to fail.

Quimpo and Shamsi [45] employed connectivity analysis strategies for prioritizing maintenance decisions. Bouchart and Goulter [46] developed a model for optimal valve locations to minimize the consequences of pipe failure events, recognizing that in reality, when a pipe fails, more customers are isolated than those situated at the pipe's two ends. Jowitt and Xu [47] proposed a micro-flow simplified distribution model to estimate the hydraulic impact of pipe failure scenarios. Fujiwara and Ganesharajah [48] explored the reliability of a water treatment plant, ground-level storage, a pumping station, and a distribution network in a series, using the expected served demand as the reliability measure. Vogel and Bolognese [49] developed a two-state Markov model for describing the overall behavior of water supply systems dominated by carry-over storage. The model quantifies the trade-offs among reservoir system storage, yield, reliability, and resilience. Schneiter et al. [50] explored the system capacity reliability (i.e., the probability that the system's carrying capacity is able to meet flow demands) for enhancing maintenance and rehabilitation decision making.

Yang et al. [51] employed the minimum cut-set method for investigating the impact of link failures on source-demand connectivity. Yang et al. [52] complemented the reliability connectivity model of [51] with Monte Carlo simulations for pipe failure impact assessments on a consumer's shortfalls. Xu and Goulter [53] developed a two stage methodology for reliability assessment of water distribution systems using a linearized hydraulic model coupled with probability distributions of nodal demands, pipe roughnesses, and reservoir/tank levels. Fujiwara and Li [54] suggested a goal programming model for flow redistribution during failure events for meeting customers' equity objectives. Tanyimboh et al. [55] used pressure-driven simulation to compute the reliability of single-source networks under random link failures. Ostfeld et al. [56] applied stochastic simulation to quantify the reliability of multi-quality water distribution systems, using the fraction of delivered volume, demand, and quality as reliability measures.

Shinstine et al. [57] coupled a cut-set method with a hydraulic steady state simulation model to quantify the reliability of two large-scale municipal water distribution networks. Ostfeld [58] classified existing reliability analysis methodologies and compared two extreme approaches for system reliability assessment: "lumped supply-lumped demand" versus stochastic simulation. Tolson et al. [59] used the same approach as [60] for optimizing the design of water distribution systems with capacity reliability constraints by linking a genetic algorithm (GA) with the first-order reliability method (FORM). Ostfeld [61] complemented the study of [4] by designing a methodology for finding the most flexible pair of operational and backup subsystems as inputs for the design of optimal reliable networks. Recently, [62] utilized first order reliability methods in conjunction with an adaptive response surface approach for analyzing the reliability of water distribution systems; and [63] compared the surrogate measures of statistical entropy, network resilience, resilience index, and the modified resilience index for quantifying the reliability of water networks.

3.2 Reliability Inclusion in Optimal Design and Operation of Water Supply Systems

Su et al. [64] were the first to incorporate reliability into least cost design of water distribution systems. Their model established a link between a steady state one loading hydraulic simulation, a reliability model based on the minimum cut-set method [65], and the general reduced gradient GRG2 [23] for system optimization. Ormsbee and Kessler [66] used a graph theory methodology for optimal reliable least cost design of water distribution systems for creating a one level system redundancy (i.e., a system design that guarantees a predefined level of service in case one of its components is out of service). Khang and Fujiwara [67] incorporated minimum pipe diameter reliability constraints into the least cost design problem of water distribution systems, showing that at most two pipe diameters can be selected for a single link. Park and Liebman [68] incorporated into the least cost design problem of water distribution systems the expected shortage of supply due to failure of individual pipes. Ostfeld and Shamir [4] used backups (i.e., subsystems of the full system that maintain a predefined level of service in case of failure scenarios) for reliable optimal design of multi-quality water distribution systems. Xu and Goulter [60] coupled the first-order reliability method (FORM), which estimates capacity reliability, with GRG2 [23] to optimize the design of water distribution systems. Ostfeld [69] developed a reliability assessment model for regional water supply systems, comprised of storage-conveyance analysis in conjunction with stochastic simulation. Afshar [70] presented a heuristic method for the simultaneous layout and sizing of water distribution systems using the number of independent paths from source nodes to consumers as the reliability criterion. Farmani et al. [71] applied for Anytown USA [72], a multi-objective evolutionary algorithm for trading off cost and the resilience index [73] as a reliability surrogate. Dandy and Engelhardt [74] used a multi-objective genetic algorithm to generate trade-off curves between cost and reliability for pipe replacement decisions. Agrawal [75] presented a heuristic iterative methodology for creating the trade-off curve between cost and reliability (measured as a one level system redundancy) through strengthening and expanding the pipe network. Reca et al. [76] compared different metaheuristic methodologies for trading off cost and reliability, quantified as the resilience index [73]. van Zyl et al. [77] incorporated reliability criteria for tank sizing. Duan et al. [78] explored the impact of system data uncertainties, such as pipe diameter and friction on the reliability of water networks under transient conditions. Ciaponi et al. [79] introduced a simplified procedure based on the unavailability of pipes for comparing design solutions with reliability considerations.

4 Optimal Operation of Water Networks

Subsequent to the well known least cost design problem of water distribution systems [1, 80, 81], optimal operation is the most explored topic in water distribution systems management. Since 1970 a variety of methods were developed to address this problem, including the utilizations of dynamic programming, linear programming, predictive control, mixed-integer, non-linear programming, metamodeling, heuristics, and evolutionary computation. Ormsbee and Lansey [82] classified to that time optimal water distribution systems control models through systems type, hydraulics, and solution methods. This section reviews the current literature on this subject.

4.1 Dynamic Programming

Dreizin [83] was the first to suggest an optimization model for water distribution systems operation through a dynamic programming (DP) scheme coupled with hydraulic simulations for optimizing pumps scheduling of a regional water supply system supplied by three pumping units. Sterling and Coulbeck [84] used a dynamic modeling approach to minimize the costs of pumps operation of a simple water supply system. Carpentier and Cohen [85] developed a decomposition-coordination methodology for partitioning a water supply system into small sub-systems which could be solved separately (i.e., decomposed) using dynamic programming, and then merged (i.e., coordinated) at the final solution. Houghtalen and Loftis [86] suggested aggregating training simulations with human operational knowledge and dynamic programming to minimize operational costs. Ormsbee et al. [87] developed a coupled dynamic programming and enumeration scheme for a single pressure zone in which the optimal tank trajectory is found using dynamic programming and the pumps scheduling using enumeration. Zessler and Shamir [88] used an iterative dynamic programming method to find the optimal scheduling of pumps of a regional water supply system. Lansey and Awumah [89] used a two level approach in which the hydraulics and cost functions of the system are generated first off-line followed by a dynamic programming model for pumps scheduling. Nitivattananon et al. [90] utilized heuristic rules combined with progressive optimality to solve a dynamic programming model for optimal pumps scheduling. McCormick and Powell [91] utilized a stochastic dynamic program framework for optimal pumps scheduling where daily demand for water are modeled as a Markov process.

4.2 Linear Programming

Olshansky and Gal [92] developed a two level linear programming methodology in which the distribution system is partitioned into sub-systems for which hydraulic simulations are run and serve further as parameters in an LP model for pumps optimal scheduling. This approach was used also by [93] who developed a linear programming model to optimize pumps scheduling in which the LP parameters are set through off-line extended period hydraulic simulation runs. Diba et al. [94] used graph-theory coupled with a linear programming scheme for optimizing the operation and planning of a water distribution system including reliability constraints.

4.3 Predictive Control

Coulbeck and Coulbeck et al. [95–97] suggested hierarchical control optimization frameworks for the optimal operation of pumps. Biscos et al. [98] used a predictive control framework coupled with mixed integer non-linear programming (MINLP) for minimizing the costs of pump operation. Biscos et al. [99] extended [98] to include the minimization of chlorine dosage.

4.4 Mixed-Integer

Ulanicki et al. [100] developed a mixed-integer model for tracking the optimal reservoirs trajectories based on the results of an initially relaxed continuous problem. Pulido-Calvo and Gutiérrez-Estrada [101] presented a model for both sizing storage and optimizing pumps operation utilizing a framework based on a mixed integer non-linear programming (MINLP) algorithm and a data driven (neural networks) scheme.

4.5 Non-Linear Programming

Chase [102] used an optimization-simulation framework coupling the general reduced gradient GRG2 [23] with a water distribution system simulation model WADISO [103] for minimizing pumps cost operation. Brion and Mays [104] developed an optimal control simulation-optimization framework for minimizing pumps operation costs in which the simulation solves the hydraulic equations and the optimization utilizes the non-linear augmented Lagrangian method [105]. Pezeshk et al. [106] linked hydraulic simulations with non-linear optimization to minimize the operation costs of a water distribution network. Cohen et al. [107–109] presented three companion papers on optimal operation of water distribution systems using non-linear programming: with water quality considerations only [107], with flow inclusion [108], and with both flow and quality [109].

4.6 Metamodeling

Broad et al. [110] used an artificial neural network (ANN) as a metamodel for optimizing the operation of a water distribution system under residual chlorine constraints. Shamir and Salomons [111] developed a framework for real-time optimal operation integrating an aggregated/reduced model, an artificial neural network, and a genetic algorithm. Broad et al. [112] extended [110] through comparing four different metamodelling scenarios and suggesting skeletonization procedures.

4.7 Heuristics

Tarquin and Dowdy [113] used heuristic analysis of pump and system head curves to identify pump combinations which reduce operation costs. Pezeshk and Helweg [114] introduced a heuristic discrete adaptive search algorithm for optimal pumps scheduling based on pressure readings at selected network nodes. Ormsbee and Reddy [115] linked a minimum-cost-constraint identification methodology with nonlinear heuristics for optimal pumps scheduling.

4.8 Evolutionary Computation

Sakarya and Mays [116] presented a simulating annealing [117] scheme for optimizing the operation of a water distribution system with water quality constraints. Cui and Kuczera [118] used a genetic algorithm (GA) [25, 119] and the shuffled complex evolution (SCE) method [120] to optimize urban water supply headworks. Ostfeld and Salomons [121, 122] minimized the total cost of pumping and water quality treatment of a water distribution system through linking a genetic algorithm with EPANET (www.epa.gov/nrmrl/wswrd/dw/epanet.html). van Zyl et al. [123] utilized a genetic algorithm (GA) linked to a hillclimber search algorithm for improving the local GA search once closed to an optimal solution. López-Ibáñez et al. [124] proposed an ant colony optimization (ACO) [125] framework for optimal pumps scheduling. Boulos et al. [126] developed the H_2ONET tool based on genetic algorithms for scheduling pump operation to minimize operation costs.

4.9 Commercial Modeling Tool

Commercial applications for energy minimization have been developed by companies such as Derceto (http://www.derceto.com/), Bentley (http://www.bentley.com/en-US/Solutions/Water+and+Wastewater/), MWH Soft (http://www.mwhglobal.com/) and others. These applications allow system design, optimal pump scheduling and system operation while minimizing system operation cost and optimizing water supply.

5 Water Quality Analysis Inclusion in Distribution Systems

Research in modeling water quality in distribution systems started in the context of agricultural usage (e.g., [127, 128]) primarily in arid regions where good water quality is limited. In 1990 the United States Environmental Protection Agency (USEPA) promulgated rules requiring that water quality standards must be satisfied at the consumer taps rather than at treatment plants. This initiated the need for water quality modeling, the development of the USEPA simulation water quantity and quality model EPANET (EPANET 2.0@2000, [129]), and raised other problems and research needs that commenced considerable research in this area to assist utilities.

Shamir and Howard [130] were the first to classify water quality models for water distribution systems. Their classification was based upon the flow conditions in the network and the contaminant concentrations in the sources: (1) *steady flow—steady concentration*. This occurs in agriculture or industry; flows are rarely steady in municipal water distribution systems, (2) *steady flow—unsteady concentration*. This appears when a pulse of contamination is distributed within the distribution system under steady flow conditions, (3) *unsteady flow—steady concentration*. Contaminant concentrations in the sources remain constant while the flow regime is unsteady, and (4) *unsteady flow—unsteady concentration*. This occurs when a pulse of contamination enters the system under unsteady flow conditions.

The above classifications set the boundary conditions for the analysis of flow and water quality in water distribution systems. These involve four major categories: simulation, optimization, chlorine control, and monitoring. This section hereafter concentrates on the optimization of multi-quality water networks.

Optimization models of water distribution systems can be classified according to their consideration of time and of the physical laws which are included explicitly [131, 132]. In time the distinction is between policy and real time models. Policy models are run off - line, in advance, and generate the operating plans for several typical and/or critical operating conditions. Real time (on-line) models are run continuously in real time, and generate an operating plan for the immediate coming period. The classification with respect to the physical laws which are considered explicitly as constraints, is: (1) *QH (discharge—head) models:* quality is not considered, and the network is described only by its hydraulic behavior; (2) *QC (discharge—quality) models:* the physics of the system are included only as continuity of water and of pollutant mass at nodes. Quality is described essentially as a transportation problem in which pollutants are carried in the pipes, and mass conservation is maintained at nodes. Such a model can account for decay of pollutants within the pipes and even chemical reactions, but does not satisfy the continuity of energy law (i.e., Kirchoff's Law no. 2), and thus there is no guarantee of hydraulic feasibility and of maintaining head constraints at nodes; and (3) *QCH (discharge—quality—head) models:* quality constraints, and the hydraulic laws, which govern the system behavior, are all considered. The QH and QC problems are relatively easier to solve than the full QCH.

Ostfeld and Shamir [131] developed a QCH policy model for optimal operation of undirected multi-quality water distribution systems under steady state conditions, which has been extended to the unsteady case in [132]. To overcome the non-smoothness problems, an approximation of the quality equations has been used, following [133]. In the unsteady case, instead of dividing each pipe into segments and tracking the movement of the quality fronts, a single approximated equation was developed, representing the average concentration of the water quality fronts in the pipes for a specific time increment. Both the steady and unsteady models were solved with GAMS [134]/MINOS [22], an on-shelf nonlinear optimization package.

Ostfeld and Shamir [4] developed a QCH methodology which integrates optimal design and reliability of a multi-quality water distribution system in a single framework. The system designed is able to sustain prescribed failure scenarios, such as any single random component failure, and still maintain a desired level of service in terms of the quantities, qualities, and pressures supplied to the consumers. In formulating and solving the model, decomposition was used. The decomposition results in an "outer" non-smooth problem in the domain of the circular flows, and an "inner" convex quadratic problem. The method of solution included the use of a non-smooth optimization technique for minimizing the "outer" problem [135], for which a member of the sub-gradient group was calculated in each iteration. The method allowed reversal of flows in pipes, relative to the direction initially assigned. The methodology was applied to Anytown USA [72] for a single loading condition, and one quality constituent. Cohen et al. [109] solved the steady state operation model of an undirected multi-quality water distribution system by decomposing the QCH problem into the QH and QC sub—problems for given water flows in the distribution system, and removal ratios at the treatment plants. The QC and QH models are solved first. The combination of their solutions serves for solving the QCH. The model has been applied to the Central Arava Network in southern Israel, which consists of 38 nodes, 39 pipes, 11 sources and 7 treatment plants.

Goldman [136] developed a simulated annealing shell linked to EPANET for solving the scheduling pumping problem of a water distribution system with water quality constraints at the consumer nodes. Sakarya and Mays [116] solved the same problem by linking the GRG2 nonlinear code [23] with EPANET. In both models, treatment facilities, valves, and varying electrical energy tariffs throughout the simulation were not considered. In addition, the unsteady water quality constraints were applied only to the last operational time period, while in reality the problem of supplying adequate water quality to consumers is a continuous operational time dependent problem.

Ostfeld et al. [56] developed a QCH application of stochastic simulation for the reliability assessment of single and multi-quality water distribution systems. The stochastic simulation framework was cast in a program entitled RAP (Reliability Analysis Program), linking Monte Carlo replications with EPANET simulations.

Three reliability measures were evaluated: the Fraction of Delivered Volume (FDV), the Fraction of Delivered Demand (FDD), and the Fraction of Delivered Quality (FDQ). Ostfeld and Salomons [121, 122] developed a genetic algorithm scheme tailored-made to EPANET, for optimizing the operation of a water distribution system under unsteady water quality conditions. The water distribution system consists of sources of different qualities, treatment facilities, tanks, pipes, control valves, and pumping stations. The objective is to minimize the total cost of pumping and treating the water for a selected operational time horizon, while delivering the consumers the required quantities, at acceptable qualities and pressures. The decision variables, for each of the time steps that encompass the total operational time horizon, included the scheduling of the pumping units, settings of the control valves, and treatment removal ratios at the treatment facilities. The constraints were domain heads and concentrations at the consumer nodes, maximum removal ratios at the treatment facilities, maximum allowable amounts of water withdraws at the sources, and returning at the end of the operational time horizon to a prescribed total volume in the tanks.

6 Water Networks Security Related Topics

Threats on a water distribution system can be partitioned into three major groups according to their resulted enhanced security: (1) a direct attack on the main infrastructure: dams, treatment plants, storage reservoirs, pipelines, etc.; (2) a cyber attack disabling the functionality of the water Supervisory Control and Data Acquisition (SCADA) system, taking over control of key components which might result water outages or insufficiently treated water, changing or overriding protocol codes, etc.; and (3) a deliberate chemical or biological contaminant injection at one of the system's nodes.

The threat of a direct attack can be minimized by improving the system's physical security (e.g., additional alarms, locks, fencing, surveillance cameras, guarding, etc.), while a cyber attack by implementing computerized hardware and software (e.g., an optical isolator between communication networks, routers to restrict data transfer, etc.).

Of the above threats, a deliberate chemical or biological contaminant injection is the most difficult to address. This is because of the uncertainty of the type of the injected contaminant and its effects, and the uncertainty of the location and injection time. Principally a contaminant can be injected at any water distribution system connection (node) using a pump or a mobile pressurized tank. Although backflow preventers provide an obstacle, they do not exist at all connections, and at some might not be functional.

The main course to enhance the security of a water distribution system against a deliberated contamination intrusion is through placing a sensor system (ASCE [137]; AWWA [138]).

In recent years there has been growing interest in the development of sensor systems with the majority of models using a single objective approach. The employment of multiobjective optimization for sensor placement started recently. This section reviews some models for sensor placement using the two approaches.

6.1 Single Objective Sensor Placement Models

Lee and Deininger [139] were the first to address the problem of sensor placement by maximizing the coverage of the demands using an integer programming model. Kumar et al. [140] improved the study of [139] by applying a greedy heuristic-based algorithm. Kessler et al. [141] suggested a set covering graph theory algorithm for the layout of sensors. Woo et al. [142] developed a sensor location design model by linking EPANET with an integer programming scheme. Al-Zahrani and Moeid [143] followed Lee and Deininger's approach using a genetic algorithm scheme [25, 119]. Ostfeld and Salomons [121, 122] extended [141, 144] to multiple demand loading and unsteady water quality propagations. Ostfeld and Salomons [145] extended [121, 122] by introducing uncertainties to the demands and the injected contamination events. Berry et al. [146] presented a mixed-integer programming (MIP) formulation for sensor placement showing that the MIP formulation is mathematically equivalent to the p-median facility location problem. Propato [147] introduced a mixed-integer linear programming model to identify sensor location for early warning, with the ability to accommodate different design objectives.

6.2 Multiobjective Sensor Placement Models

Watson et al. [148] were the first to introduce a multiobjective formulation to sensor placement by employing a mixed-integer linear programming model over a range of design objectives. The Battle of the Water Sensors [149] highlighted the multiobjective nature of sensor placement: [150] developed a constrained multi-objective optimization framework entitled the Noisy Cross-Entropy Sensor Locator (nCESL) algorithm based on the Cross Entropy methodology proposed by [151, 152] proposed a multiobjective solution using an "Iterative Deepening of Pareto Solutions" algorithm; [153] suggested a predator-prey model applied to multiobjective optimization, based on an evolution process; [154] proposed a multiobjective genetic algorithm framework coupled with data mining; [155, 156] used the multiobjective Non-Dominated Sorted Genetic Algorithm–II (NSGA-II) [157] scheme; [158] used a multiobjective optimization formulation, which was solved using a genetic algorithm, with the contamination events randomly generated using a Monte Carlo procedure.

7 A Look into the Future

Traditionally, water distribution networks were designed, operated, and maintained through utilizing offline small discrete datasets. Those were the governing and limiting constraints imposed on modeling challenges and capabilities. This situation is dramatically changing: from a distinct framework of data collection to a continuous transparent structure. With multiple types of sensor data at multiple scales, from embedded real-time hydraulic and water quality sensors to airborne and satellite-based remote sensing, how can those be efficiently integrated into new tools for decision support for water distribution networks is a major challenge.

This new reality is expected to limit all current modeling efforts capabilities and require new thinking on approaches for managing water distribution networks: from a state of lack of data to a situation of overflowing information. New tools for data screening, algorithms and metamodeling constructions, as well as computational efficiency are anticipated to govern all future developments for water distribution networks analysis.

References

1. Alperovits, E., Shamir, U.: Design of optimal water distribution systems. Water Resour. Res. **13**(6), 885–900 (1977)
2. Eiger, G., Shamir, U., Ben-Tal, A.: Optimal design of water distribution networks. Water Resour. Res. **30**(9), 2637–2646 (1994)
3. Kessler, A., Shamir, U.: Analysis of the linear programming gradient method for optimal design of water supply networks. Water Resour. Res. **25**(7), 1469–1480 (1989)
4. Ostfeld, A., Shamir, U.: Design of optimal reliable multiquality water supply systems. J. Water Resour. Planning Manage. Div. ASCE **122**(5), 322–333 (1996)
5. Quindry, G.E., Brill, E.D., Liebman, J.C., Robinson, A.R.: Comment on "Design of optimal water distribution systems" by Alperovits E. and Shamir U. Water Resour. Res. **15**(6), 1651–1654 (1979)
6. Quindry, G.E., Brill, E.D., Liebman, J.C.: Optimization of looped water distribution systems. J. Environ. Eng. ASCE **107**(EE4), 665–679 (1981)
7. Lansey, K.E., Mays, L.W.: Optimization models for design of water distribution systems. In: Reliability Analysis of Water Distribution Systems, Mays L. W. Ed., pp. 37–84 (1989)
8. Ormsbee, L.E., Contractor, D.N.: Optimization of hydraulic networks. In: Proceedings, International Symposium on Urban Hydrology, Hydraulics, and Sediment Control, pp. 255–261. Kentucky, Lexington KY (1981)
9. Taher, S.A., Labadie, J.W.: Optimal design of water-distribution networks with GIS. J. Water Resour. Planning Manage. Div. ASCE **122**(4), 301–311 (1996)
10. Shamir, U.: Optimal design and operation of water distribution systems. Water Resour. Res. **10**(1), 27–36 (1974)
11. Watanatada, T.: Least-cost design of water distribution systems. J. Hydraul. Div. ASCE **99**(HY9), 1497–1513 (1973)
12. Salomons, E.: Optimal design of water distribution systems facilities and operation. MS Thesis, Technion, Haifa, Israel (In Hebrew) (2001)
13. Savic, D., Walters, G.: Genetic algorithms for least cost design of water distribution networks. J. Water Resour. Planning Manage. Div. ASCE **123**(2), 67–77 (1997)

14. Simpson, A.R., Dandy, G.C., Murphy, L.J.: Genetic algorithms compared to other techniques for pipe optimization. J. Water Resour. Planning Manage. Div. ASCE **120**(4), 423–443 (1994)
15. Vairavamoorthy, K., Ali, M.: Pipe index vector: a method to improve genetic-algorithm-based pipe optimization. J. Hydraul. Eng. ASCE **131**(12), 1117–1125 (2005)
16. Wu, Z.Y., Walski, T.: Self-adaptive penalty approach compared with other constraint-handling techniques for pipeline optimization. J. Water Resour. Planning Manage. Div. ASCE **131**(3), 181–192 (2005)
17. Loganathan, G.V., Greene, J.J., Ahn, T.J.: Design heuristic for globally minimum cost water-distribution systems. J. Water Resour. Planning Manage. Div. ASCE **121**(2), 182–192 (1995)
18. Eusuff, M.M., Lansey, K.E.: Optimization of water distribution network design using the shuffled frog leaping algorithm. J. Water Resour. Planning Manage. Div. ASCE **129**(3), 210–225 (2003)
19. Maier, H.R., Simpson, A.R., Zecchin, A.C., Foong, W.K., Phang, K.Y., Seah, H.Y., Tan, C.L.: Ant colony optimization for design of water distribution systems. J. Water Resour. Planning Manage. Div. ASCE **129**(3), 200–209 (2003)
20. Singh, R.P., Mahar, P.S.: Optimal design of multidiameter, multioutlet pipelines. J. Water Resour. Planning Manage. Div. ASCE **129**(3), 226–233 (2003)
21. Samani, M.V., Mottaghi, A.: Optimization of water distribution networks using integer linear programming. J. Hydraul. Eng. ASCE **132**(5), 501–509 (2006)
22. Murtagh, B.A., Saunders, M.A.: A projected lagrangian algorithm and its implementation for sparse nonlinear constraints. Math. Program. Study **16**, 84–117 (1982)
23. Lasdon, L.S., Waren, A.D.: GRG2 user's guide. University of Texas, USA (1986). 50p
24. Abadie, J.: Application of the GRG method to optimal control problems. In: Abadie, J. (ed.) Integer and Nonlinear Programming, pp. 191–211. North Holland Publishing, Amsterdam (1970)
25. Holland, J.H.: Adaptation in natural and artificial systems. The University of Michigan Press, Ann Arbor, Ann Arbor (1975)
26. Gessler, J.: Pipe network optimization by enumeration. In: Proceedings Computer Applications for Water Resources, pp. 572–581. ASCE, New York, N. Y. (1985)
27. Fujiwara, O., Khang, D.B.: A two-phase decomposition method for optimal design of looped water distribution networks. Water Resour. Res. **26**(4), 539–549 (1990)
28. Schaake, J.C., Lai, D.: Linear programming and dynamic programming application to water distribution network design. Report No. 116, Department of Civil Engineering, Massachusetts Institute of Technology, Cambridge, Massachusetts (1969)
29. Dorigo, M., Maniezzo, V., Colorni, A.: Ant system: optimization by a colony of cooperating agents. IEEE Trans. Syst. Man Cybern. Part B **26**(1), 29–41 (1996)
30. Stützle, T., Hoos, H.H.: MAX–MIN ant system. Future Gener. Comput. Syst. **16**, 889–914 (2000)
31. Halhal, D., Walters, G.A., Savic, D.A., Ouazar, D.: Scheduling of water distribution system rehabilitation using structured messy genetic algorithms. Evol. Comput. **7**(3), 311–329 (1999)
32. Kapelan, Z.S., Savic, D.A., Walters, G.A.: Multi-objective sampling design for water distribution model calibration. J. Water Resour. Planning Manage. Div. ASCE **129**(6), 466–479 (2003)
33. Keedwell, E.C., Khu, S.T.: More choices in water distribution system optimization. In: Advances in Water Supply Management, Proceedings of Computers and Control in the Water Industry, pp. 257–265. London (2003)
34. Prasad, T.D., Park, N.-S.: Multi-objective genetic algorithms for design of water distribution networks. J. Water Resour. Planning Manage. Div. ASCE **130**(1), 73–82 (2004)
35. Farmani, R., Savic, D.A., Walters, G.A.: Evolutionary multi-objective optimization in water distribution network design. Eng. Optim. **37**(2), 167–183 (2005)

36. Vamvakeridou-Lyroudia, L.S., Walters, G.A., Savic, D.A.: Fuzzy multi-objective optimization of water distribution networks. J. Water Resour. Planning Manage. Div. ASCE **131**(6), 467–476 (2005)
37. Babayan, A., Savic, D.A., Walters, G.A.: Multi-objective optimization for the least-cost design of water distribution systems under correlated uncertain parameters. In: Proceedings of the EWRI/ASCE World Water and Environmental Resources Congress, Anchorage, Alaska, published on CD (2005)
38. Shamir, U., Howard, C.D.: Water supply reliability theory. J. Am. Water Works Assoc. **37**(7), 379–384 (1981)
39. Vogel, R.M.: Reliability indices for water supply systems. J. Water Resour. Planning Manage. Div. ASCE **113**(4), 563–579 (1987)
40. Wagner, J.M., Shamir, U., Marks, D.H.: Water distribution reliability: analytical methods. J. Water Resour. Planning Manage. Div. ASCE **114**(3), 253–275 (1988)
41. Wagner, J.M., Shamir, U., Marks, D.H.: Water distribution reliability: simulation methods. J. Water Resour. Planning Manage. Div. ASCE **114**(3), 276–294 (1988)
42. Bao, Y., Mays, L.W.: Model for water distribution system reliability. J. Hydraul. Eng. **116**(9), 1119–1137 (1990)
43. Duan, N., Mays, L.W.: Reliability analysis of pumping systems. J. Hydraul. Eng. **116**(2), 230–248 (1990)
44. Jacobs, P., Goulter, I.: Estimation of maximum cut-set size for water network failure. J. Water Resour. Planning Manage. Div. ASCE **117**(5), 588–605 (1991)
45. Quimpo, R.G., Shamsi, U.M.: Reliability-based distribution system maintenance. J. Water Resour. Planning Manage. Div. ASCE **117**(3), 321–339 (1991)
46. Bouchart, F., Goulter, I.: Reliability improvements in design of water distribution networks recognizing valve location. Water Resour. Res. **27**(12), 3029–3040 (1991)
47. Jowitt, P.W., Xu, C.: Predicting pipe failure effects in water distribution networks. J. Water Resour. Planning Manage. Div. ASCE **119**(1), 18–31 (1993)
48. Fujiwara, O., Ganesharajah, T.: Reliability assessment of water supply systems with storage and distribution networks. Water Resour. Res. **29**(8), 2917–2924 (1993)
49. Vogel, R., Bolognese, R.: Storage—reliability—resilience—yield relations for over—year water supply systems. Water Resour. Res. **31**(3), 645–654 (1995)
50. Schneiter, C.R., Haimes, Y.Y., Li, D., Lambert J.H.: Capacity reliability of water distribution networks and optimum rehabilitation decision making. Water Resour. Res. 32(7):2271–2278 (1996)
51. Yang, S.-I., Hsu, N.-S., Louie, P.W.F., Yeh, W.-G.Y.: Water distribution network reliability: connectivity analysis. J. Infrastruct. Syst. **2**(2), 54–64 (1996)
52. Yang, S.-I., Hsu, N.-S., Louie, P.W.F., Yeh, W.-G.Y.: Water distribution network reliability: stochastic simulation. J. Infrastruct. Syst. **2**(2), 65–72 (1996)
53. Xu, C., Goulter, I.C.: Probabilistic model for water distribution reliability. J. Water Resour. Planning Manage. Div. ASCE **124**(4), 218–228 (1998)
54. Fujiwara, O., Li, J.: Reliability analysis of water distribution networks in consideration of equity, redistribution, and pressure—dependent demand. Water Resour. Res. **34**(7), 1843–1850 (1998)
55. Tanyimboh, T.T., Tabesh, M., Burrows, R.: Appraisal of source head methods for calculating reliability of water distribution networks. J. Water Resour. Planning Manage. Div. ASCE **127**(4), 206–213 (2001)
56. Ostfeld, A., Kogan, D., Shamir, U.: Reliability simulation of water distribution systems—single and multiquality. Urban Water, Elsevier Sci. **4**(1), 53–61 (2002)
57. Shinstine, D.S., Ahmed, I., Lansey, K.E.: Reliability/availability analysis of municipal water distribution networks: case studies. J. Water Resour. Planning Manage. Div. ASCE **128**(2), 140–151 (2002)
58. Ostfeld, A.: Reliability analysis of water distribution systems. J. Hydroinformatics **6**(4), 281–294 (2004)

59. Tolson, B.A., Maier, H.R., Simpson, A.R., Lence, B.J.: Genetic algorithms for reliability—based optimization of water distribution systems. J. Water Resour. Planning Manage. Div. ASCE **130**(1), 63–72 (2004)
60. Xu, C., Goulter, I.C.: Reliability-based optimal design of water distribution networks. J. Water Resour. Planning Manage. Div. ASCE **125**(6), 352–362 (1999)
61. Ostfeld, A.: Water distribution systems connectivity analysis. J. Water Resour. Planning Manage. Div. ASCE **131**(1), 58–66 (2005)
62. Torii, A.J., Lopez, R.H.: Reliability analysis of water distribution networks using the adaptive response surface approach. J. Hydraul. Eng. (2011), posted ahead of print. doi:10.1061/(ASCE)HY.1943-7900.0000504
63. Tanyimboh, T.T., Tietavainen, M.T., Saleh, S.: Reliability assessment of water distribution systems with statistical entropy and other surrogate measures. Water Sci.Technol. Water Supply **11**(4), 437–443 (2011)
64. Su, Y.C., Mays, L.W., Duan, N., Lansey, K.E.: Reliability-based optimization model for water distribution systems. J. Hydraul. Eng. **114**(12), 1539–1556 (1987)
65. Tung, Y.K.: Evaluation of water distribution network reliability. In: Hydraulics and Hydrology in the Small Computer Age, Proceedings of the Specialty Conference, American Society of Civil Engineers Hydraulics Division, vol. 1, pp. 1–6. Orlando, FL (1985)
66. Ormsbee, L., Kessler, A.: Optimal upgrading of hydraulic-network reliability. J. Water Resour. Planning Manage. Div. ASCE **116**(6), 784–802 (1990)
67. Khang, D., Fujiwara, O.: Optimal adjacent pipe diameters in water distribution networks with reliability constraints. Water Resour. Res. **28**(6), 1503–1505 (1992)
68. Park, H., Liebman, J.C.: Redundancy-constrained minimum-cost design of water-distribution nets. J. Water Resour. Planning Manage. Div. ASCE **119**(1), 83–98 (1993)
69. Ostfeld, A.: Reliability analysis of regional water distribution systems. Urban Water **3**, 253–260 (2001)
70. Afshar, M.H., Akbari, M., Mariño, M.A.: Simultaneous layout and size optimization of water distribution networks: engineering approach. J. Infrastruct. Syst. **11**(4), 221–230 (2005)
71. Farmani, R., Walters, G.A., Savic, D.A.: Trade—off between total cost and reliability for Anytown water distribution network. J. Water Resour. Planning Manage. Div. ASCE **131**(3), 161–171 (2005)
72. Walski, T.M., Brill, D., Gessler, J., Goulter, I.C., Jeppson, R.M., Lansey, K.E., Lee, H.L., Liebman, J.C., Mays, L., Morgan, D.R., Ormsbee, L.: Battle of the network models: epilogue. J. Water Resour. Planning Manage. Div. ASCE **113**(2), 191–203 (1987)
73. Todini, E.: Looped water distribution networks design using a resilience index based heuristic approach. Urban Water **2**(3), 115–122 (2000)
74. Dandy, G.C., Engelhardt, M.O.: Multi—objective trade—offs between cost and reliability in the replacement of water mains. J. Water Resour. Planning Manage. Div. ASCE **132**(2), 79–88 (2006)
75. Agrawal, M.L., Gupta, R., Bhave, P.R.: Reliability—based strengthening and expansion of water distribution networks. J. Water Resour. Planning Manage Div. ASCE **133**(6), 531–541 (2007)
76. Reca, J., Martínez, J., Baños, R., Gil, C.: Optimal design of gravity - fed looped water distribution networks considering the resilience index. J. Water Resour. Planning Manage. Div. ASCE **134**(3), 234–238 (2008)
77. van Zyl, J.E., Piller, O., Gat, Y.: Sizing municipal storage tanks based on reliability criteria. J. Water Resour. Planning Manage. Div. ASCE **134**(6), 548–555 (2008)
78. Duan, H.-F., Tung, Y.-K., Ghidaoui, M.S.: Probabilistic analysis of transient design for water supply systems. J. Water Resour. Planning Manage. Div. ASCE (2010). doi:10.1061/(ASCE)WR.1943-5452.0000074
79. Ciaponi, C., Franchioli, L., Papiri, S.: A simplified procedure for water distribution networks reliability assessment. J. Water Resour. Planning Manage. Div. ASCE (2011), posted ahead of print. doi:10.1061/(ASCE)WR.1943-5452.0000184

80. Jacoby, S.: Design of optimal hydraulic networks. J. Hydraul. Div. ASCE **94**(HY3), 641–661 (1968)
81. Karmeli, D., Gadish, Y., Meyers, S.: Design of optimal water distribution networks. J. Pipeline Div. ASCE **94**(1), 1–9 (1968)
82. Ormsbee, L.E., Lansey, K.E.: Optimal control of water supply pumping systems. J. Water Resour. Planning Manage. Div. ASCE **120**(2), 237–252 (1994)
83. Dreizin, Y.: Examination of possibilities of energy saving in regional water supply systems. M.Sc. Thesis, Technion—Israel Institute of Technology (1970), 85p
84. Sterling, M.J.H., Coulbeck, B.: A dynamic programming solution to optimization of pumping costs. Proc. Inst. Civil Eng. **59**(4), 813–818 (1975)
85. Carpentier, P., Cohen, G.: Decomposition, coordination and aggregation in the optimal control of a large water supply network. In: Proceedings of IFAC World Congress, Budapest, Hungary, July. Proceedings of the 9th Triennial IFAC World Congress, Budapest, pp 3207–3212 (1984)
86. Houghtalen, R.J., Loftis, J.C.: Improving water delivery system operation using training simulators. J. Water Resour. Planning Manage. Div. ASCE **115**(5), 616–629 (1989)
87. Ormsbee, L.E., Walski, T.M., Chase, D.V., Sharp, W.W.: Methodology for improving pump operation efficiency. J. Water Resour. Planning Manage. Div. ASCE **115**(2), 148–164 (1989)
88. Zessler, U., Shamir, U.: Optimal operation of water distribution systems. J. Water Resour. Planning Manage. Div. ASCE **115**(6), 735–752 (1989)
89. Lansey, K.E., Awumah, K.: Optimal pump operations considering pump switches. J. Water Resour. Planning Manage. Div. ASCE **120**(1), 17–35 (1994)
90. Nitivattananon, V., Sadowski, E.C., Quimpo, R.G.: Optimization of water supply system operation. J. Water Resour. Planning Manage. Div. ASCE **122**(5), 374–384 (1996)
91. McCormick, G., Powell, R.S.: Optimal pump scheduling in water supply systems with maximum demand charges. J. Water Resour. Planning Manage. Div. ASCE **129**(5), 372–379 (2003)
92. Olshansky, M., Gal, S.: Optimal operation of a water distribution system." IBM—Israel, Technical Report 88.239 (1988), 52p
93. Jowitt, P.W., Germanopoulos, G.: Optimal pump scheduling in water-supply networks. J. Water Resour. Planning Manage. Div. ASCE **118**(4), 406–422 (1992)
94. Diba, A., Louie, P.W.F., Mahjoub, M., Yeh, W.W.-G.: Planned operation of large-scale water-distribution system. J. Water Resour. Planning Manage. Div. ASCE **121**(3), 260–269 (1995)
95. Coulbeck, B.: Optimal operations in non-linear water networks. Optimal Control Appl. Methods **1**(2), 131–141 (1980)
96. Coulbeck, B., Brdys, M., Orr, C.H., Rance, J.P.: A hierarchical approach to optimized control of water distribution systems: part I decomposition. Optimal Control Appl. Methods **9**(1), 51–61 (1988)
97. Coulbeck, B., Brdys, M., Orr, C.H., Rance, J.P.: A Hierarchical approach to optimized control of water distribution systems: part II. Lower-level algorithm. Optimal Control Appl. Methods **9**(2), 109–126 (1988)
98. Biscos, C., Mulholland, M., Le Lann, M.V., Brouckaert, C.J., Bailey, R., Roustan, M.: Optimal operation of a potable water distribution network. Water Sci. Technol. **46**(9), 155–162 (2002)
99. Biscos, C., Mulholland, M., Le Lann, M.-V., Buckley, C.A., Brouckaert, C.J.: Optimal operation of water distribution networks by predictive control using MINLP. Water SA **29**(4), 393–404 (2003)
100. Ulanicki, B., Kahler, J., See, H.: Dynamic optimization approach for solving an optimal scheduling problem in water distribution systems. J. Water Resour. Planning Manage. Div. ASCE **133**(1), 23–32 (2007)
101. Pulido-Calvo, I., Gutiérrez-Estrada, J.C.: Selection and operation of pumping stations of water distribution systems. Environ. Res. J. **5**(3), 1–20 (2011)

102. Chase, D.V.: A computer program for optimal control of water supply pump stations: development and testing. USACERL TECHNICAL REPORT N-90/14, US Army Corps of Engineers, Construction Engineering Research Laboratory (1990), 98p
103. Gessler, J., Walski, T.M.: Water distribution system optimization. Technical Report EL-85-11, U.S. Army Engineer Waterways Experiment Station, Vicksburg MS., NTIS No. AD A163 493 (1985)
104. Brion, L.M., Mays, L.W.: Methodology for optimal operation of pumping stations in water distribution systems. J. Hydraul. Eng. ASCE **117**(11), 1551–1569 (1991)
105. Brion, L.M.: Methodology for optimal operation of pumping stations in water distribution systems. PhD thesis, University of Texas at Austin, Texas (1990)
106. Pezeshk, S., Helweg, O.J., Oliver, K.E.: Optimal operation of ground-water supply distribution systems. J. Water Resour. Planning Manage. Div. ASCE **120**(5), 573–586 (1994)
107. Cohen, D., Shamir, U., Sinai, G.: Optimal operation of multi-quality networks-I: introduction and the Q-C model. Eng. Optim. **32**(5), 549–584 (2000)
108. Cohen, D., Shamir, U., Sinai, G.: Optimal operation of multi-quality networks-II: the Q-H model. Eng. Optim. **32**(6), 687–719 (2000)
109. Cohen, D., Shamir, U., Sinai, G.: Optimal operation of multi-quality networks-III: the Q-C-H model. Eng. Optim. **33**(1), 1–35 (2000)
110. Broad, D.R., Dandy, G.C., Maier, H.R.: Water distribution system optimization using metamodels. J. Water Resour. Planning Manage. Div. ASCE **131**(3), 172–180 (2005)
111. Shamir, U., Salomons, E.: Optimal real-time operation of urban water distribution systems using reduced models. J. Water Resour. Planning Manage. Div. ASCE **134**(2), 181–185 (2008)
112. Broad, D.R., Maier, H.R., Dandy, G.C.: Optimal operation of complex water distribution systems using metamodels. J. Water Resour. Planning Manage. Div. ASCE **136**(4), 433–443 (2010)
113. Tarquin, A.J., Dowdy, J.: Optimal pump operation in water distribution. J. Hydraul. Eng. ASCE **115**(2), 158–168 (1989)
114. Pezeshk, S., Helweg, O.J.: Adaptive search optimization in reducing pump operating costs. J. Water Resour. Planning Manage. Div. ASCE **122**(1), 57–63 (1996)
115. Ormsbee, L.E., Reddy, S.L.: Nonlinear heuristic for pump operations. J. Water Resour. Planning Manage. Div. ASCE **121**(4), 302–309 (1995)
116. Sakarya, B.A., Mays, L.W.: Optimal operation of water distribution pumps considering water quality. J. Water Resour. Planning Manage. Div. ASCE **126**(4), 210–220 (2000)
117. Kirkpatrick, S., Galett, C.D., Vecchi, M.P.: Optimization by simulated annealing. Science **220**, 621–630 (1983)
118. Cui, L.-J., Kuczera, G.: Optimizing urban water supply headworks using probabilistic search methods. J. Water Resour. Planning Manage. Div. ASCE **129**(5), 380–387 (2003)
119. Goldberg, D.E.: Genetic algorithms in search, optimization, and machine learning. Addison-Wesley, New York (1989)
120. Duan, Q., Sorooshian, S., Gupta, V.: Effective and efficient global optimization for conceptual rainfall-runoff models. Water Resour. Res. **28**(4), 1015–1031 (1992)
121. Ostfeld, A., Salomons, E.: Optimal operation of multiquality water distribution systems: unsteady conditions. Eng. Optim. **36**(3), 337–359 (2004)
122. Ostfeld, A., Salomons, E.: Optimal layout of early warning detection stations for water distribution systems security. J. Water Resour. Planning Manage. Div. ASCE **130**(5), 377–385 (2004)
123. van Zyl, J.E., Savic, D.A., Walters, G.A.: Operational optimization of water distribution systems using a hybrid genetic algorithm. J. Water Resour. Planning Manage. Div. ASCE **130**(2), 160–170 (2004)
124. López-Ibáñez, M., Prasad, T.D., Paechter, B.: Ant colony optimization for optimal control of pumps in water distribution networks. J. Water Resour. Planning Manage. Div. ASCE **134**(4), 337–346 (2008)

125. Dorigo, M.: Optimization, learning and natural algorithms. Ph.D. thesis, Politecnico di Milano, Milan, Italy (1992)
126. Boulos, P.F., Wu, Z., Orr, C.H., Moore, M., Hsiung, P., Thomas, D.: Optimal pump operation of water distribution systems using genetic algorithms. www.rbfconsulting.com/papers/genetic_algo.pdf (2011)
127. Liang, T., Nahaji, S.: Managing water quality by mixing water from different sources. J. Water Resour. Planning Manage. Div. ASCE **109**, 48–57 (1983)
128. Sinai, G., Koch, E., Farbman, M.: Dilution of brackish waters in irrigation networks—an analytic approach. Irrig. Sci. **6**, 191–200 (1985)
129. EPANET 2.0 (2002). http://www.epa.gov/nrmrl/wswrd/dw/epanet.html
130. Shamir, U., Howard, C.D.D.: Topics in modeling water quality in distribution systems. In: Proceedings of the AWWARF/EPA Conference on Water Quality Modeling in Distribution Systems, pp. 183–192. Cincinnati, Ohio (1991)
131. Ostfeld, A., Shamir, U.: Optimal operation of multiquality distribution systems: steady state conditions. J. Water Resour. Planning Manage. Div. ASCE **119**(6), 645–662 (1993)
132. Ostfeld, A., Shamir, U.: Optimal operation of multiquality distribution systems: unsteady conditions. J. Water Resour. Planning Manage. Div. ASCE **119**(6), 663–684 (1993)
133. Cohen, D.: Optimal operation of multi-quality networks. D.Sc. Thesis, Faculty of Agricultural Engineering, Technion—Israel (in Hebrew), 400 p (1992)
134. Brooke, A., Kendrick, D., Meeraus, A.: GAMS: a user's guide, Scientific Press, USA (1988), 289p
135. Shor, N.Z.: Minimization Methods for Non-Differentiable Functions. Springer, New York (1985), 159p
136. Goldman, E.F.: The application of simulated annealing for optimal operation of water distribution systems. PhD dissertation, Arizona State University, 242p (1998)
137. American Society of Civil Engineers (ASCE): Guidelines for designing an online contaminant monitoring system. http://www.asce.org/static/1/wise.cfm#Monitoring (2004). Accessed 2 Aug 2007
138. American Water Works Association (AWWA): Security guidance for water utilities. http://www.awwa.org/science/wise (2004). Accessed 2 Aug 2007
139. Lee, B., Deininger, R.: Optimal locations of monitoring stations in water distribution system. J. Environ. Eng. ASCE **118**(1), 4–16 (1992)
140. Kumar, A., Kansal, M.L., Arora, G.: Identification of monitoring stations in water distribution system. J. Environ. Eng. ASCE **123**(8), 746–752 (1997)
141. Kessler, A., Ostfeld, A., Sinai, G.: Detecting accidental contaminations in municipal water networks. J. Water Resour. Planning Manage. Div. ASCE **124**(4), 192–198 (1998)
142. Woo, H.M., Yoon, J.H., Choi, D.Y.: Optimal monitoring sites based on water quality and quantity in water distribution systems. In: Bridging the Gap: Meeting the World's Water and Environmental Resources Challenges, Proceedings of the ASCE EWRI annual conference, Orlando, Florida, published on CD (2001)
143. Al-Zahrani, M., Moied, K.: Locating optimum water quality monitoring stations in water distribution system. In: Bridging the Gap: Meeting the World's Water and Environmental Resources Challenges, Proceedings of the ASCE EWRI annual conference, Orlando, Florida, published on CD (2001)
144. Ostfeld, A., Kessler, A., Goldberg, I.: A contaminant detection system for early warning in water distribution networks. Eng. Optim. **36**(5), 525–538 (2004)
145. Ostfeld, A., Salomons, E.: Securing water distribution systems using online contamination monitoring. J. Water Resour. Planning Manage. Div. ASCE **131**(5), 402–405 (2005)
146. Berry, J.W., Hart, W.E., Phillips, C.A., Uber, J.G., Watson, J.P.: Sensor placement in municipal water networks with temporal integer programming models. J. Water Resour. Planning Manage. Div. ASCE **132**(4), 218–224 (2006)
147. Propato, M.: Contamination warning in water networks: general mixed-integer linear models for sensor location design. J. Water Resour. Planning Manage. Div. ASCE **132**(4), 225–233 (2006)

148. Watson, J.P., Greenberg, H.J., Hart, W.E.: A multiple-objective analysis of sensor placement optimization in water networks. In: Critical Transitions in Water and Environmental Resources Management, Proceedings of the ASCE EWRI annual conference, Salt Lake City, Utah, published on CD (2004)
149. Ostfeld, A., Uber, J., Salomons, E.: Battle of the water sensor networks (BWSN): a design challenge for engineers and algorithms. In: 8th Annual Water Distribution System Analysis Symposium Cincinnati, Ohio, USA, published on CD (2006)
150. Dorini, G., Jonkergouw, P., Kapelan, Z., di Pierro, F. Khu, S.T., Savic, D: An efficient algorithm for sensor placement in water distribution systems. In: 8th Annual Water Distribution System Analysis Symposium Cincinnati, Ohio, USA, published on CD (2006)
151. Eliades, D., Polycarpou, M: Iterative deepening of Pareto solutions in water sensor Networks. In: 8th Annual Water Distribution System Analysis Symposium Cincinnati, Ohio, USA, published on CD (2006)
152. Rubinstein, R.Y.: The simulated entropy method for combinatorial and continuous optimization. Methodol Comput. Appl. Probab. **2**, 127–190 (1999)
153. Gueli, R.: Predator—prey model for discrete sensor placement. In: 8th Annual Water Distribution System Analysis Symposium Cincinnati, Ohio, USA, published on CD (2006)
154. Huang, J.J., McBean, E.A., James, W.: Multi-objective optimization for monitoring sensor placement in water distribution systems. In: 8th Annual Water Distribution System Analysis Symposium Cincinnati, Ohio, USA, published on CD (2006)
155. Ostfeld, A., Salomons, E.: Sensor network design proposal for the battle of the water sensor networks (BWSN). In: 8th Annual Water Distribution System Analysis Symposium Cincinnati, Ohio, USA, published on CD (2006)
156. Preis, A., Ostfeld, A.: Multiobjective sensor design for water distribution systems security. In: 8th Annual Water Distribution System Analysis Symposium Cincinnati, Ohio, USA, published on CD (2006)
157. Deb, K., Agrawal, S., Pratap, A., Meyarivan, T.: A fast elitist non-dominated sorting genetic algorithm for multi-objective optimization: NSGA-II. In: Proceedings of the Parallel Problem Solving from Nature VI Conference, pp. 849–858. Paris, France (2000)
158. Wu, Z.Y., Walski, T.: Multi objective optimization of sensor placement in water distribution systems. In: 8th Annual Water Distribution System Analysis Symposium Cincinnati, Ohio, USA, published on CD (2006)

Transportation Systems: Monitoring, Control, and Security

Stelios Timotheou, Christos G. Panayiotou and Marios M. Polycarpou

Abstract Transportation is one of the main cornerstones of human civilization which facilitates the movement of people and goods from one location to another. People routinely use several transportation modes, such as road, air, rail and water for their everyday activities. However, the continuous global population increase and urbanization around the globe is pushing transportation systems to their limits. Unquestionably, the road transportation system is the one mostly affected because it is difficult and costly to increase the capacity of existing infrastructure by building or expanding new roads, especially in urban areas. Towards this direction, Intelligent Transportation Systems (ITS) can have a vital role in enhancing the utilization of the existing transportation infrastructure by integrating electronic, sensing, information and communication technologies into a transportation system. However, such an integration imposes major challenges in the monitoring, control and security of transportation systems. This chapter surveys the state of the art and the challenges for the implementation of ITS in road transportation systems with a special emphasis on monitoring, control and security.

Keywords Road transport · Intelligent transportation systems (ITS) · Vehicle/network monitoring · Vehicle/cooperative/network control · Cyber-physical security · Survey

---

This work is partially funded by the European Research Council Advanced Grant FAULT-ADAPTIVE (ERC-2011-AdG-291508).

S. Timotheou (✉) · C.G. Panayiotou · M.M. Polycarpou
KIOS Research Center for Intelligent Systems and Networks,
University of Cyprus, Nicosia, Cyprus
e-mail: timotheou.stelios@ucy.ac.cy

C.G. Panayiotou
e-mail: christosp@ucy.ac.cy

M.M. Polycarpou
e-mail: mpolycar@ucy.ac.cy

E. Kyriakides and M. Polycarpou (eds.), *Intelligent Monitoring, Control, and Security of Critical Infrastructure Systems*, Studies in Computational Intelligence 565,
DOI 10.1007/978-3-662-44160-2_5

1 Introduction

Transportation systems are an indispensable part of human activity which facilitates the movement of people and goods from one location to another. In recent years, people have become so dependent on transportation modes, such as road, air, rail and water that different transportation systems face not only opportunities, but also challenges. The most widely used and the one most affected is the road transportation system.

The ever increasing urbanization and motorization overcrowds cities with vehicles, leading to undesired everyday phenomena, such as congestion and accidents. Inefficient management of urban mobility causes several adverse effects, such as long travel times and increased travel costs, degraded quality of life for the travellers, as well as high CO_2 emissions and huge waste of fossil-fuel energy. In the USA, transportation systems account for about 25–30 % of the total energy consumption and 75 % of petroleum fuel consumption [1]. In addition, the growing road traffic increases the accident risk which results in a large number of injuries and casualties, further increases congestion and costs economies huge amounts of money. In the USA there are on average around 40,000 fatalities a year, while congestion causes travellers 5.5 billion h, 2.9 billion gallons of fuel and a congestion cost of $121 billion [2].

To alleviate these societal, economic, energy and environmental problems caused by the increasing use of transportation systems drastic actions need to be taken. One direction, is to increase the capacity of the transportation infrastructure by constructing new roads. However, it is both difficult and costly to expand roads in urban areas, so this is unlikely to happen on a large-scale. Another direction, is to take actions that will reduce traffic either by managing traffic demand or by using alternative transportation modes. Traffic demand management can be achieved either explicitly by restricting specific vehicles off the roads or implicitly through different pricing schemes, such as congestion charging, high occupancy toll lanes and vehicle-travelled miles fees [3]. For example, during the Beijing Olympics of 2008, an alternating restriction was imposed to the private car owners according to the vehicle license plate number (odd/even), which reduced traffic in half [4]. Also, in several large cities, such as London and Stockholm, congestion charging schemes have resulted in the reduction of traffic volume, travel time and pollution by about 15–20 % [5, 6]. The wide adoption of public transportation modes could result in huge traffic reductions but this is not always convenient while certain countries lack the cultural background for such adoptions. A third direction, is towards the integration of intelligent transportation systems (ITS) technologies into the existing infrastructure which increases capacity without building new roads. For example, implementing ITS to achieve real-time optimal operation of traffic lights in the USA could reduce travel time by 25 % and fuel consumption by 10 %.

ITS solutions can reduce travel-time, ease inconvenience and congestion, improve safety, and reduce pollutant emissions by integrating electronic, sensing, information and communication technologies (ICT) into the transportation system to optimally manage traffic. ITS also play a vital role in enabling the enforcement and operation of strategies aiming at reducing traffic. Additionally, ITS can increase the speed, reliability, convenience and safety of public transportation means, giving strong incentives to travellers for adoption. These goals can be realized by monitoring traffic conditions (e.g., travel times, queue lengths, accidents, incidents, construction works) for the short/long-term management and real-time control of traffic flows, as well as through ICT systems that facilitate the operation of transportation systems, assist the drivers, provide information/ guidance to the travellers, and provide the capability for vehicles to communicate with other entities (referred to as V2X), including vehicle-to-vehicle (V2V) and vehicle-to-infrastructure (V2I) communication.

Nevertheless, the monitoring and control of traffic through ITS is very challenging for several reasons. Firstly, it is difficult to deal with the large-scale nature of a transportation system which involves millions of travellers/cars and a vast network of roads. For example, monitoring a transportation system requires the deployment of numerous different sensors that provide real-time information; nevertheless, aggregating raw-data to a central operations centre and extracting useful information is very challenging due to the large volume, velocity and variety of the incoming data. Secondly, traffic dynamics are nonlinear and often exhibit chaotic behaviour which makes the control of traffic difficult. Besides, high uncertainty in traffic dynamics due to the human driving style, various road incidents, such as construction works, accidents or even a slowly-moving vehicle complicates the situation even further making traffic modelling more challenging. Another challenge regards the proactive handling of the inherent human behaviour; for example, disseminating information about a congested highway, might have a negative impact in the sense that all drivers might adapt their travel plan resulting in the congestion of other areas than the particular highway. Finally, providing a high-level of security for the transportation system is also of vital importance to allow the normal operation of ITS under different security threats.

This aim of this chapter is to describe the state of the art in ITS and the challenges associated with the implementation of ITS solutions in road transportation systems with a special emphasis on monitoring, control and security. The structure of the chapter is the following. Section 2, provides an introduction to ITS. Sections 3–5, discuss in detail the state of the art in the monitoring, control and security of ITS, respectively, and attempt to capture the main challenges and open issues of these areas. Finally, Sect. 6 provides concluding remarks.

2 Introduction to Intelligent Transportation Systems

The introduction of ICT technologies in our everyday life has significantly improved our quality of life by transforming many industries, such as education, health-care and governmental services. Currently, ICT is in the process of transforming transportation systems by integrating electronic, sensing, information and communication technologies, an area called Intelligent Transportation Systems (ITS). ITS are capable of monitoring and controlling transportation systems, developing advanced traveller/operator applications, and providing assistive technologies for the drivers. To better understand ITS and their vital role in road transportation systems, this section describes the main applications areas, the enabling technologies and the benefits of ITS.

2.1 ITS Application Areas

The meaning and potential of a technology can be better understood through its capacity to provide numerous applications. Here, we outline several applications of ITS through a taxonomy of application categories based on functional intent. ITS have the following five main application areas [7]:

- Advanced Traffic Management Systems (ATMS)
- Advanced Traveller Information Systems (ATIS)
- Advanced Public Transportation Systems (APTS)
- ITS-Enabled Transportation Pricing Systems (ETPS)
- Advanced Vehicle Control Systems (AVCS)

ATMS concerns the intelligent control of traffic-related devices, such as traffic signals, variable message signs and ramp meters to improve the utilization of transportation systems. Traffic signal control, concerns the optimal scheduling of signal indications in order to optimize some appropriate metric, such as average speed delay, vehicle throughput or average number of stops for each vehicle. To achieve this, it is not only necessary to have sophisticated control algorithms to achieve the specific goal but also reliable real-time information about the condition of the transportation network. Variable message signs are electronic panel signs that provide travellers with information about special events in the road network allowing them to take appropriate actions. Ramp metering is another powerful strategy for increasing the flow of vehicles in highways by regulating the number of vehicles entering a highway, to avoid congested states that might lead to the deterioration of traffic flows. ATMS are also concerned with the development of holistic monitoring systems that collect a variety of data from different sources, such as detector loops, cameras and probe vehicles. Data are aggregated and translated to meaningful information by traffic management centres, which then disseminate the information to the travellers.

ATIS are devices or telematics services that provide to the travellers real-time travel and traffic information, such as directions towards specific destinations based on traffic, congestion-levels, road works, weather conditions etc. Information dissemination can take place via variable message signs, radio announcements, and the Internet. Individual in-car assistive devices for navigation and telematics services are also included in this category. Navigation devices provide personalized navigation directions for transit routes based on the minimization of different metrics, such as delay, distance or energy. Telematics devices provide location-specific information to the drivers regarding congestion, accidents, weather conditions or even parking which can be very beneficial in allowing the drivers to take appropriate actions in advance so as to minimize the effect of any evolving situation.

APTS help to make public transport a more attractive option for commuters by improving the efficiency, convenience and safety of public transportation by means of systems that provide real-time status information of different public transport modes (e.g., bus, subway, rail), or devices that allow electronic fare payment, easing public transportation commuting and reducing travel times. APTS can play a crucial role in improving the overall operation of the transportation system by reducing road traffic, easing congestion and significantly decreasing the amount of energy consumed in transportation systems.

ETPS enable road-user charging methods to alleviate congestion. Charging methods include automatic toll collection when entering highways, congestion charging when entering high-occupancy parts of a city, fee-based express lanes, vehicle-miles travelled usage fees and variable parking fees. Such charging methods can incentivise travellers to share cars, use public transportation means, or travel through less-congested routes. Despite the efficiency of charging methods, there are several implementation challenges regarding their automatic operation, enforcement, accurate monitoring and nation-wide deployment, which can only be surpassed via ITS technologies.

Finally, AVCS use advanced electronics and techniques to improve vehicles' behaviour and assist the driver, with the dominant concern being safety. Such systems can be used to provide safety warnings (e.g., rear end/front end collision, roadway departure, intersection collision, vehicle distance), increase highway capacity via "platooning", a group of cars traveling in an organized manner, or provide the means for advanced cruise control. AVCS also provide the means for information exchange either through Vehicle-to-Infrastructure (V2I) communication to monitor the state of the transportation network or disseminate useful information to vehicles via road-side units (RSU), and through V2V communication for collaborative real-time decision making between vehicles. Nowadays, fully autonomous vehicles (AVs) are also becoming a reality [8]; many large manufacturers including Mercedes Benz, Nissan, Toyota, Volvo and Google have already developed AVs which are currently testing. AVs have already driven successfully thousands of miles in real-life conditions and have even acquired driving licenses by certain U.S. states [9]. Therefore, it is not long before our roads fill with AVs; once the technology matures, each vehicle will be easily and cheaply converted into an AV. An abstract representation of the different ITS functionalities in road transport is provided in Fig. 1.

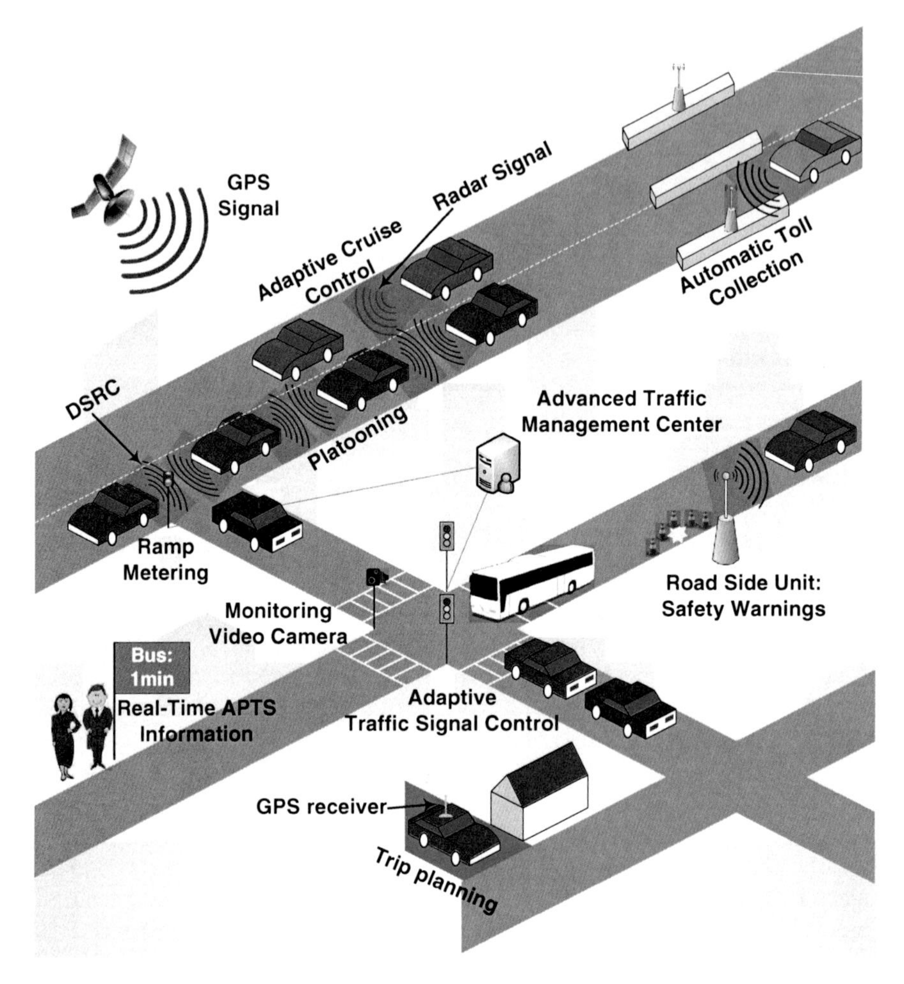

Fig. 1 Abstract representation of different ITS functionalities

2.2 Enabling ITS Technologies

The full potential of ITS services is made possible through technological advancements in several fields, such as electronics, communications, positioning systems and information systems. The most important technologies that enable ITS applications are the following [7]:

- Location-related technologies
- In-car sensors
- Network monitoring devices
- Communication solutions
- Computing platforms
- Software advancements

Location awareness is fundamental towards the implementation of ATIS applications, e.g., for navigation and warning systems, location-based services in APTS, estimation of vehicle-miles travelled for ETPS and traffic data collection for ATMS. A vehicle location can be estimated using global positioning system (GPS) receivers based on the absolute position of several satellites (at least 3–4) and the relative distance between the vehicle and each of the satellites. GPS receivers can achieve an accuracy of 10–30 m and require line-of-sight with the satellites; hence, the presence of skyscrapers in downtown environments can inhibit the operation of GPS devices. Geographical Information Systems (GIS) are also fundamental in the implementation of ITS applications through the construction, analysis, manipulation, and display of geographic information. GIS provide detailed digital maps for routing and scheduling in ATIS, assist APTS in planning and operations, and are valuable in the analysis, monitoring and management of traffic in ATMS.

In-car sensors are fundamental for the implementation of AVCS technologies. On-board sensors not only provide critical information for the normal operation of the car, such as temperature, airbag status, velocity and direction, but also enable assistive technologies related to parking, warnings and advanced control. For example, front and rear radars and ultrasonic sensors measure distance from surrounding objects to warn about slowly moving vehicles, prevent collisions and assist in parking. On-board cameras are essential in AVs for the visual perception of the surrounding environment, traffic sign recognition and pedestrian detection. Lidars are remote sensing devices that illuminate surrounding objects with a laser and analyse the reflected light to generate a precise 3-D map of an AV's surrounding environment. GPS devices also enable assistive control functionalities through warnings regarding potential collisions, lane departure and hazardous locations.

There are also several technologies that can be used to monitor the road network, such as detector loops, roadside cameras, probe vehicles, bluetooth devices, seismic sensors and wireless systems. These devices measure different traffic characteristics, such as traffic volume and density, vehicle speed, vehicle category, travel times, etc., which are essential in ATMS for the management and control of the network's traffic flows. Monitoring technologies are described in more detail in Sect. 3.1.

Communication solutions allow the deployment of cooperative technologies that operate beyond the line-of-sight of in-car sensors. Through V2X cooperation dangerous situations can be perceived and avoided, data collection can be far more detailed compared to network monitoring equipment, while information dissemination to travellers can be more accurate and useful. There are several ITS communication solutions with different characteristics (communication range, data rate, frequency bands, etc.). Dedicated Short Range Communications (DSRC) enable V2X communications for distances up to 1,000 m, operating at 5.8 GHz or 5.9 GHz; it is the main technology for several ITS applications, such as ETPS services, information provision and V2V cooperation. The range of DSRC can be extended by relaying information in the vehicular network until the information reaches the desired destination. ATIS and APTS significantly benefit from mobile 3G and 4G (4th generation) data transfer technologies that provide medium to high data-rate, long range Internet connectivity to travellers.

In spite of these enabling technologies, ITS applications would remain unrealizable without advances in software (e.g., intelligent processing, database management) and computer hardware (e.g., processors, personal digital assistants (PDAs), memory). According to Moore's law the processing power and memory of computing platforms is doubling every 18 months [10]; these have enabled the development of advanced techniques for the processing of large volumes of data for the extraction of useful information in reasonable time. Towards this direction, cloud computing is another driving force which provides the illusion of infinite computing resources on demand, according to the user needs and without long-term commitments [11]; ATMS can significantly benefit from cloud computing by storing and processing their data on the cloud [12]. PDAs also have a prominent role in several areas of ITS; the emergence of smartphones and navigation devices is enabling several traveller services, such as route guidance, traffic information retrieval, and driver assistive applications [13], while they can be used for the collection of good quality traffic data from individual travellers.

Despite the aforementioned hardware innovations, intelligence in ITS emanates from software. Intelligent algorithms constitute the main ingredient of ITS applications. They are responsible for the processing of noisy, incomplete and even misleading raw data for the extraction of useful information, for the modelling, management and optimal control of traffic in ATMS, for the derivation of optimal navigation routes in ATIS, for the management of APTS and for the intelligent control of AVCS. The intelligent approaches for the solution of problems arising in ITS are discussed in more detail in Sect. 4.

2.3 ITS Benefits

ITS delivers safety, performance, environmental and economic benefits. Most importantly, ITS increases the safety of travellers and pedestrians. In 2010, there were 1.24 million fatalities [14] associated with road accidents around the world, of which about 75 % were attributed to human error [15]. Departing from the traditional philosophy of protecting passengers in the event of an accident, AVCS applications aim at preventing accidents e.g., through systems for collision avoidance, pedestrian detection, emergency breaking and lane changing warning. For example, IntelliDrive, an initiative of the U.S. Department of Transportation to provide an intelligent infrastructure with V2I capabilities for the identification and warning of drivers about threats and hazards on the roadway [16], could potentially cut accidents to one fifth [17]. Other ITS areas can assist in implicitly reducing accidents through the reduction of congestion. For instance, accidents were reduced by approximately 25 % by introducing ramp metering in the freeways of Twin Cities, Minnesota [18].

In terms of performance, all ITS areas contribute towards the reduction of congestion, which is the key for safer, greener, more cost-efficient and comfortable road travel. Towards this direction, intelligent management and control of traffic, as well as the exploitation of advanced navigation systems, increase the capacity of existing infrastructure; the introduction of APTS and ETPS reduces the number of road vehicles, while AVCS minimize unnecessary travel delays by reducing the number of crashes. Quantifying congestion reduction from the simultaneous deployment of various ITS solutions may be difficult, but there are numerous studies that examined the performance benefits from individual ITS solutions. For instance, a pilot project that introduced the heuristic ramp metering coordination system (HERO) at 6 consecutive inbound on-ramps on the Monash freeway in Melbourne, Australia, increased average flow and speed by 8.4 and 58.6 % respectively [19], while congestion charging in London and Stockholm have resulted in the reduction of traffic volume, travel time and pollution by about 15–20 % [5, 6].

ITS solutions also deliver significant environmental benefits. Alleviating congestion reduces energy waste and air pollution; this is particularly important, as the sector of transportation systems is a main contributor towards the consumption of petroleum fuels and greenhouse gas emissions. In addition, ATIS navigation devices can provide guidance towards routes that minimize energy or CO_2 emissions. APTS and ETPS solutions also contribute towards these directions, by contributing towards the reduction of vehicles in the roads.

Finally, ITS provide several economic benefits. Firstly, ITS have an excellent benefit-cost ratio; for example, it is estimated that a real-time traffic management program across the US would yield a benefit-cost ratio of 25–1 [7]. Another example is the HERO system introduced in the Monash freeway in Melbourne, which had an economic payback period of just 11 days [19]. Secondly, ITS can boost productivity by transporting people and goods faster and safer. Thirdly, ITS can diminish the economic impact of accidents due to vehicle damages and loss of life. According to a study from Cambridge Systematics Inc., prepared for AAA in 2011, the annual cost of congestion and accidents in the US alone is around 100 and 300$ billion, respectively, which accounts for approximately 2.5 % of the GDP [20].

3 Monitoring Road Transportation Systems

Monitoring is the act of collecting and managing information concerning the characteristics and status of network or vehicular resources of interest, such as traffic flows, road violations and driver status. Monitoring is inextricably interwoven with ITS, providing valuable information for the implementation of prediction, estimation, classification and event-detection applications in all areas of ITS. It is also a very active research area due to the challenges associated with the collection and processing of transportation data given their high velocity, volume, and variety, as well as the requirement for scalable, extensible and secure solutions.

The network monitoring process involves several stages. Initially, monitoring devices measure generated events from various sources and encode data to a given format; at this or any later stage, data processing may take place, such as filtering or event detection. Then, collected data and events are disseminated to a transportation operations centre, where data from multiple sources are jointly processed to identify important traffic events (e.g., traffic congestion and accidents) and estimate or predict the network state. Finally, processed data need to be properly presented to the end-users for interpretation and further actions.

Vehicular monitoring may regard certain functions of the vehicle, the surrounding environment or the driver. The monitoring process in this case is similar, but information is disseminated to the driver in the form of notifications, warnings or driving assistance. Vehicular monitoring systems may be used among others for the pedestrian detection, lane keeping, traffic sign detection and driver inattention detection.

This section reviews the monitoring devices, explains the monitoring parameters and traffic characteristics of interest, outlines some main methodologies used for the analysis of collected data and describes the main monitoring applications.

3.1 Monitoring Technologies

There are numerous devices for monitoring individual vehicles and the road transportation network, each capable of measuring different characteristics.

Apart from the traditional in-car sensors that monitor various operational characteristics of the car, several more advanced vehicle monitoring devices (VMDs) are emerging. Radars and ultrasonic sensors are used to prevent collision by measuring the distance from surrounding objects (including pedestrians or vehicles) and warning the driver if something is in short-range. GPS devices are exploited to monitor the position of the vehicle and provide lane departure warnings to the driver. Video cameras are also becoming a main VMD due to the importance of visual information, the progress in computer-vision and their price-to-performance improvement. Video cameras are mostly used for driver assistance or autonomous driving, due to their ability to detect, identify, track other vehicles and perceive the environment [21]. Another key VMD are lidars, used in AVs to generate high-resolution 3-D maps of the surrounding environment.

Network monitoring devices (NMDs) are intrusive (installed in the road or pavement) or non-intrusive (mounted on posts, unmanned aerial vehicles (UAVs) or satellites) technologies that measure and report different traffic characteristics (e.g., concentration, flow, speed, travel times, counts, turning movements, vehicle type, weight, etc.). According to the spatiotemporal measurement characteristics NMDs report different traffic characteristics, yielding five measurement categories [22]:

- at a point (measurement of flow, speed, time/distance headway, count, weight, vehicle type),

- over a short section of the road (10 m) (aforementioned traffic characteristics and occupancy),
- over a length of road (0.5 km) (density),
- using moving-observers (speeds, travel times),
- wide-area sampling from multiple ICT sources (density, turning movements, speeds, vehicle type).

Measurements at a point are mostly performed using induction loop detectors and video cameras, but other technologies, such as microwave beams, radars, photocells and ultrasonics are also in operation. Point detectors can sense the passage of vehicles and provide traffic volume counts, flow rates (vehicles per unit time per lane) and time headways (time distance between adjacent vehicles). Speed measurements can also be obtained by microwave and radar point detectors. Piezoelectric sensors can measure vertical forces, estimating a vehicle's weight per axle [23]. Vehicles can also be categorized (such as motorcycles, cars, minivans, trucks, and buses) by different point detectors [24], by exploiting measured vehicle characteristics, such as length, size and weight.

Short-section measurements are performed by installing point detectors in pairs closely spaced apart. In addition to the traffic characteristics measured using point detectors, short-section measurements can yield speeds (by calculating the time needed for a vehicle to pass both detectors) for all detector types, as well as occupancy. Occupancy is an essential traffic parameter characterizing the concentration of cars, through the percentage of time that a vehicle passes from the detection zone.[1]

Another fundamental traffic parameter that needs to be measured is density. Density is dual to occupancy, measuring the number of vehicles per unit length for a specific time instance. Hence, density can be measured over a long section of the road, typically around 0.5–1 km, either from aerial photography (e.g., satellites and UAVs) or from cameras installed on high structures. The importance of occupancy or density is emphasized by the fact that, along with flow rate and speed, they are the three fundamental traffic parameters characterizing the traffic dynamics.

Moving observers are also emerging as important NMDs, especially for the monitoring of urban transportation networks. Moving observers are floating cars, equipped with special equipment, such as GPS and speedometer, that record their speeds and travel times for different road segments over time. The advent of V2X, combined with smartphones' capability to record position and speed, imply that every car has the potential to be transformed into a moving observer that sends data to the infrastructure. Having a large number of floating data-collecting cars means that wide-area traffic characteristics can be captured in real-time which could have a major impact on the monitoring and control of road transportation systems.

---

[1] The fact that occupancy is measured over a short road section is a practical limitation; ideally, occupancy measurements can be performed with point detectors.

Recently, advances in ICT have allowed the development of novel, wide-area, low-cost data collecting methods [25]. One approach exploits the abundance of mobile phones to anonymously estimate their position and speed; this information can be fused with other data for the real-time monitoring of traffic characteristics [26]. Wireless sensor networks (WSN) consisting of inexpensive nodes equipped with magnetometers, can also be used for vehicle detection, which is essential for inferring vehicle counts, turning movements and speeds and vehicle length [27]. Finally, bluetooth detectors are also gaining widespread use as NMDs because they are non-invasive, low cost and simple solutions. The goal of bluetooth detectors is mainly the estimation of travel times of specific road segments by matching the Medium Access Control (MAC) address of mobile devices passing from the end-points of these segments [28].

Apart from providing a variety of traffic measurements, video cameras can employ advanced computer vision algorithms to deliver advanced functionalities, such as automatic license plate recognition and scene analysis, which cannot be performed by other NMDs. These capabilities enable a wealth of applications for automatic incident detection and traffic policy enforcement, which are particularly important in ITS. A summary of the various NMDs along with the main principles of operation, applications, advantages and disadvantages can be found in [29].

3.2 Applications

Monitoring has numerous applications in transportation systems. The most fundamental application of monitoring solutions regards the identification of different road users, which enables more advanced applications. Traditional NMDs and vision-based solutions can count vehicles, and often can classify vehicles according to their type. Depending on the capabilities of an NMD, vehicle classification (VCL) can be accomplished using several different metrics, such as weight, length and speed, vehicle silhouette, etc. Using vision-based or mobile phone solutions the detection of pedestrians is also possible [30, 31].

Measuring and aggregating spatiotemporal data regarding fundamental traffic parameters, such as count and classification of vehicles, concentration (density or occupancy), flow, speed and travel times are essential in the management and planning of transportation systems (MPTS). Examination and analysis of these data allows transportation engineers to identify main operational problems, such as dangerous or highly congested network regions and plan accordingly to solve those issues by constructing new or upgrading existing infrastructure, or by adopting policies that will reduce congestion and accident risk. In addition, these data allows the optimization and generalisation of certain functions of the transportation system; for example, offline optimization of traffic signal scheduling requires estimates about the queue lengths at each intersection and the travel time between intersections, which can be estimated using collected traffic data. Infrastructure health monitoring can also be achieved automatically through the collection and processing of acoustic sensor data that capture the noise generated by the interaction between car tires and the road surface [32].

Another important application of monitoring is traffic state estimation (TSE). The high cost and installation difficulty of NMDs hinders their widespread deployment, which implies that a large portion of the transportation network remains unmonitored. For this reason, several model-based techniques have been developed to address this issue by finding the state of the roads of the considered transportation region that minimizes a certain cost measure, while respecting the known traffic information. In this context, traffic state can mean either density or speed as a function of time and space for highway traffic, or it can mean queue lengths at intersections in urban areas. Even if there is abundance of measurements, TSE is still important due to the noisy, incomplete or even erroneous nature of measurements. For example, the recent emergence of data collection from floating user cars equipped with GPS receivers, allows the collection of a large volume of relatively low quality data that need appropriate processing to extract good estimates of the traffic state. Closely-related and equally important problems to TSE are the short-term prediction (STP) of the traffic state which is utilized in ATIS (e.g., link travel times and congestion levels), as well as the estimation of the origin-destination matrix (EODM), which is an indicator of the traffic flows in different paths of the transportation network [33]. Traffic data can also be exploited to detect anomalies in the flow of traffic; closer examination of that area can determine if there is an incident, such as accidents, road blockages or congestion, that needs to be dealt with.

Automatic incident detection (AID) is another important application of monitoring, mostly achieved using vision-based solutions. AID is more complex than simply measuring certain traffic parameters, and often requires several processing stages such as the detection, classification, identification and tracking of multiple moving objects, as well as the analysis of the scene. Detection is required to distinguish moving targets of interest from the background and from each other and monitor their behaviour. To fully identify moving targets, classification of road users (e.g., pedestrians, cars, buses, tracks) and license plate recognition can be performed. In addition, it is often necessary to track the trajectory of individual vehicles over time and infer whether a specific incident is taking place. For example, when monitoring a scene to detect illegal turning, each vehicle has to be detected, identified and tracked, to infer whether it is following an illegal trajectory. Apart from illegal turning, there are several other monitoring applications related to the enforcement of transportation policies, such as illegal stopping or parking, bus lane or one-way street violation, speeding, red light running and stolen vehicle recovery. Vision-based solutions can also be used for easy-pass toll charging, parking monitoring to provide guidance to travellers via ATIS and accident detection.

The main driving force of VMDs are video cameras as they enable key technologies in driver assistive systems and AVs. The ability of cameras to detect, recognize and monitor the trajectories of other vehicles is essential for crash avoidance. Cameras can also detect pedestrians, so as if a pedestrian is discovered in a dangerous region with respect to the moving vehicle, then certain proactive measures can be taken to protect them from colliding. Apart from the monitoring of moving objects, cameras are essential in perceiving the surrounding environment to assist the driver in avoiding hazards (e.g., pedestrians detection, collision avoidance, car following, traffic sign identification and lane departure warning [34]). Video or infrared cameras can also monitor the driver for inattention, taking appropriate actions in case of distraction, drowsiness or fatigue which impair the driver's performance and increase the risk for accidents. Figure 2, summarizes the different monitoring technologies and the main applications emanating from each technology.

3.3 Techniques

The development of vehicle and network monitoring applications has given rise to numerous techniques from various areas of engineering, such as statistical and probabilistic analysis, computer vision, machine learning, control systems and computational intelligence. In this section, the main techniques of monitoring applications are discussed.

A prominent application of traffic monitoring regards TSE and STP. The challenging nature of these problems (due to the incompleteness, heterogeneity and inaccuracy of measurements, the dynamically changing environmental conditions, as well as the unpredictable nature of human behaviour) has given rise to two main technique classes. The first, makes use of traffic flow physics to construct appropriate models and efficiently incorporate traffic data to obtain estimation and/or prediction of traffic. The most common approach utilizes first or second order density-flow or velocity models, combined with different Kalman-filtering approaches (linear [35], extended [36], ensemble [37], localized-extended [38]) to optimize estimation performance against noise. Technique variations occur from the various measurement sources (loop detectors, floating vehicles, mobile phone probes) that provide data for different traffic parameters and hence require models modifications to better incorporate information; fusing data from multiple sources can also improve performance, especially in urban environments where traffic estimation is more challenging [39]. The second class of techniques is purely data-driven, without considering a specific model from traffic engineering, so that the estimation is based on statistical or machine learning techniques. Data-driven techniques have the advantage of being able to incorporate, apart from standard traffic measurements, other useful information, such as meteorological data, vehicle mix and driver mix. Data-driven techniques mostly rely on parametric methods that assume an underlying general model (e.g., multivariate time-series [40, 41]) or non-parametric methods where estimation and prediction is achieved by implicitly constructing the relationship between independent and dependent variables (e.g., through clustering [42], neural networks [43], SVM classifiers [44]). A recent survey on short-term traffic prediction methods can be found in [45].

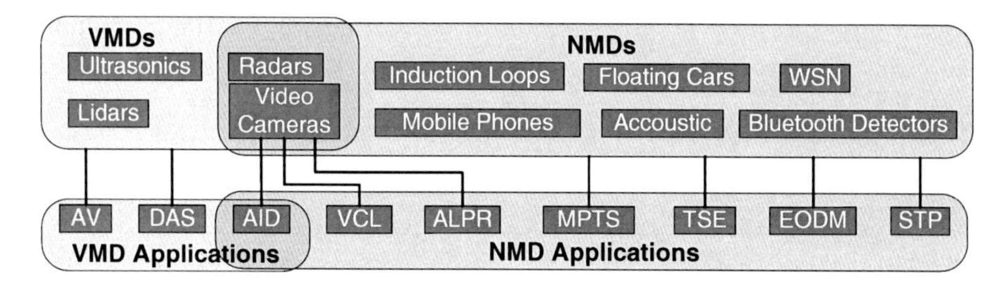

Fig. 2 Monitoring devices and applications

A plethora of techniques have been developed for detecting and classifying vehicles into categories (e.g., passenger cars, vans, cars with trailer, buses, trucks, motorcycles), due to the variety of NMDs deployed in the road infrastructure, each providing different kinds of measurement and having its own strengths and shortfalls. In [46] the variation rate of the natural frequency resulting from the passage of cars and the magnetic profile of an inductive dual-loop detector were used as input to a multilayer feed-forward neural network to classify vehicles into five categories. Inductive single-loop detectors were used in [47], to estimate the speed and length of each vehicle and classify cars using a simple rule-based method. The authors of [48] perform VCL using a network of acoustic and seismic sensors with wireless communication capabilities in a distributed two stage procedure. The first stage involved local classification from various acoustic and seismic signal features through machine learning classifiers (k-nearest neighbors (kNN), support vector machines (SVM), maximum likelihood (ML) and linear vector quantization (LVQ)); in the second stage, local decisions were centrally fused using weighted majority voting exploiting the fact that multiple sensors classified each vehicle. A feature extraction low complexity classification algorithm was also developed in [49], based on the length, height and height profile of each vehicle obtained from a microwave radar detector. In [50] vibration sensors were employed for the categorization of trucks according to the received signal profile and the time of arrival and departure of a vehicle using a signal processing procedure to extract the number of vehicle axles. Computer vision techniques have also been developed to tackle the problem of VCL using still images or video sequences. Such techniques can extract specific features from each vehicle, including size, linearity properties [24] and edges [51], or use general extraction methods [52] to extract features that are then compared to a library of associated features for each vehicle class. In this way similarity scores can be established between target vehicles and different vehicle categories; different classifiers can then be employed according to the sequence of similarity scores for each vehicle, such as maximum likelihood [24], Bayesian inference [53] and ensembles of classifiers [52].

Apart from the detection and classification of vehicles, computer vision techniques are also fundamental in AID [54]. AID is comprised of the following stages: vehicle detection and classification (already discussed), identification through ALPR, tracking and analysis of the scene based on the tracking of multiple vehicles to detect whether a significant event has occurred (e.g., two vehicles very close to each other is a potential accident [55]).

ALPR is usually achieved through detection of the license plate, segmentation into individual characters and recognition of each character, using still images or video sequences. This task is quite challenging due to the diversity in license plate formats, the dynamically changing outdoor illumination conditions and the low resolution of image regions due to dirt and physical damage. For this reason, several methodologies for ALPR have been developed ranging from monochrome and colour image processing techniques (e.g., based on morphological characteristics, component analysis, hierarchical representations, image transforms, histogram processing) to machine learning (e.g., cascade classifiers, SVM, hidden Markov Models) and computational intelligence (e.g., artificial neural networks, genetic algorithms). A review of the various techniques used in the three stages of ALPR can be found in [56, 57].

Vehicle tracking allows the observation of the path of each vehicle, which is important for traffic policy enforcement in traffic surveillance. Tracking involves two main phases; in the first phase, features are extracted from the vehicle or the foreground, and used in the second phase by a dynamic model to predict the trajectory of the moving vehicle. The most commonly used dynamic model is the Kalman filter which can optimally estimate the state of a linear time invariant systems using measurements with Gaussian noise [24, 58]. Nevertheless because the motion model is in general nonlinear, and with non-Gaussian noise, particle filters have also been developed for this task [59]. Other techniques that have been proposed for vehicle tracking include spatiotemporal Markov Random fields [60] and graph correspondence [61]. A comprehensive presentation of computer vision techniques for urban environments can be found in [62].

Apart from network monitoring, computer-vision is also of paramount importance in AVs and DAS in intelligent vehicles. The implementation of such systems requires an integrated monitoring framework that incorporates the vehicle, the surrounding environment and the driver [63, 64]. Monitoring the behaviour of the vehicle is perhaps the easiest components, as it is a process that can be performed using multiple on-board sensors in a controlled manner. Contrary to vehicle monitoring, perceiving the surrounding environment using vision-based systems is quite challenging and requires among others to detect, track and analyse the behaviour of other moving-objects (pedestrians and vehicles), as well as to observe the road-way infrastructure (e.g., road lanes, traffic lights, traffic signs). Early object detection systems relied on simple image features, such as shape, symmetry, texture and edges, however, recent detectors use more general and robust features, such as histogram of oriented gradients (HOG) [65], Haar wavelets [66] and Scale Invariant Feature Transform (SIFT) [67] features. For the classification of image segments into objects of interest, machine learning classifiers are usually employed, such as linear and latent SVMs, Adaboost cascade classification, hidden Markov Models and Gaussian Mixture Models [68]. The abundance of research works on this topic is evidenced by the multiple survey papers that have been conducted for pedestrian [30, 69–71] and vehicle detection [72, 73]. Variations of Kalman and particle filtering are the most widely used techniques to provide robust position estimations of the detected objects; in these techniques the state vector is usually comprised of the coordinates and velocity of the pixels that parameterize the tracked object [72]. Tracking of pedestrians and vehicles is vital for the early diagnosis and prevention of collisions by assessing the risk that the predicted trajectories of two moving objects coincide [70, 71]. Detection and tracking are also prerequisites of behaviour analysis, which is also essential for AVs or a complete DAS that enhances safety and driving quality by providing information, warnings and autonomous control functionality. Characterization of on-road vehicle behaviour refers to the identification of specific driving behaviour from sets of predefined behaviours related to: (a) driving environment (urban, freeway, intersection, non-intersection), (b) performed manoeuvres (e.g., overtaking, merging, lane changing, lane keeping, turning), (c) long-term vehicle trajectory prediction, and (d) driving style (e.g., erratic, aggressive, normal). Due to the predefined categorization of vehicle behaviours, characterization is usually achieved based on unsupervised (e.g., clustering, template matching scores), supervised (e.g., SVM, neural networks), and probabilistic (Dynamic Bayesian Networks, hidden Markov Models, particle filters) techniques. A review of vehicle behaviour studies can be found in [72].

Perceiving the road-way infrastructure using computer vision is also fundamental in AVCS. In [74], various studies on lane detection and tracking were reviewed, identifying five common steps of the developed algorithms: (a) road modelling, (b) road marking extraction, (c) pre-processing, (d) vehicle modelling, (e) position tracking. More recently, supervised learning techniques based on neural networks and SVM were employed for lane detection from a collection of images, while particle filtering was used for lane tracking [75]. Detecting traffic signs is also an important in DASs and AVs. Traffic sign detection usually consists of three stages: (a) segmentation, used to obtain roughly the segment of the image containing the sign, (b) feature extraction (such as edges, colour, HOG, Haar wavelets), and (c) detection. For shape-based features, the most common detection techniques are radial symmetry voting or Hough detection, while for HOG and Haar wavelet features the methods of choice are SVM and cascade classifiers. A survey of traffic sign detection methods can be found in [34].

An integrated DAS needs to monitor not only the environment and the vehicle but also the driver. Monitoring the inattention of the driver using in-car cameras can enhance safety, by warning the driver for potentially hazardous situations. The two main factors of driver inattention are fatigue (physical and mental fatigue, sleepiness) and distraction (auditory, visual, cognitive, biochemical). Hence, techniques have been develop to monitor the driver and the vehicle in order to infer such inattention events. Fatigue can be detected from physical measures (e.g., eye closure duration, frequency, duration and energy of blinks, head nodding), as well as from driving performance measures (e.g., vehicle following distance, dynamics of steering-wheel motion). Distraction can also be detected based on physical and driving performance measures, but the quantities measured in this case are different (gaze direction, head motions, in-vehicle audio-visual operation, driving data, lane boundaries). Using different sets of these inattention characteristics several machine learning and computation intelligence techniques have been proposed to classify normal from distracted driving, which are summarized in [76].

In summary, the wealth of technologies and multiplicative applications of monitoring have resulted in the development of numerous techniques from the fields of computer-vision, machine learning, computational intelligence, statistical and probabilistic analysis, which manipulate data from single sources. Nevertheless, improved results can be obtained by combining the advantages of multiple heterogeneous sources using data fusion techniques [77].

3.4 Challenges and Open Issues

Monitoring transportation systems is a challenging task due to their large-scale, dynamic, uncertain and heterogeneous nature with several open issues that require further investigation. In terms of monitoring technologies, challenges arise in advancing existing monitoring devices and developing novel low-cost wide-area measurement units. Despite the wealth of intrusive and non-intrusive NMDs, all technologies suffer from specific disadvantages, such as high capital cost (e.g., video-cameras), high cost and disruption of traffic when installing/maintaining devices (e.g., inductive loops, magnetometers), low accuracy and limited number of supported measurement types (e.g., ultrasonics), as well as performance deterioration in adverse weather conditions (e.g., infrared, video-cameras, ultrasonics) [29]. Hence, new advancements are required to eliminate the drawbacks of these technologies. Similarly, there is a need to develop low-cost VMDs, such as video-cameras, differential GPS, radars and lidars, in order to improve the economic viability of DAS in passenger vehicles. In addition, the emergence of smartphones and V2X communications allows the deployment of novel wide-area measurement solutions that can boost the collection of high-quality traffic measurements. Towards this direction, the exploration of cooperative vehicle sensing approaches could significantly add value to the monitoring ability of the network.

Apart from the collection of data, smartphones can be exploited to deliver novel applications that enhance the safety and performance of travellers. Although some smartphone apps have already been developed, such as WalkSafe [78] and CarSafe [79], that enhance pedestrian and driver safety respectively, there is room for more research and development in this area that could have a major impact on ITS. The development of applications that allow user-system interaction is also quite challenging. Currently, TSE relies solely on measurements collected from NMDs deployed in road networks; nevertheless, crowdsourcing information directly from users can add considerable value especially for the detection and identification of traffic incidents.

Open issues also exist in processing data collected from VMDs and NMDs. In VMDs, significant challenges arise in processing images collected from computer vision equipment for the detection and tracking of vehicles and pedestrians, as well as for scene analysis to automatically infer incidents and vehicle behaviours. These challenges arise from dynamically changing scenes that involve different viewing angles and poses, variable illumination and lighting conditions, various distances between objects and the cameras, cluttered background and partial occlusion. Although significant advancements have been achieved towards the detection of pedestrians and vehicles under normal conditions, performance rapidly deteriorates for partially occluded objects or under night-light conditions [30]. Hence, better feature extraction techniques, exploitation of inter-frame motion features and more robust classifiers are needed to deal with the detection problem under adverse conditions. In addition, as the building of a perfect vehicle or pedestrian detector is impossible, systems should leverage knowledge from other sensing devices or even the driver.

On the other hand, the processing of data collected from NMDs face challenges related to the large-scale nature of the network, the abundance of information acquired from multiple heterogeneous sources, as well as the noisy and often faulty nature of the collected data. Collecting large volume, high velocity and rich variety data (referred to as "big-data") may be desirable, but the aggregation and processing of big-data is challenging. The obvious solution to deal with the incoming data is to increase the communication bandwidth and memory/computation capabilities of processing servers at transportation operation centers which, however, is not scalable. Hence, other solutions need to be explored, such as processing streaming data at the sources to either compress input data (via appropriate signal processing techniques based on dimensionality reduction or sparse learning) or only report information on important events. Another potential solution is to perform distributed cooperative processing among sources to avoid the need of a central processing station. Dealing with missing or faulty data still constitutes a significant challenge; towards this direction, robust techniques need to be employed from fields, such as statistical data analysis and fault tolerance. Finally, fusion techniques need further exploration to exploit the unique characteristics of heterogeneous data sources for better monitoring.

4 Control of Road Transportation Systems

Despite the fact that monitoring is important for improved operation of the transportation system, the key ingredient towards this direction is the real-time control of the system's operation. By actively controlling the transportation system both at the vehicle and network level, several key ITS benefits can be delivered including congestion alleviation, safety improvement, CO_2 emissions and energy consumption reduction. It should be emphasized that the word "control" in this context, is not simply indicative of control theory techniques; it is used in a general sense to imply the active, intelligent management and operation of various transportation entities (vehicles, traffic lights, traffic flows, navigation devices) to deliver individual or system benefits. In this section, a review of the main strategies used for controlling the transportation system and of the different techniques employed to maximize the impact of each control strategy is provided.

4.1 Control Strategies

Traditionally, control strategies are implemented at the vehicle and network level. At the vehicle level, control strategies aim primarily towards safety and performance and appear in the form of car-centric or driver-centric measures. The former category includes semi-autonomous DASs, which take partial control of the vehicle both according to the driver's commands and in critical situations, and AVs that have full control leaving the driver completely out of the loop. Driver-centric control measures aim at assisting the driver's decision making process e.g., by providing advice or navigation instructions. At the network level, control measures mostly aim at reducing congestion in urban or freeway networks; main strategies include traffic signal control, gating, ramp metering, variable speed limits, dynamic congestion pricing and route guidance systems. The emergence of V2X communications is giving rise to novel control strategies relying on V2X cooperation, such as vehicle platoons, cooperative situational awareness or decision making and intelligent vehicle highway systems.

4.1.1 Vehicle Control Strategies

Car-centric control strategies are associated with semi-autonomous DAS and AVs. Semi-autonomous DAS monitor certain functionalities of the vehicle and take partial control when needed. Measures included in this category are intelligent speed adaptation (ISA), adaptive cruise control (ACC), forward collision avoidance and autonomous parking [80]. An ISA system monitors the speed of the vehicle in accordance to the road state (geometry, speed limit) and accordingly provides feedback to the driver or automatically adjusts the speed. An ACC system controls the vehicle relative to the distance and speed difference compared to the predecessor car; in urban areas, stop-and-go ACC (operates at low speeds) adjust the distance from the front car, while high-speed ACC controls the speed of the vehicle relative to the front car. Forward collision avoidance systems also sense the relative speed and distance from the predecessor vehicle, but their goal is to mitigate imminent collisions by breaking the vehicle in emergency situations. Autonomous parking systems temporarily take full control of the maneuvering to park the vehicle at a location indicated by the driver.

AVs have permanent full control of the vehicle with the driver completely out of the loop [8]. Apart from the advanced environmental monitoring functionalities needed to raise situational awareness controlling the vehicle's motion and interacting with other actors of the transportation network (vehicles, pedestrians, bicycles) are also quite challenging to accomplish a safe and optimal journey. Motion is comprised of several tasks, such as path planning, obstacle avoidance and maneuvering control. Global path planning provides the optimal path that needs to be followed to reach a certain destination, while local path planning provides the immediate motion trajectory for obstacle avoidance and route keeping. Based on the local trajectory, the maneuvering of the vehicle must be controlled for stable and safe navigation, in accordance to the road conditions and the states of neighboring vehicles. Interaction of an AV with other actors of the transportation network is also quite important. Towards this direction, the trajectory of the AV has to be decided according to the predicted trajectories of other actors to minimize the collision risk, or consensus decisions need to be reached to allow the right-of-way for a specific actor e.g., at an unsignalized intersection.

Driver-centric control measures aim at assisting the decision making of the driver in order to improve performance and safety. With respect to safety, advisory systems may inform the driver of potentially hazardous situations or provide advice on the safety of specific maneuvers, such as lane changing and intersection crossing [81]. Regarding performance, the most important advisory systems provide instructions for optimal route navigation according to a predefined metric including distance, time, energy consumption and CO_2 emissions. Route navigation is either static when road state information is based on historical data, or dynamic when real-time information is utilized [82].

4.1.2 Network Control Strategies

At the network level, the aim is the introduction of strategies and measures that control the flow, speed and density of vehicles within certain parts of the transportation infrastructure (e.g., urban areas and freeways) towards reducing congestion. This is achieved either by increasing the throughput of cars, reducing the total time spent in the network or even by regulating the number of vehicles within congested parts of the network, which has the effect of maintaining high network capacity compared to the congestion capacity, a phenomenon called capacity drop.[2] Suitable network control strategies mainly fall within the ATMS area of ITS and include traffic signal control and gating for urban networks, ramp metering, link control and roadside route guidances systems for freeway networks, as well as integrated approaches that combine multiple measures for mixed networks. Nevertheless, appropriate strategies can be found in other ITS areas, such as smart parking systems in ATIS, bus priority and schedule control in APTS and dynamic pricing in ETPS.

---

[2] It has been found that in transportation networks, the congestion capacity is lower than the maximum capacity of the network by about 5–15 % [83].

The installation of traffic lights at intersections is the main control measure in ATMS for urban networks [84]. Traffic lights were originally introduced to allow safe crossing of vehicles and other actors from an intersection which can be accomplished fairly easy by designing signal sequences with no conflicts. The challenging part is the determination of an efficient signal sequence that in addition to safety, optimizes an appropriate traffic metric, such as total delay, number of vehicle stops and vehicle throughput. Apart from managing isolated intersections, traffic signal control can also be used in a coordinated manner to optimize the performance of a network of intersections, e.g., by allowing green waves for long arterials to minimize the number of vehicle stops as they travel across multiple intersections or by coordinating signal sequences to avoid underutilized signal phases. To prevent over-saturation of sensitive urban road-links and hence avoid capacity drop, gating is also used as a means to hold-back traffic via prolong red phases that restrain the vehicles from reaching the protected regions [85].

Ramp metering, applied at on-ramps in the form of green/red signals, is a major control measure that regulates the entrance in freeway networks. In this way, the traffic flow of a freeway is maintained at a desired level which prevents congestion and capacity drop [86]. This results in a somewhat counter-intuitive outcome, as by restricting vehicles from entering the freeway, the system-wide performance is improved so that even the temporarily restricted vehicles may reach their destination faster. Even better performance can be achieved using coordinated ramp metering, where the traffic flows at multiple on-ramps are simultaneously regulated [87]. In addition, to increasing the throughput of the mainline, coordinated ramp metering can better manage the on-ramp queues to avoid traffic spill-back into the urban network and also increases the total volume of traffic served by preventing the blockage of off-ramps.

Apart from regulating the traffic entering a freeway, measures can be taken to control the traffic on the freeway [86]. When the mainline's traffic flow is close to capacity, simple disturbances, such as lane changing and sudden braking can cause backwards-propagating shock-waves that eventually cause stop-and-go congestion, a phenomenon called phantom jam[3] [88]. Hence, to avoid congestion in such circumstances, several link control measures can be applied to homogenize traffic, such as lane keeping, prohibited lane upstream use (due to high use of associated on-ramps) and variable speed limits. Although, the effect of these measures is difficult to be accurately captured using validated mathematical models, studies have shown that these measures can be quite effective if used properly (an example study on variable speed limits can be found in [89]).

While ramp metering and link control regulate traffic flows that enter or are present on freeway networks, roadside route guidance systems have an impact on the volume of traffic that is directed towards certain routes, by influencing the drivers' decisions at bifurcation nodes of the network [86]. This is achieved by

[3] It is called phantom congestion, because severe congestion is caused with no obvious reason, such as lane closure, merging and accident.

providing real-time information (e.g., travel times, congestion levels of certain links) to the drivers via appropriate en-route devices, such as radio services (RDS-TMC), variable message signs (VMS) or in-car devices. Using this strategy, control can be achieved measuring in real-time the state of the network and altering the guidance instructions accordingly so as to achieve the desired control objectives. Effectively implementing such strategies is quite challenging because there is an inherent risk that drivers might over-react to certain guidelines which will create congestion at the suggested links; if this occurs on a regular basis, eventually drivers will lose their trust, making the guidance system useless.

Apart from strategies implemented in ATMS, control measures introduced in other areas of ITS can help improve the transportation system. In ATIS, apart from in-car and roadside advisory and guidance systems, smart parking systems can help alleviate congestion[4] and increase driver convenience in urban networks. This is achieved by deploying smart systems that can manage parking by proving advice to travelers on available spaces and also reserving spaces for specific users according to their needs [90]. In APTS, control measures can help increase the efficiency and reliability of public means e.g., by controlling the headway of buses to help them meet their schedule or by giving priority to public means at traffic signals when they fall behind schedule [91].

4.1.3 Cooperative Strategies

V2X communications allow the cooperation of different actors in the transportation network, which is a key enabler for novel strategies to control the operation of individual vehicles, groups of vehicles or even the entire transportation network.

At the vehicle level, V2V communication allows the collection of information regarding the state of the road actors and the infrastructure beyond the line-of-sight of the driver which enables cooperative advisory safety applications, such as slow vehicle warning and hazardous location warning. In addition, the exchange of information in a V2V fashion, is a key enabler of cooperative safety and efficiency control strategies. For instance, by cooperating with the surrounding actors, a vehicle can perform ACC safer and more efficiently, while it can mitigate collisions more reliably by knowing the state of surrounding cars faster. Another important example regards uncontrolled intersection management. Through negotiation and cooperative control with other vehicles that want to traverse an uncontrolled intersection, vehicles can pass much faster and safer by appropriate speed control, without the need to stop [92].

The coordination of a group of vehicles can also be achieved using V2V communication. A fundamental concept in this area, is vehicle platooning which allows the movement of a team of vehicles in an optimal formation to improve the efficiency and safety [93]. Platoons consist of a leader vehicle and several follower

---

[4] The search for parking comprises 30 % of total congestion in downtown areas [90].

vehicles; the leader is responsible for providing information to the followers on the road state and on the maneuvers that need to be coordinately performed. Due to the inherent control of vehicles in platoon formation, intra-platoon space can be kept constant and very small, while inter-platoon space can be higher for improved safety; hence, platooning is considered safer than ad hoc movement, and can increase efficiency of both the grouped vehicles (less fuel consumption due to improved aerodynamics and better maneuvering) and the infrastructure (higher capacity due to small intra-platoon space and smoother driving behavior).

Cooperative control strategies have also been developed at the network level. Closely related to vehicle platooning is the concept of Intelligent Vehicle Highway Systems (IVHS), which provide assistance for intelligent vehicles and vehicle platoons and coordinate vehicles for improved system performance through V2I communication [80]. Apart from providing information on the road state, the infrastructure can facilitate platoon maneuvering, control the formation of platoons to locally mitigate congestion, e.g., by computing the maximum platoon size and velocity, or even indicate the path of each platoon to alleviate congestion at the system level. Although, IVHS have not been implement in practice yet, several control frameworks have been proposed including PATH, Dolphin, Auto21 CDS, CVIS and Safespot; a comparative survey of these projects can be found in [80]. Cooperative network level control strategies can also be implemented mainly using V2V cooperation with a degree of infrastructure support. In this regard, several control frameworks have been suggested that treat vehicles as a network of autonomous agents that have communication, computation and control capabilities, so that they can exchange information with other agents, make decisions and implement these decisions autonomously, with the infrastructure having a supporting role in reserving resources for each agent and facilitating negotiation between agents [94]. Such approaches can transform traditional control strategies for network management, e.g., by eliminating the need for traffic lights in urban networks or ramp-metering in highways. Figure 3, summarizes the control strategies discussed in this section.

4.2 Design Control Methodologies

The design of appropriate methodologies for the implementation of various control strategies, described in the previous section, is quite challenging due to the complex, large-scale, real-time, dynamic and uncertain behavior of the underlying problems. As a result, for each control strategy numerous design methodologies have been suggested emanating from diverse fields of engineering and physical sciences, such as control theory, computational intelligence, operational research and multi-agent systems. This section attempts to briefly outline the main methodological approaches that have been used for the design of different strategies for vehicle and network level control; the aim is not to provide a complete survey of these methods, but to illustrate the main concepts and provide pointers for further reading.

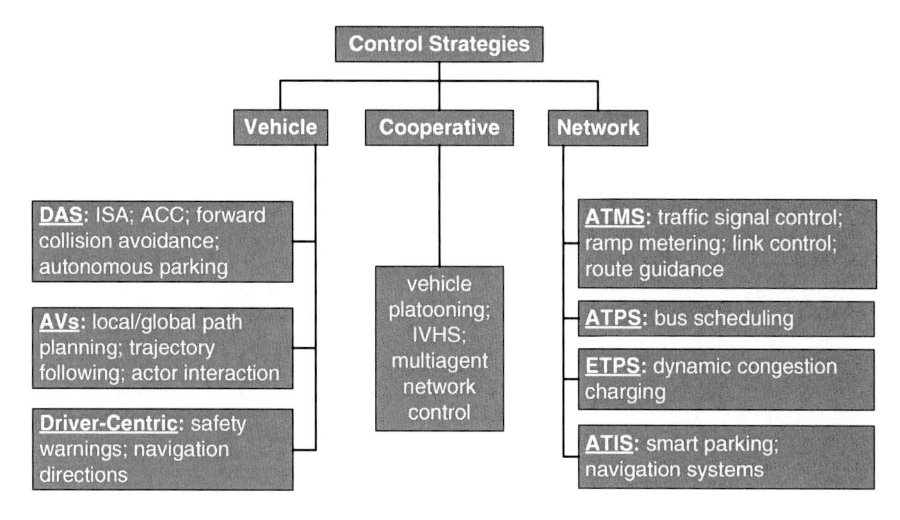

Fig. 3 Summary of control strategies

4.2.1 Vehicle-Oriented

At the vehicle level, a large body of design methodologies is concentrated towards the development of longitudinal and lateral maneuvering control of the vehicle using model-based control theory. The idea is to incorporate different models that describe the lateral and longitudinal dynamics of the vehicle, for the design of controllers that guarantee the satisfaction of control objectives.

Lateral motion systems are designed for steering control for lateral lane keeping, yaw stability[5] and rollover prevention. Lateral systems usually employ kinematic models which describe the system based on geometric relationships between the vehicle and the road (valid at low-speeds, e.g., for automatic parking) or dynamic models with different states, such as the inertial lateral position and yaw angle of the vehicle, or the lateral position and yaw errors in relation to a desired trajectory (suitable for high speeds). Different design methodologies are used depending on the available information; if full information is available, controller design is usually based on state-feedback, else if information is associated with the look-ahead lateral position obtained using computer vision, then output-feedback control is preferred.

Longitudinal systems are used to control parameters associated with the forward/backward movement of the vehicle, such as speed and acceleration, relative distance from preceding vehicles, slip ratio and braking forces. Longitudinal control strategies include anti-lock brake systems (ABS), ACC, collision avoidance and vehicle platooning. To control longitudinal motion two major dynamical models are necessary for the vehicle and driveline dynamics. Vehicle dynamics

[5] Yaw stability prevents vehicles from skidding and spinning out.

include external forces that affect the motion of the vehicle e.g., from tires, aerodynamic drag, rolling resistance and gravity, while driveline dynamics model internal factors that affect motion, such as the internal combustion engine, the torque converter, the transmission and the wheels. ABS employ logic based threshold algorithms and nonlinear control algorithms to prevent wheels from locking or to maintain an optimal slip ratio. Collision avoidance and ACC can be achieved using two-level hierarchical control. At the higher level, usually linearized, fuzzy and nonlinear feedback control algorithms [93] are used for regulating the acceleration of the vehicle to achieve the desired objective (e.g., vehicle following by maintaining constant distance/time-gap from the preceding vehicle, speed control, collision avoidance); at the lower level, nonlinear control algorithms are employed for calculating the throttle input required to track the desired acceleration. In ACC, transitional controllers are also used to achieve smooth transition from vehicle following mode to speed control mode and vice versa. Adaptive control and reinforcement learning is also often employed to account for unknown, dynamically varying, and partially observable model parameters, such as the vehicle mass, aerodynamic drag coefficient and rolling resistance. A comprehensive description of models and design control methodologies for lateral and longitudinal motion systems can be found in [95] and references therein.

In vehicle platooning, not only individual vehicle stability needs to be maintained, but also string stability which requires that the spacing error does not amplify as it propagates across the platoon [93]. If constant distance is the objective, then cooperative control via V2V communications is necessary to achieve string stability, while if constant time-gap is sought, stability can be achieved by autonomously controlling each vehicle. Due to the nonlinearity of the string stability problem, different control approaches have been developed including approaches based on linearization or fuzzification of the model, as well as nonlinear and adaptive control.

4.2.2 Network-Oriented

The large variety of road control strategies and the different design specifications have resulted in a wealth of solution methodologies emanating from diverse areas, such as control theory, optimization and computational intelligence [80, 86]. The applicability of such methodological families to different control strategies is usually indicated by the design specifications rather than the type of control strategy, as they have common traits in terms of computational complexity, constraint handling, incorporation of traffic models, scalability and inclusion of future inputs. Hence, the discussion of different solution methods focuses on their ability to handle different design specifications, such as responsiveness to real-time traffic, road network size, and solution architecture.

Methods based on mixed-integer linear programming (MILP) arise when binary decisions have to be made, e.g., to capture green/red alternations in traffic signal control [96] or space allocation in smart parking [90]. In traffic signal control,

MILP methods usually employ a linear traffic model to capture the behavior of the underlying network, but the ability of the method to handle constraints allows the incorporation of different design specifications, such as minimum and maximum green/red time. MILP methods can be applied to system-wide problems and produce optimal solutions, but they suffer from exponential complexity, so that they are only suitable for fixed-time control, where a schedule for traffic lights is centrally pre-specified for different periods of the days based on historical data. Although MILP can solve system-wide problems, their exponentially growing complexity hinders the solution of large-scale problems.

To overcome the issue of exponential complexity, integer variables are often relaxed into continuous variables, so that the problem can be handled using linear programming (LP). Although, the resulting solution is worse than the MILP one, appropriate rounding heuristics can be used to provide near-optimal integer solutions. Meta-heuristic methods, such as genetic algorithms (GA), particle swarm optimization (PSO) and differential evolution (DE), can also be used in fixed-time strategies to deal with the curse of dimensionality of integer problems [97]. Meta-heuristic methods imitate natural processes, such as genetic evolution, human thinking and social behavior of organisms in swarms or colonies to provide solutions to complex problems. In this regard, they are usually used for the solution of integer, nonlinear complex optimization problems by iteratively adjusting the solution according to the mechanisms of such a natural process and evaluating the objective to move towards better solutions. Meta-heuristics can even be applied to cases with no explicit mathematical model available; in these cases, the objective is evaluated through simulation. Despite the popularity of meta-heuristic methods due to their generality and good performance, they suffer from poor handling of hard problem constraints, offer no guarantee on the quality of the obtained solutions, and take substantial time to run so that their applicability is restricted to fixed-time system-wide strategies.

Traffic responsive strategies make use of real-time measurements to compute online a suitable control plan. In these strategies the plan is periodically updated, so that a new solution is derived in each time period. Hence, control methodologies employed for responsive strategies need to be of low computational complexity. For isolated traffic signal and ramp metering control, approaches usually rely on static control, dynamic programming, fuzzy or neural network controllers and reinforcement learning.

In static control, actions are taken based on measurements about the current state and often on the output (feedback) of the considered system, towards the satisfaction of the control objective (e.g., ALINEA uses desired output flow downstream an on-ramp [98]). Simple proportional-integral feedback control has also been applied for route guidance at isolated freeway bifurcations and gating in urban networks [85].

Dynamic programming is usually employed for traffic signal control to provide online optimal signal schedules based on real-time measurements, by constructing and solving a realistic state-transition model. Although such methods (e.g., PRODYN [99], RHODES [100]) provide excellent results for isolated intersections, the solution of multiple intersection problems is limited due to the exponentially growing size of the state-transition model. The inherent complexity and uncertainty involved in the real-time solution of road control strategies has resulted in the introduction of methods that explicitly handle such situations, such as fuzzy logic and reinforcement learning.

Fuzzy controllers have been proposed for handling imprecise or missing data. Using fuzzy sets input and output traffic parameters can be modelled, such as speed, flow, occupancy, metering rate and red/green signal ratio, and fed into an inference engine to provide crisp values to the fuzzy sets. Inference is accomplished using well designed rules obtained from expert knowledge and offline simulations or online experience. The low complexity and intuitive nature of fuzzy controllers made them popular for both ramp metering and traffic signal control. Neurofuzzy controllers have also been developed where the decision making is done by the fuzzy controller, while the fine tuning of fuzzy membership functions is done by a neural network using reinforcement learning for both isolated [101] and system-wide problems [102].

Another important methodology for isolated traffic control is reinforcement learning (RL). RL relies on self-learning from direct interaction with the environment according to the reward gained from different actions—both from offline simulations and online experience—and requires no explicit model of the environment, contrary to fuzzy systems and dynamic programming respectively. Similar to dynamic programming, RL experiences exponential state-space but it has low computational requirements, which allows the handling of a larger space. For this reasons, RL have been used for isolated ramp-metering and traffic signal control [103], as well as for distributed traffic signal control using multi-agent reinforcement learning techniques [104].

Optimal control (OC) and model predictive control (MPC) methodologies have also been employed for different system-wide road control strategies (e.g., traffic signal control, ramp metering and variable speed limits [105]). Although both methodologies rely on linear and nonlinear models to determine optimal control actions based on predicted future demands, OC is essentially open-loop while MPC is closed-loop. This means that the determined control actions are applied for the whole time window considered in OC or for a few time-units in a rolling horizon fashion in MPC. While MPC is suboptimal under accurate modelling and perfect demand prediction, a rolling horizon MPC approach is robust to disturbances and model mismatch errors. Problems arising in OC and MPC are usually solved using linear and quadratic programming when a linear model (e.g., CMT) is used [106], while nonlinear programming is the method of choice for higher order models (e.g., METANET) [87]. One drawback of OC and MPC is computational complexity, as these problems are centrally solved, and hence dealing with large networks is difficult.

4.3 Challenges and Open Issues

The introduction of ITS technologies raise a number of challenging questions related to the control of vehicles and the road network that need to be dealt with for the successful implementation of advanced control strategies associated with the actors of transportation systems.

At the vehicle-level, there are several open issues and questions associated with car-centric and driver-centric control, such as the introduction of passenger AVs and multi-criteria dynamic navigation e.g., based on fuel-consumption and CO_2 emissions. Although, AVs are currently tested in public roads, there are still a number of issues than need to be addressed related to reliability, cost, fault-tolerance and liability, before they can be realised for sale. Despite the fact that AVs have already autonomously driven thousands of miles, reliability remains a major issue because any situation ignored may lead to an error with devastating consequences. Another challenging issue is the prediction of the behaviour and cooperation with non-autonomous entities; for example, humans can easily cooperative to reach consensus on the right-of-way at an uncontrolled intersection, but the presence of an AV can make the situation difficult and tricky. Currently, AVs rely on costly equipment to perform their functionalities; hence, commercial deployment of cheap passenger AVs remains a challenge. Other important issues that arise in AVs are related to liability and legislation. At present, AVs require driver presence, but AVs should also benefit people with no driving license or people with disabilities. Nevertheless, in these cases a question arises with respect to who is responsible for an accident if there is no qualified driver in the AV or if the accident occurred due to system error. These are questions that need to be addressed with proper legislation.

At the network level, challenges arise from the large-scale, dynamic and uncertain nature of the system. Traffic control methodologies proposed so far may solve some issues, but suffer from others; for example, dynamic methods are not scalable, computational intelligence methods are usually suboptimal and of large computational complexity, while mathematical programming techniques usually rely on traffic models that may not closely capture the behaviour of the network. Important challenges in network traffic control also arise due to the imminent emergence of intelligent vehicles. If control strategies are not properly adapted to account for the evolving transportation environment, intelligent vehicles may even lead to performance deterioration. For example, if intelligent vehicles acquire information about the low congestion level of a specific route, then exploitation of this information by the majority of vehicles will lead to congestion of the specific route. Towards this direction, it is important to develop network control methodologies that proactively take into account the reaction of travellers to a certain control measure.

Several open issues also exist with respect to cooperative control strategies. One issue regards the implementation of platooning. Firstly, there is an interoperability issue for platoons formed by heterogeneous vehicles, as these vehicles will be characterized by different dynamics and will carry equipment from different manufacturers. Secondly, the fact that platooned vehicles travel at high speeds and in close distances might create a strong psychological resistance from the drivers to adopt this congestion alleviation strategy. Thirdly, investigations need to be conducted to find the best platooning strategy, as there are different variations of platooning. Another important challenge is associated with the interaction of cooperative V2V strategies and network control strategies. Challenges within this area include the development of completely decentralised network control strategies based only on V2V cooperation (e.g., using self-organization techniques), the development of new network control strategies (e.g., variable "real-valued" speed limits) that exploit the unique characteristics of new transportation paradigms, and the investigation of potential implications on traffic flow behaviour.

5 Security of Road ITS

Security in transportation systems concerns the detection, identification, mitigation and protection against physical and cyber-physical threats towards users or the infrastructure. The large concentration of people using mass transit systems, the fundamental economic role and the network character of transportation systems, as well as the emergence of V2X communications in ITS have made them particularly vulnerable against physical and cyber security threats [107]. This section discusses major threats, solutions and challenges associated with transportation security with special emphasis towards cyber-security in ITS.

5.1 Security Threats

Security threats can be categorized into physical and cyber-physical. Physical security threats directly affect users, infrastructure or operations and include—most importantly—terrorist attacks, natural disasters, crimes in public modes, robberies of valuable cargos and the movement of hazardous materials. A recent report produced by the Mineta Transportation Institute's National Transportation Security Center (MTI/NTSC), of the Department of Homeland Security, USA, estimates that between 1970 and 2010 there were 1,633 attacks on all public surface transport (passenger trains/buses, stations, highway infrastructure) in the world, resulting in more than 6,000 fatalities and 18,000 injuries [108]. To prevent and reduce the effect of terrorist attacks, different security measures are taken, such as frequent police patrolling of potentially targeted areas, random security checks at busy transit stations, acquisition and evaluation of intelligence information at an international level, as well as frequent evaluation of the system's resilience towards the identification and protection of vulnerable areas (e.g., tunnels and bridges). Although resilience is also important in natural disasters, creating a resilient infrastructure against terrorism is fundamentally different because terrorists actively change their targets and tactics based on new technologies and the current state of the infrastructure so as to avoid reinforced targets and capitalize on new weaknesses [107]. Criminal acts can be dealt with more easily with different local security measures, such as police patrolling and the installation of surveillance cameras in public means. The transport of hazardous materials (hazmats) (e.g., flammable, explosive, poisonous, corrosive, or radioactive) is also an important security issue, as incidents during the transportation process can cause public health and safety, environmental and economic issues [109]. This problem can be dealt with using different measures, such as safety regulations, analysis of accidents involving hazmat shipments, hazard mitigation, risk assessment of transportation segments and selection of appropriate routes [110].

Cyber-security threats emanate from the emergence of V2X communications in ITS. Although, V2X cooperation can have a significant impact if exploited properly, their security constitutes a significant challenges that needs to be properly addressed. The challenge stems from the multi-facet nature of security that requires solutions satisfying several objectives, such as confidentiality and integrity of information, availability of services, as well as accountability and authenticity of users. Besides, cyber-security threats should not be underestimated; the reason is that such attacks can be remotely launched and orchestrated in a way that could affect not only individual vehicles but large regions of the transportation system with paramount safety, economic, operational and societal implications. Next, the main cyber attack methods that compromise different security objectives are outlined including malware infection, information inconsistency, denial of service (DoS), impersonation, privacy disclosure, and eavesdropping [111, 112].

Internet malwares, such as viruses and worms, can affect the normal operation of computers of transportation operation centers and cause disruptions of different services. In addition, if remote communication with a system is possible, an attacker might infect it and take control. For example, in 2008 a teenager took control of the tram system in Lodz, Poland, causing four train derails [113], while in 2010, researchers showed that it is possible to infect a car via Bluetooth and take control of the brakes and engine [114].

Information inconsistency is an attack that affects the integrity of information by injecting wrong information into the network with selfish or malicious goals. For example, a vehicle might deceive other vehicles to change direction by disseminating wrong information about the state of the road or its position. Wrong information about the network state can also be propagated by replaying old messages or by modifying messages en-route from the source to the destination. It is also possible for an attacker to make legitimate users estimate wrong information e.g., about their location using GPS spoofing, an attack that counterfeits GPS signals [115].

Denial of service (DoS) attacks aim towards the deprivation of service availability, either by hampering communications or overloading services available to the legitimate users. Examples of attacks were communications are disturbed include flooding and jamming. In flooding, the attacker overloads the vehicular network with a high volume of meaningless data traffic in order to overload the communication channel and hinder nodes from decoding important messages in time. In jamming, the attacker deliberately transmits strong communication signals that interfere with transmitted signals of legitimate users, so that the error rate of received information unacceptably increases. Furthermore, by sending spam messages towards a particular ITS service an attacker can overwhelm its servers with requests and deprive users from accessing the service.

A malicious user can also impersonate a legitimate user by MAC or IP spoofing, if address-based authentication is used, or masquerade into a legitimate user by using a stolen identity to be granted access to her privileges. Moreover, by launching a Sybil attack, a malicious user can simultaneously send messages using different identities to mislead other vehicles and alter the distribution of traffic. Sybil attacks can even defeat majority checks used to validate the correctness of an information received from multiple neighbouring vehicles, in case the malicious messages are more than the benign ones.

Important security threats also exist against information and user confidentiality, such as eavesdropping and privacy disclosure. In eavesdropping, the attacker intercepts messages of a private communication between legitimate users to illegally gain access and utilize private information, e.g., the password of a commercial transaction. Privacy disclosure is also a very important threat in which an attacker may gain access to the identity of the user along with other sensitive personal information, or collect location information through the analysis of message traffic to construct the profile of a driver, breaching her privacy.

Finally, the safety implications of V2X cooperation, require that users can be held accountable for misbehaviours, e.g., intentional provision of wrong information to other users or not adherence to decisions taken jointly with other road users. Such liability issues require that users cannot deny the transmission or reception of critical communication messages. A summary of security threats is provided in Fig. 4.

5.2 ITS Cyber-Security Solutions

The development of solutions for ITS cyber security threats is fundamental for the implementation of V2X communications in ITS. In this section we outline the main methodologies proposed towards the satisfaction of different security objectives, such as malware protection, information integrity, service availability, as well as user confidentiality and accountability.

Protecting transportation operation centers against malware is essential to avoid disruptions; these can be achieved using traditional web security countermeasures, such as firewalls and anti-malware software, as well as by disconnecting from the Internet computers whose operation is highly critical [116]. Regarding the infection of a vehicle's on-board computers, a solution would be to completely

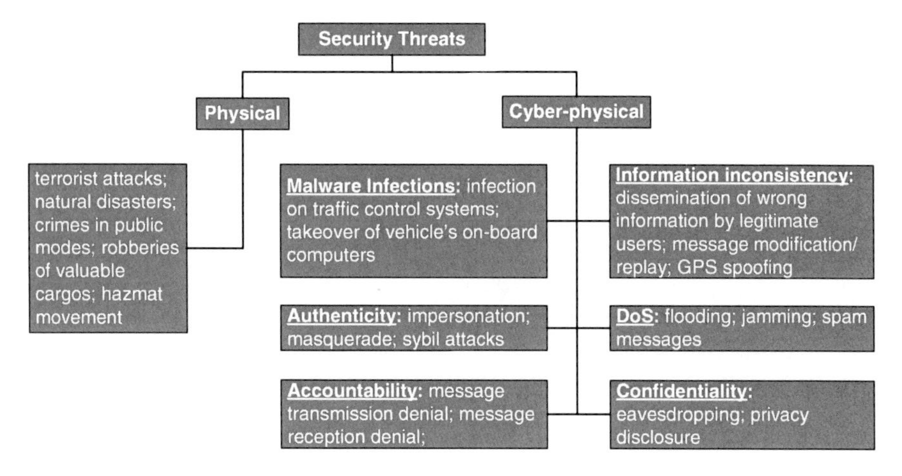

Fig. 4 Categorization of security threats

lock-down remote control capabilities, such as memory reflashing of engine control units (ECUs) and diagnostic testing. Nevertheless, this would ignore the needs of several stakeholders, such as the ability of car owners or small auto shops for repairs. A proposed solution to this problem is to require certain validation and physical access to issue such commands [114].

Protection from certain attacks against information integrity, authentication, confidentiality (e.g., eavesdropping) and accountability can be achieved through traditional solutions from secret and public-key cryptography (also called symmetric and asymmetric cryptography respectively) [117]. For example, for protection against message alternation, a secret checksum algorithm is required that computes the parity of different message blocks and a shared secret key so that if an attacker alters the message the checksum will be wrong. Eavesdropping can be eliminated by encrypting transmitted messages so that the attacker will need to know the secret/private key of the receiving entity to decrypt the message. Authenticity and accountability can be achieved through the use of asymmetric cryptography supported by an intermediary trusted public key infrastructure (PKI) comprised of several authorities for certificate management (e.g., for user certificate registration, generation, revocation and distribution). Certificates are PKI-signed messages that contain the identity and public-key of different users. Hence if user A wants to communicate with user B then s/he only needs to know a priori the public-key of PKI; then by obtaining the certificate of B from PKI, s/he can initiate a procedure using B's public-key (stored in the certificate) for mutual authentication (user B also needs to obtain A's certificate from the PKI). Source authentication and accountability can also be accomplished using asymmetric cryptography. To achieve this, the transmitting node, A, signs the message with its private key so that successful decryption of the message by a receiving node, B, ensures transmission from A. Nevertheless, traditional security methods cannot solve all security threats in ITS and novel security algorithms have been developed specifically in the context of ITS.

In the context of information integrity, information inconsistency due to deliberate dissemination of wrong information by legitimate users, can be addressed through data validation or user trust management. In data validation, each vehicular node cross-validates received information (e.g., about an accident) against its own situational awareness model; one such approach [118] ranks possible explanations from data it has collected using a developed heuristic and selects the highest ranked explanation. On the other hand, trust management methodologies attempt to detect and prevent information inconsistency by building a metric for the trust level (usually based on honesty and discernment) of each user or specific messages. Apart from relying on its own experiences to evaluate trustworthiness, users can also collect opinions from nodes that forward a message [119], passively wait for other nodes to report on the same event [120], or even proactively ask her neighbours [121] (for a survey on trust management see [122]). The replay of old messages can be dealt with by incorporating timestamps that show the generation time so that old messages can be rejected. Solution methods against GPS spoofing attacks have also been developed which rely on monitoring and analysing the amplitude of GPS signals, as spoofed signals tend to be larger that original satellite signals [123].

To limit the impact of DoS attacks, alleviation measures can be taken against different security threats [124]. Firstly, it is important to be able to detect and remotely deactivate misbehaving devices (e.g., devices that launch flooding or spamming attacks). Secondly, alternative communication paths need to be maintained to avoid sending messages through routes that are under attack. Thirdly, communication units should support frequency agility, such as frequency hopping, to provide protection against jamming.

The nature of ITS requires both encryption/authentication of communications and user privacy which are conflicting objectives; hence, conditional privacy-preserving authentication (CPPA) protocols have been proposed. CCPA protocols perform anonymous message authentication, but maintain the user/vehicle identity at a trusted authority in order to reveal it in case of a traffic event dispute. Examples of different proposed CCPA protocols include:

- RSU-based where short-lived anonymous keys are generated from RSUs for communication with vehicles.
- Group-oriented signature-based where entities are organized in groups so that if a group member signs and sends a message, signature verification can be performed by all users while anonymity is maintained within the group.
- Pseudonym-based where each vehicle installs a depository of certificates each containing a different vehicle pseudo-identify and a corresponding private key, so that each time a message has to be transmitted it is signed using a randomly selected certificate.

More information on anonymous authentication in vehicular networks can be found in [125] and references therein.

Apart from providing anonymous authentication, CCPA protocols may also provide user accountability and location concealment. User accountability is achieved by signing the anonymous messages so that in case of a dispute the identities of the users responsible for the dispute can be revealed so that users cannot repudiate the sending of a critical message. Additionally, the use of randomly selected pseudonyms can prevent adversaries from tracking the location of users.

5.3 Challenges and Open Issues

ITS cyber-physical security significantly differs from traditional network security, due to the challenging nature of ITS services and applications, which are characterized by real-time constraints, low-tolerance to errors, large volume of secure messages and high-mobility. Besides ITS security is a relatively new research domain with few research contributions and practically developed solutions. As a result, there are several challenges that need to be further addressed for the successful deployment of ITS security solutions.

To begin with, to guarantee the authenticity, non-repudiation and integrity of transmitted messages, ITS security solutions assume encryption and signing of the messages. Nevertheless, the high frequency of message transmission and the large number of communicating entities implies that each vehicle will have to perform a large number of signature verifications and decryptions which are considered computationally expensive operations. For example, the authors of [126] estimated that each vehicle has to perform 8,800 such operations per second, which can be prohibitive for real-time ITS applications. Hence, it is essential to develop faster algorithms for verification and decryption or reconsider the cases in which these operations are essential to be performed.

Toward this direction, the security requirements of different ITS services should be examined based on the intended usages [126]. This emanates from the fact that ITS involve primarily control applications so that the effect of each security threat should be examined in accordance to its consequences on the safety and efficiency of the system. For instance, in real-time collision avoidance, a message deletion attack might have no effect as the driver will continue to rely on her own situational awareness, while a message modification attack could result in immediate braking of the car which can even increase the probability of collision. Furthermore, robust security solutions need to be developed so that the system can automatically detect and mitigate attacks in real-time, without significant performance deterioration; in the worst case scenario, the performance and safety of the attacked system should not degrade more than a legacy system with no ITS services deployed.

The multi-stakeholder and cross-domain nature of ITS services constitutes another major challenge. The implementation and operation of ITS services involves several stakeholders, such as manufacturers, network operators, governments and customers with often conflicting objectives. Thus, the successful deployment of ITS security solutions depends on providing the right incentives to these stakeholders. Additionally, vehicles can travel across multiple administrative domains with different certification authorities; this requires the development of cross-certified relationships between these authorities which can be challenging as one legal entity may not accept credential issued by another authority.

Finally, there is a need to develop and practically test holistic architectures that integrate solutions for all security problems. Such architectures should even be able to account for seemingly conflicting security requirements, such as authenticity and privacy.

6 Conclusions

Advancements in electronics, communications, positioning and information systems have led to the introduction of ITS in road transportation systems with the potential to revolutionize everyday travel. The incorporation of ITS technologies into different road-actors (i.e., vehicles, travellers, infrastructure) is transforming transportation systems from infrastructure-centric to vehicle-centric with the potential to drastically improve safety and comfort, alleviate congestion, as well as reduce energy consumption and pollutant emissions.

This chapter has introduced ITS in road transport with emphasis towards three main areas: monitoring, control and security. Monitoring allows the collection and management of information enabling several applications and informing road-actor control processes. Control allows improved operation of different road-actors; vehicle, network and cooperative control strategies has been explained, along with different design control methodologies highlighting their strengths and weaknesses. The security of road transportation systems has also been discussed, with special emphasis on cyber-physical security threats and proposed solutions. The main challenges and open issues associated with each ITS area have also been discussed.

References

1. U.S. Energy Information Administration: Total Energy Flow, 2011 (2011)
2. U. D. o. T. Federal Highway Administration: Urban Congestion Trends—Operations: The Key to Reliable Travel. U.S. Department of Transport, Federal Highway Administration, Washington, DC, USA (2013)
3. Anas, A., Lindsey, R.: Reducing urban road transportation externalities: road pricing in theory and in practice. Rev. Environ. Econ. Policy **5**(1), 66–88 (2011)
4. Zhang, J., Wang, F.-Y., Wang, K., Lin, W.-H., Xu, X., Chen, C.: Data-driven intelligent transportation systems: A survey. IEEE Trans. Intell. Transp. Syst. **12**(4), 1624–1639 (2011)
5. Eliasson, J.: A cost-benefit analysis of the Stockholm congestion charging system. Transp. Res. Part A Policy Pract. **43**(4), 468–480 (2009)
6. Leape, J.: The London congestion charge. J. Econ. Perspect. **20**(4), 157–176 (2006)

7. Ezell, S.: Intelligent Transportation Systems. The Information Technology & Innovation Foundation (ITIF), Washington, DC, USA (2010)
8. Ibanez-GuzmÃ¡n, J., Laugier, C., Yoder, J.-D., Thrun, S.: Autonomous driving: context and state-of-the-art. In: Eskandarian, A. (ed.) Handbook of Intelligent Vehicles, pp. 1271–1310. Springer, London (2012)
9. Pinto, C.: How autonomous vehicle policy in California and Nevada addresses technological and non-technological liabilities. Intersect Stanf. J. Sci. Technol. Soc. **5**, 1 – 16 (2012)
10. Mack, C.A.: Fifty years of Moore's law. IEEE Trans. Semicond. Manuf. **24**(2), 202–207 (2011)
11. Armbrust, M., Fox, A., Griffith, R., Joseph, A.D., Katz, R., Konwinski, A., Lee, G., Patterson, D., Rabkin, A., Stoica, I., et al.: A view of cloud computing. Commun. ACM **53**(4), 50–58 (2010)
12. Li, Q., Zhang, T., Yu, Y.: Using cloud computing to process intensive floating car data for urban traffic surveillance. Int. J. Geogr. Inf. Sci. **25**(8), 1303–1322 (2011)
13. Lane, N.D., Miluzzo, E., Lu, H., Peebles, D., Choudhury, T., Campbell, A.T.: A survey of mobile phone sensing. IEEE Commun. Mag. **48**(9), 140–150 (2010)
14. World Health Organization: Global status report on road safety 2013: supporting a decade of action (2013)
15. Malta, L., Miyajima, C., Takeda, K.: A study of driver behavior under potential threats in vehicle traffic. IEEE Trans. Intell. Transp. Syst. **10**(2), 201–210 (2009)
16. Amanna, A.: Overview of IntelliDrive/Vehicle Infrastructure Integration (VII). Technical report. Virginia Tech Transportation Institute (2009)
17. Frequency of Target Crashes for IntelliDrive Safety Systems. U.S. Department of Transport, National Highway Traffic Safety Administration, Washington, DC, USA (2010)
18. Cambridge Systematics: Twin cities ramp meter evaluation. Prepared for Minnesota Department of Transportation (2001)
19. Papamichail, I., Papageorgiou, M., Vong, V., Gaffney, J.: Heuristic ramp-metering coordination strategy implemented at Monash freeway, Australia. Transp. Res. Rec. J. Transp. Res. Board **2178**(1), 10–20 (2010)
20. Cambridge Systematics: Crashes versus Congestion—What's the Cost to Society? (2011)
21. Bertozzi, M., Broggi, A., Fascioli, A.: Vision-based intelligent vehicles: State of the art and perspectives. Robot. Auton. Syst. **32**(1), 1–16 (2000)
22. Hall, F.L.: Traffic stream characteristics. In: Lieu, H., (ed.) Traffic Flow Theory. US Federal Highway Administration (1996)
23. Labry, D., Dolcemascolo, V., Jacob, B., Stanczyk, D.: Piezoelectric sensors for weigh-in-motion systems: sensor behaviour analysis and recommendations. In: 4th International Conference on Weigh-In-Motion, Taipei, Taiwan (2005)
24. Hsieh, J.-W., Yu, S.-H., Chen, Y.-S., Hu, W.-F.: Automatic traffic surveillance system for vehicle tracking and classification. IEEE Trans. Intell. Transp. Syst. **7**(2), 175–187 (2006)
25. Antoniou, C., Balakrishna, R., Koutsopoulos, H.N.: A synthesis of emerging data collection technologies and their impact on traffic management applications. Eur. Trans. Res. Rev. **3**(3), 139–148 (2011)
26. Calabrese, F., Colonna, M., Lovisolo, P., Parata, D., Ratti, C.: Real-time urban monitoring using cell phones: a case study in Rome. IEEE Trans. Intell. Transp. Syst. **12**(1), 141–151 (2011)
27. Coleri, S., Cheung, S.Y., Varaiya, P.: Sensor networks for monitoring traffic. In: Allerton Conference on Communication, Control and Computing, pp. 32–40 (2004)
28. Ahmed, H., El-Darieby, M., Abdulhai, B., Morgan, Y.: Bluetooth-and Wi-Fi-based mesh network platform for traffic monitoring. In: Transportation Research Board 87th Annual Meeting (2008)
29. Mimbela, L., Klein, L., Kent, P., Hamrick, J., Luces, K., Herrera, S.: Summary of Vehicle Detection and Surveillance Technologies used in Intelligent Transportation Systems. Federal Highway Administration's (FHWA) Intelligent Transportation Systems Program Office, August 2007

30. Dollar, P., Wojek, C., Schiele, B., Perona, P.: Pedestrian detection: an evaluation of the state of the art. IEEE Trans. Pattern Anal. Mach. Intell. **34**(4), 743–761 (2012)
31. Mohan, P., Padmanabhan, V.N., Ramjee, R.: Nericell: rich monitoring of road and traffic conditions using mobile smartphones. In: Proceedings of the 6th ACM Conference on Embedded Network Sensor Systems, SenSys'08, New York, NY, USA, pp. 323–336. ACM (2008)
32. Paje, S., Bueno, M., Terán, F., Viñuela, U.: Monitoring road surfaces by close proximity noise of the tire/road interaction. J. Acoust. Soc. Am. **122**, 2636 (2007)
33. Bera, S., Rao, K.V.K.: Estimation of origin-destination matrix from traffic counts: the state of the art. Eur. Trans. Trasporti Europei **49**, 2–23 (2011)
34. Mogelmose, A., Trivedi, M., Moeslund, T.: Vision-based traffic sign detection and analysis for intelligent driver assistance systems: perspectives and survey. IEEE Trans. Intell. Transp. Syst. **13**(4), 1484–1497 (2012)
35. Herrera, J.C., Bayen, A.M.: Incorporation of Lagrangian measurements in freeway traffic state estimation. Transp. Res. Part B Methodol. **44**(4), 460–481 (2010)
36. Wang, Y., Papageorgiou, M.: Real-time freeway traffic state estimation based on extended Kalman filter: a general approach. Transp. Res. Part B Methodol. **39**(2), 141–167 (2005)
37. Work, D.B., Blandin, S., Tossavainen, O.-P., Piccoli, B., Bayen, A.M.: A traffic model for velocity data assimilation. Appl. Math. Res. Express **2010**(1), 1–35 (2010)
38. van Hinsbergen, C.P., Schreiter, T., Zuurbier, F.S., Van Lint, J., van Zuylen, H.J.: Localized extended Kalman filter for scalable real-time traffic state estimation. IEEE Trans. Intell. Transp. Syst. **13**(1), 385–394 (2012)
39. Kong, Q.-J., Li, Z., Chen, Y., Liu, Y.: An approach to urban traffic state estimation by fusing multisource information. IEEE Trans. Intell. Transp. Syst. **10**(3), 499–511 (2009)
40. Stathopoulos, A., Karlaftis, M.G.: A multivariate state space approach for urban traffic flow modeling and prediction. Transp. Res. Part C Emerg. Technol. **11**(2), 121–135 (2003)
41. Bar-Gera, H.: Evaluation of a cellular phone-based system for measurements of traffic speeds and travel times: a case study from Israel. Transp. Res. Part C Emerg. Technol. **15**(6), 380–391 (2007)
42. Antoniou, C., Koutsopoulos, H.N., Yannis, G.: Dynamic data-driven local traffic state estimation and prediction. Transp. Res. Part C Emerg. Technol. **34**, 89–107 (2013)
43. Vlahogianni, E.I., Karlaftis, M.G., Golias, J.C.: Optimized and meta-optimized neural networks for short-term traffic flow prediction: a genetic approach. Transp. Res. Part C Emerg. Technol. **13**(3), 211–234 (2005)
44. Tyagi, V., Kalyanaraman, S., Krishnapuram, R.: Vehicular traffic density state estimation based on cumulative road acoustics. IEEE Trans. Intell. Transp. Syst. **13**(3), 1156–1166 (2012)
45. Zhang, Y.: How to provide accurate and robust traffic forecasts practically? In: Abdel-Rahim, A. (ed.) Intelligent Transportation Systems. InTech (2012)
46. Ki, Y.-K., Baik, D.-K.: Vehicle-classification algorithm for single-loop detectors using neural networks. IEEE Trans. Veh. Technol. **55**(6), 1704–1711 (2006)
47. Coifman, B., Kim, S.: Speed estimation and length based vehicle classification from freeway single-loop detectors. Transp. Res. Part C Emerg. Technol. **17**(4), 349–364 (2009)
48. Duarte, M.F., Hen Hu, Y.: Vehicle classification in distributed sensor networks. J. Parallel Distrib. Comput. **64**(7), 826–838 (2004)
49. Urazghildiiev, I., Ragnarsson, R., Ridderstrom, P., Rydberg, A., Ojefors, E., Wallin, K., Enochsson, P., Ericson, M., Lofqvist, G.: Vehicle classification based on the radar measurement of height profiles. IEEE Trans. Intell. Transp. Syst. **8**(2), 245–253 (2007)
50. Bajwa, R., Rajagopal, R., Varaiya, P., Kavaler, R.: In-pavement wireless sensor network for vehicle classification. In: 10th International Conference on Information Processing in Sensor Networks (IPSN), pp. 85–96. IEEE (2011)
51. Ma, X., Grimson, W.E.L.: Edge-based rich representation for vehicle classification. In: Tenth IEEE International Conference on Computer Vision, vol. 2, pp. 1185–1192. IEEE (2005)
52. Zhang, B.: Reliable classification of vehicle types based on cascade classifier ensembles. IEEE Trans. Intell. Transp. Syst. **14**(1), 322–332 (2013)

53. Mei, X., Ling, H.: Robust visual tracking and vehicle classification via sparse representation. IEEE Trans. Pattern Anal. Mach. Intell. **33**(11), 2259–2272 (2011)
54. Shehata, M.S., Cai, J., Badawy, W.M., Burr, T.W., Pervez, M.S., Johannesson, R.J., Radmanesh, A.: Video-based automatic incident detection for smart roads: the outdoor environmental challenges regarding false alarms. IEEE Trans. Intell. Transp. Syst. **9**(2), 349–360 (2008)
55. Veeraraghavan, H., Masoud, O., Papanikolopoulos, N.P.: Computer vision algorithms for intersection monitoring. IEEE Trans. Intell. Transp. Syst. **4**(2), 78–89 (2003)
56. Anagnostopoulos, C.-N., Anagnostopoulos, I.E., Psoroulas, I.D., Loumos, V., Kayafas, E.: License plate recognition from still images and video sequences: A survey. IEEE Trans. Intell. Transp. Syst. **9**(3), 377–391 (2008)
57. Du, S., Ibrahim, M., Shehata, M., Badawy, W.: Automatic License Plate Recognition (ALPR): a state-of-the-art review. IEEE Trans. Circ. Syst. Video Technol. **23**(2), 311–325 (2013)
58. Coifman, B., Beymer, D., McLauchlan, P., Malik, J.: A real-time computer vision system for vehicle tracking and traffic surveillance. Transp. Res. Part C Emerg. Technol. **6**(4), 271–288 (1998)
59. Gao, T., Wang, P., Wang, C., Yao, Z.: Feature particles tracking for moving objects. J. Multimedia **7**(6), 408–414 (2012)
60. Kamijo, S., Matsushita, Y., Ikeuchi, K., Sakauchi, M.: Traffic monitoring and accident detection at intersections. IEEE Trans. Intell. Transp. Syst. **1**(2), 108–118 (2000)
61. Gupte, S., Masoud, O., Martin, R.F., Papanikolopoulos, N.P.: Detection and classification of vehicles. IEEE Trans. Intell. Transp. Syst. **3**(1), 37–47 (2002)
62. Buch, N., Velastin, S.A., Orwell, J.: A review of computer vision techniques for the analysis of urban traffic. IEEE Trans. Intell. Transp. Syst. **12**(3), 920–939 (2011)
63. Trivedi, M.M., Gandhi, T., McCall, J.: Looking-in and looking-out of a vehicle: computer-vision-based enhanced vehicle safety. IEEE Trans. Intell. Transp. Syst. **8**(1), 108–120 (2007)
64. Apostoloff, N., Zelinsky, A.: Vision in and out of vehicles: integrated driver and road scene monitoring. Int. J. Robot. Res. **23**(4–5), 513–538 (2004)
65. Dalal, N., Triggs, B.: Histograms of oriented gradients for human detection. In: Proceedings of IEEE Conference on Computer Vision and Pattern Recognition (CVPR), vol. 1, pp. 886–893. IEEE (2005)
66. Viola, P., Jones, M.: Rapid object detection using a boosted cascade of simple features. In: Proceedings of IEEE Conference on Computer Vision and Pattern Recognition (CVPR), vol. 1, pp. I-511–I-518. IEEE (2001)
67. Lowe, D.G.: Object recognition from local scale-invariant features. In: Proceedings of the 17th IEEE Conference on Computer vision, vol. 2, pp. 1150–1157. IEEE (1999)
68. Bishop, C.M., Nasrabadi, N.M.: Pattern recognition and machine learning, vol. 1. springer, New York (2006)
69. Geronimo, D., Lopez, A.M., Sappa, A.D., Graf, T.: Survey of pedestrian detection for advanced driver assistance systems. IEEE Trans. Pattern Anal. Mach. Intell. **32**(7), 1239–1258 (2010)
70. Gandhi, T., Trivedi, M.M.: Pedestrian collision avoidance systems: a survey of computer vision based recent studies. In: IEEE Intelligent Transportation Systems Conference (ITSC), pp. 976–981. IEEE (2006)
71. Gandhi, T., Trivedi, M.M.: Pedestrian protection systems: issues, survey, and challenges. IEEE Trans. Intell. Transp. Syst. **8**(3), 413–430 (2007)
72. Sivaraman, S., Trivedi, M.: Looking at vehicles on the road: a survey of vision-based vehicle detection, tracking, and behavior analysis. IEEE Trans. Intell. Transp. Syst. **14**(4), 1773–1795 (2013)
73. Sun, Z., Bebis, G., Miller, R.: On-road vehicle detection: a review. IEEE Trans. Pattern Anal. Mach. Intell. **28**(5), 694–711 (2006)

74. McCall, J.C., Trivedi, M.M.: Video-based lane estimation and tracking for driver assistance: survey, system, and evaluation. IEEE Trans. Intell. Transp. Syst. **7**(1), 20–37 (2006)
75. Kim, Z.: Robust lane detection and tracking in challenging scenarios. IEEE Trans. Intell. Transp. Syst. **9**(1), 16–26 (2008)
76. Dong, Y., Hu, Z., Uchimura, K., Murayama, N.: Driver inattention monitoring system for intelligent vehicles: a review. IEEE Trans. Intell. Transp. Syst. **12**(2), 596–614 (2011)
77. Faouzi, N.-E.E., Leung, H., Kurian, A.: Data fusion in intelligent transportation systems: progress and challenges—a survey. Inf. Fusion **12**(1), 4–10 (2011)
78. Wang, T., Cardone, G., Corradi, A., Torresani, L., Campbell, A.T.: WalkSafe: a pedestrian safety app for mobile phone users who walk and talk while crossing roads. In: Proceedings of the Twelfth Workshop on Mobile Computing Systems and Applications, p. 5. ACM (2012)
79. You, C.-W., Lane, N.D., Chen, F., Wang, R., Chen, Z., Bao, T.J., Montes-de Oca, M., Cheng, Y., Lin, M., Torresani, L., et al.: CarSafe app: alerting drowsy and distracted drivers using dual cameras on smartphones. In: Proceeding of the 11th annual international conference on Mobile systems, applications, and services, pp. 13–26. ACM (2013)
80. Baskar, L.D., De Schutter, B., Hellendoorn, J., Papp, Z.: Traffic control and intelligent vehicle highway systems: a survey. IET Intell. Trans. Syst. **5**(1), 38–52 (2011)
81. Tideman, M., van Der Voort, M., van Arem, B., Tillema, F.: A review of lateral driver support systems. In: Proceedings of the 10th IEEE Intelligent Transportation Systems Conference (ITSC), Seattle, Washington, USA, pp. 992–999. IEEE (2007)
82. Kim, S., Lewis, M.E., White III, C.C.: Optimal vehicle routing with real-time traffic information. IEEE Trans. Intell. Transp. Syst. **6**(2), 178–188 (2005)
83. Chung, K., Rudjanakanoknad, J., Cassidy, M.J.: Relation between traffic density and capacity drop at three freeway bottlenecks. Transp. Res. Part B Methodol. **41**(1), 82–95 (2007)
84. Koonce, P., Rodegerdts, L., Lee, K., Quayle, S., Beaird, S., Braud, C., Bonneson, J., Tarnoff, P., Urbanik, T.: Traffic signal timing manual. Technical report (2008)
85. Keyvan-Ekbatani, M., Kouvelas, A., Papamichail, I., Papageorgiou, M.: Exploiting the fundamental diagram of urban networks for feedback-based gating. Transp. Res. Part B Methodol. **46**(10), 1393–1403 (2012)
86. Papageorgiou, M., Diakaki, C., Dinopoulou, V., Kotsialos, A., Wang, Y.: Review of road traffic control strategies. Proc. IEEE **91**(12), 2043–2067 (2003)
87. Kotsialos, A., Papageorgiou, M., Mangeas, M., Haj-Salem, H.: Coordinated and integrated control of motorway networks via non-linear optimal control. Transp. Res. Part C Emerg. Technol. **10**(1), 65–84 (2002)
88. Flynn, M.R., Kasimov, A.R., Nave, J.-C., Rosales, R.R., Seibold, B.: Self-sustained nonlinear waves in traffic flow. Phys. Rev. E **79**(5), 056113 (2009)
89. Papageorgiou, M., Kosmatopoulos, E., Papamichail, I.: Effects of variable speed limits on motorway traffic flow. Transp. Res. Rec. J. Transp. Res. Board **2047**(1), 37–48 (2008)
90. Geng, Y., Cassandras, C.: New "Smart Parking" system based on resource allocation and reservations. IEEE Trans. Intell. Transp. Syst. **14**, 1129–1139 (2013)
91. Xuan, Y., Argote, J., Daganzo, C.F.: Dynamic bus holding strategies for schedule reliability: optimal linear control and performance analysis. Transp. Res. Part B Methodol. **45**(10), 1831–1845 (2011)
92. Fajardo, D., Au, T.-C., Waller, S.T., Stone, P., Yang, D.: Automated intersection control: performance of a future innovation versus current traffic signal control. Transp. Res. Rec. J. Transp. Res. Board **2259**(1), 223–232 (2011)
93. Kavathekar, P., Chen, Y.: Vehicle platooning: a brief survey and categorization. In: Proceedings of the ASME 2011 International Design Engineering Technical Conferences and Computers and Information in Engineering Conference (2011)
94. Chen, B., Cheng, H.H.: A review of the applications of agent technology in traffic and transportation systems. IEEE Trans. Intell. Transp. Syst. **11**(2), 485–497 (2010)
95. Rajamani, R.: Vehicle Dynamics and Control. Springer, Berlin (2011)

96. Timotheou, S., Panayiotou, C.G., Polycarpou, M.M.: Towards distributed online cooperative traffic signal control using the cell transmission model. In: Proceedings of the 16th International IEEE Annual Conference on Intelligent Transportation Systems, The Hague, Netherlands, pp. 1737–1742. IEEE (2013)
97. Zhao, D., Dai, Y., Zhang, Z.: Computational intelligence in urban traffic signal control: A survey. IEEE Trans. Syst. Man Cybern. Part C Appl. Rev. **42**(4), 485–494 (2012)
98. Papageorgiou, M., Hadj-Salem, H., Blosseville,J.-M.: ALINEA: a local feedback control law for on-ramp metering. Transp. Res. Rec. **1320**, 58–64 (1991)
99. Farges, J.L., Henry, J.J., Tufal, J.: The PRODYN real time traffic algorithm. In: Proceedings of the 4th IFAC/IFIP/IFORS Symposium on Control in Transportation Systems, Baden-Baden, Germany (1983)
100. Mirchandani, P., Head, L.: A real-time traffic signal control system: architecture, algorithms, and analysis. Transp. Res. Part C Emerg. Technol. **9**(6), 415–432 (2001)
101. Bingham, E.: Reinforcement learning in neurofuzzy traffic signal control. Eur. J. Oper. Res. **131**(2), 232–241 (2001)
102. Srinivasan, D., Choy, M.C., Cheu, R.L.: Neural networks for real-time traffic signal control. IEEE Trans. Intell. Transp. Syst. **7**(3), 261–272 (2006)
103. Abdulhai, B., Pringle, R., Karakoulas, G.J.: Reinforcement learning for true adaptive traffic signal control. J. Transp. Eng. **129**(3), 278–285 (2003)
104. El-Tantawy, S., Abdulhai, B., Abdelgawad, H.: Multiagent reinforcement learning for integrated network of adaptive traffic signal controllers (MARLIN-ATSC): methodology and large-scale application on downtown Toronto. IEEE Trans. Intell. Transp. Syst. **14**(3), 1140–1150 (2013)
105. Hegyi, A., De Schutter, B., Hellendoorn, H.: Model predictive control for optimal coordination of ramp metering and variable speed limits. Transp. Res. Part C Emerg. Technol. **13**(3), 185–209 (2005)
106. Gomes, G., Horowitz, R.: Optimal freeway ramp metering using the asymmetric cell transmission model. Transp. Res. Part C Emerg. Technol. **14**(4), 244–262 (2006)
107. Cox, A., Prager, F., Rose, A.: Transportation security and the role of resilience: a foundation for operational metrics. Transp. Policy **18**(2), 307–317 (2011)
108. Jenkins, B.M., Butterworth, B.R.: Explosives and Incendiaries Used in Terrorist Attacks on Public Surface Transportation: A Preliminary Empirical Examination. Technical Report. Mineta Transportation Institute, College of Business, San Jose State University (2010)
109. Abkowitz, M., List, G., Radwan, A.E.: Critical issues in safe transport of hazardous materials. J. Transp. Eng. **115**(6), 608–629 (1989)
110. Kara, B.Y., Verter, V.: Designing a road network for hazardous materials transportation. Transp. Sci. **38**(2), 188–196 (2004)
111. Gillani, S., Shahzad, F., Qayyum, A., Mehmood, R.: A survey on security in vehicular ad hoc networks. Communication Technologies for Vehicles, pp. 59–74, Springer, Berlin (2013)
112. Tchepnda, C., Moustafa, H., Labiod, H., Bourdon, G.: Security in vehicular networks. In: Moustafa, H., Zhang, Y. (eds.), Vehicular Networks, Techniques, Standards and Applications, pp. 331–353. CRC Press, Boca Raton (2009)
113. Loukas, G., Gan, D., Vuong, T.: A review of cyber threats and defence approaches in emergency management. Future Internet **5**(2), 205–236 (2013)
114. Koscher, K., Czeskis, A., Roesner, F., Patel, S., Kohno, T., Checkoway, S., McCoy, D., Kantor, B., Anderson, D., Shacham, H., et al.: Experimental security analysis of a modern automobile. In: Proceedings of the IEEE Symposium on Security and Privacy (SP), pp. 447–462. IEEE (2010)
115. O'Hanlon, B.W., Psiaki, M.L., Bhatti, J.A., Shepard, D.P., Humphreys, T.E.: Real-time GPS spoofing detection via correlation of encrypted signals. Navigation **60**(4), 267–278 (2013)
116. Jensen, M., Gruschka, N., Herkenhöner, R.: A survey of attacks on web services. Comput. Sci.-Res. Dev. **24**(4), 185–197 (2009)
117. Kaufman, C., Perlman, R., Speciner, M.: Network Security: Private Communication in a Public World. Prentice Hall Press, Upper Saddle River (2002)

118. Golle, P., Greene, D., Staddon, J.: Detecting and correcting malicious data in VANETs. In Proceedings of the 1st ACM international workshop on Vehicular ad hoc networks, pp. 29–37. ACM (2004)
119. Chen, C., Zhang, J., Cohen, R., Ho, P.-H.: A trust modeling framework for message propagation and evaluation in VANETs. In: Proceedings of the 2nd International Conference on Information Technology Convergence and Services (ITCS), pp. 1–8. IEEE (2010)
120. Patwardhan, A., Joshi, A., Finin, T., Yesha, Y.: A data intensive reputation management scheme for vehicular ad hoc networks. In; Proceedings of the Third Annual International Conference on Mobile and Ubiquitous Systems: Networking and Services, pp. 1–8. IEEE (2006)
121. Minhas, U.F., Zhang, J., Tran, T., Cohen, R.: Towards expanded trust management for agents in vehicular ad hoc networks. Int. J. Comput. Intell. Theory Pract. **5**(1), 3–15 (2010)
122. Ma, S., Wolfson, O., Lin, J.: A survey on trust management for intelligent transportation systems. In: Proceedings of the 4th ACM SIGSPATIAL International Workshop on Computational Transportation Science, pp. 18–23. ACM (2011)
123. Warner, J.S., Johnston, R.G.: GPS spoofing countermeasures. Homel. Secur. J.**10**, 22–30 (2003)
124. ETSI: Intelligent Transport Systems (ITS) security: threat, vulnerability and risk analysis. Technical report, European Telecommunications Standards Institute, Sophia, Antipolis, France (2010)
125. Xiong, H., Guan, Z., Hu, J., Chen, Z.: Anonymous authentication protocols for vehicular ad hoc networks: an overview. In: Sen, J. (ed.) Applied Cryptography and Network Security, p. 53. Intech, Rijeka, Croatia (2012)
126. Zhao, M., Walker, J., Wang, C.-C.: Security challenges for the intelligent transportation system. In: Proceedings of the First International Conference on Security of Internet of Things, pp. 107–115. ACM (2012)

Algorithms and Tools for Intelligent Monitoring of Critical Infrastructure Systems

Cesare Alippi, Romolo Camplani, Antonio Marullo and Manuel Roveri

Abstract Critical Infrastructure Systems (CIS) are essential services to sustain both society and economy. In fact, CIS can be considered as vital systems for a geographic area or a country. Such valuable assets have to be carefully monitored because their partial or complete failure (caused by natural hazards or criminal acts) could produce severe costs in terms of environment, economy and, in the worst scenario, human lives. The need to protect and maintain CIS and the surrounding environment is pushing the research for the development of intelligent monitoring systems, able to detect anomalies and events and to adapt autonomously to the changes in the system under investigation. In this chapter, we describe an intelligent hardware-software architecture for CIS monitoring, specifically designed for asynchronous events detection, remote configurability and diagnosis. In particular, this monitoring system is based on a novel hybrid architecture, in which different sensors, architectures and physical phenomena under monitoring coexist and cooperate to provide different views of the same physical phenomenon. In fact, the proposed monitoring system is able to gather both high frequency signals (microscopic level), such as accelerometer signals, and low-dynamic signals (macroscopic level), such as temperature and inclination. The monitoring system is connected to a remote data center, which collects, interprets and forwards them to the stakeholders in the desired format. The design principles driving the monitoring system are introduced. As a practical application will be shown a CIS monitoring system employed to monitor the Rialba's tower, a

C. Alippi · R. Camplani · M. Roveri
Politecnico Di Milano, Milan, Italy
e-mail: alippi@elet.polimi.it

R. Camplani
e-mail: camplani@elet.polimi.it

M. Roveri
e-mail: roveri@elet.polimi.it

A. Marullo (✉)
Altran Italia, Milan, Italy
e-mail: antonio.marullo@altran.com

E. Kyriakides and M. Polycarpou (eds.), *Intelligent Monitoring, Control, and Security of Critical Infrastructure Systems*, Studies in Computational Intelligence 565,
DOI 10.1007/978-3-662-44160-2_6

rock tower-like limestone complex overlooking an area of strategic importance connecting the Lecco and Como provinces in north Italy. The rock tower is indeed exposed to a rock toppling risk, thus menacing an area characterized by the presence of a freeway, a railway line and gas and power distribution pipelines.

Keywords Environmental monitoring · Critical infrastructure systems · Sensor networks · Energy harvesting · Embedded systems · Adaptive sensing

1 Introduction

Critical Infrastructure Systems (CIS) are a pivotal element for the correct working of societies and economies. Since a failure, either partial or complete, would mean a high price in terms of environmental, economic or even human losses, it is justified to place such valuable assets under careful monitoring. This need is leading to the development of intelligent monitoring systems, capable of detecting anomalies and dangers, adapting themselves to the changes in the monitored system.

Unfortunately, there are not many published works about the monitoring of CIS. Reference will mainly be made to works in the areas of structural health monitoring and sensor networks. The main kinds of sensor networks employed for structure health monitoring, including traditional systems as well as wireless and hybrid sensor networks, are described in [1]. In [2] a discussion of the principles behind a smart monitoring of structures is presented, underlining how it is necessary to collect not just the data from the sensors, but also the metadata about the proper working of the sensors. In [3] there is a discussion regarding the issues and challenges arising by the application of wireless sensor networks to CIS protection, with a focus for network security. In [4] and [5] the focus is on the usage of the information collected by the CIS monitoring, without requiring a specific CIS monitoring system implementation.

The monitoring system described in this chapter is designed to integrate different kinds of sensor networks, operating at different levels. At the lowest level there are the sensing subsystems, constituted by sensor networks collecting local data, while at the higher level there's a set of web services that analyze and holistically integrate together the sensing subsystems to perform a smart monitoring of the whole CIS.

In Sect. 2 the application of this monitoring system is described. This application, the Rialba's tower, is a natural system surrounded by CIS, such as railroads and power lines. In this application the proposed system has been deployed and has performed its monitoring operations for more than 2 years, in total autonomy.

In Sect. 3 the general monitoring system is described, followed by the description of two sensor networks that constitute the monitoring subsystems on the field. In Sect. 4 the chapter is concluded by summarizing the experiences obtained by the long term operation of the proposed system.

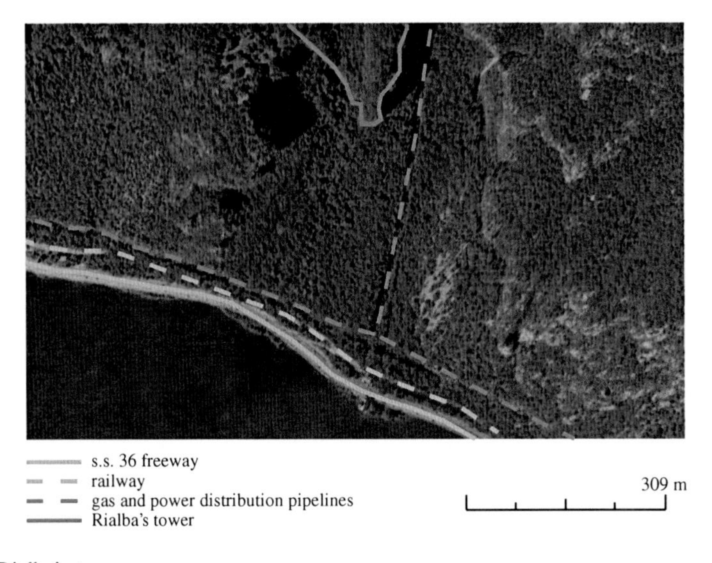

Fig. 1 Rialba's tower area

2 Application Scenario

As applicative scenario for employing the principles behind the monitoring of CISs, the Rialba's tower has been chosen: a rock tower-like limestone complex overlooking an area of strategic importance connecting the Lecco and Como provinces in north Italy.

The area overlooked by Rialba's Tower is of strategical importance, containing infrastructures, such as the s.s.36 freeway, a railway line and gas and power distribution pipelines. Figure 1 shows the position of the Tower and of the CIS surrounding it. The risk represented by the Rialba's Tower for the surrounding infrastructures, is that of a rock topple with the consequence of several tons of rock falling down on the nearby structures and the consequent damages.

The Rialba's Tower is crossed by four big fractures dividing it in five towers of rock, as shown in the illustration in Fig. 2. Of these towers, just the most external two are exposed to a rock toppling risk and thus need to be monitored. The risk of a rock topple of the Rialba's Tower is real, although remote. Studies performed by archeologists in the area show how in the past another portion of the Rialba's Tower, next to the one delimited by fracture F4, existed in the area and how it ended in the nearby lake subsequently to a rock toppling event.

The rock toppling phenomenon would be generated by a weakening of the base of the structure, due to the worsening of existing fractures in the rock and the birth of new ones. Such damage to the base would trigger a loss of equilibrium leading to the rock falling.

To keep the external two towers properly surveilled, several geophysical quantities are under observation. To monitor the insurgence of new fractures at the

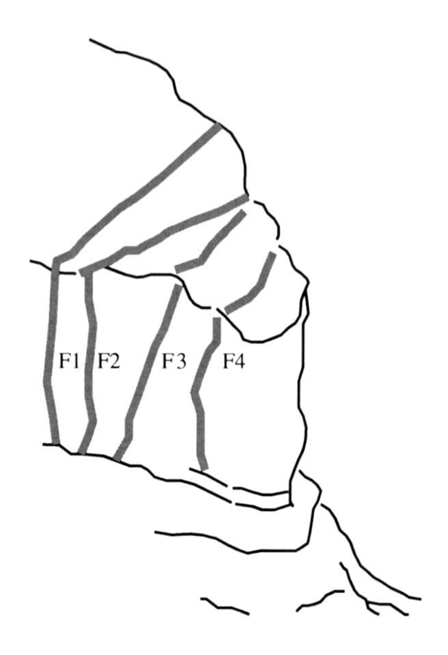

Fig. 2 Rialba's tower main fractures

basis, sensors have been employed, capable of detecting the acoustic signals deriving from the development of micro-cracks in the rock. At the same time, the top of the towers is monitored through wired strain gauge and high precision inclinometers. The impact of water is also monitored with a pluviometer on the top of the Rialba's Tower measuring the precipitation in the area, and two piezometers at the basis measuring the rate of the water flowing down along the fractures F3 and F4.

3 Monitoring System

The monitoring system in its entirety can be seen as a central server in the control room (but could also be a virtualized service in the cloud) to which the different monitoring subsystems in the field send their data, following a publish-subscriber paradigm. More precisely the monitoring system is constituted by a set of restful web services designed to process the data coming from the monitoring subsystems and send new parameters to the same subsystems.

The services exposed to the subsystems are the data insert service and the parameters fetching service. The monitoring system also provided to its human users a web application for data visualization and a web application for parameter setting. The data insert service is accessed asynchronously by the various subsystems. After the systems identify themselves, they send the aggregated data regarding a single sensorial and processing unit (SPU) as a binary stream. Thanks to xml configuration files [6] describing the data coming from each unit and their

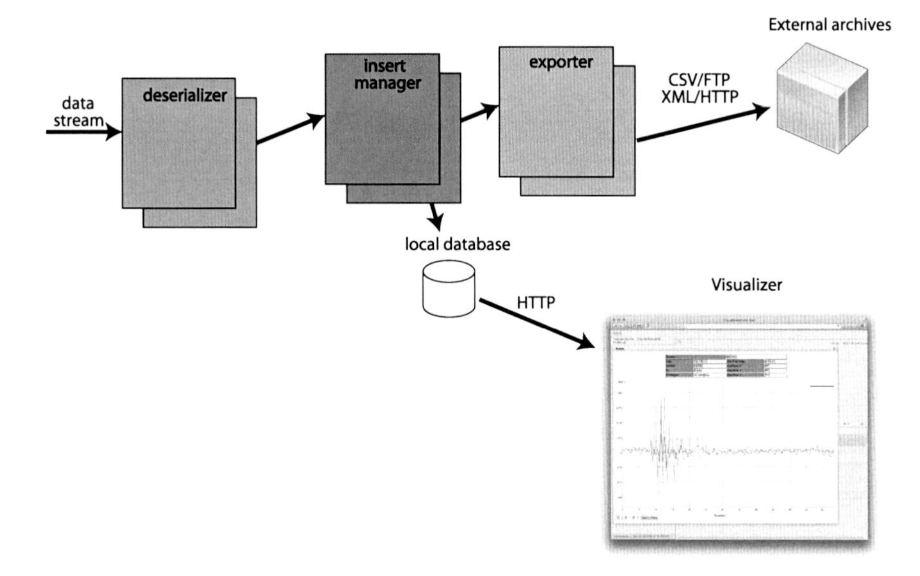

Fig. 3 Data paths

characteristics, the binary object gets partitioned and deserialized to follow the data path shown in Fig. 3.

The deserialized data get indeed passed to an insert manager that calibrates their values and inserts them together with their metadata in a local database. This database is then used by the data visualization web application to present the acquired data to users. At the same time the insert manager evaluates the data notifying through emails any stakeholder configured in the xml files to receive any alarm concerning the data and turns the deserialized data to an exporter module.

The exporter module, always following the xml configuration specific for the received data, exports the data in the desired form (usually xml or text file containing comma separated values) and sends them to all the registered consumers (usually decision systems supported by web services or ftp archives that would subsequently expose the data to legacy systems).

Such a data management chain allows to manage and analyze in real time data coming from heterogeneous sources in a coherent and unified manner. There's no constraint on the data, whether they must be periodic or asynchronous, scalar or array.

On the opposite, to request a specific periodicity from the local monitoring subsystems or to change the behavior of a specific SPU, the monitoring system offers a web interface (like the one in Fig. 4) where the user can conveniently set the desired values. Once the parameter values are committed, they get serialized, following the rules specified in an xml configuration file specific for each monitoring subsystem component, and saved in a local database. When the remote systems ask new parameters for their SPU, the parameters fetching service selects

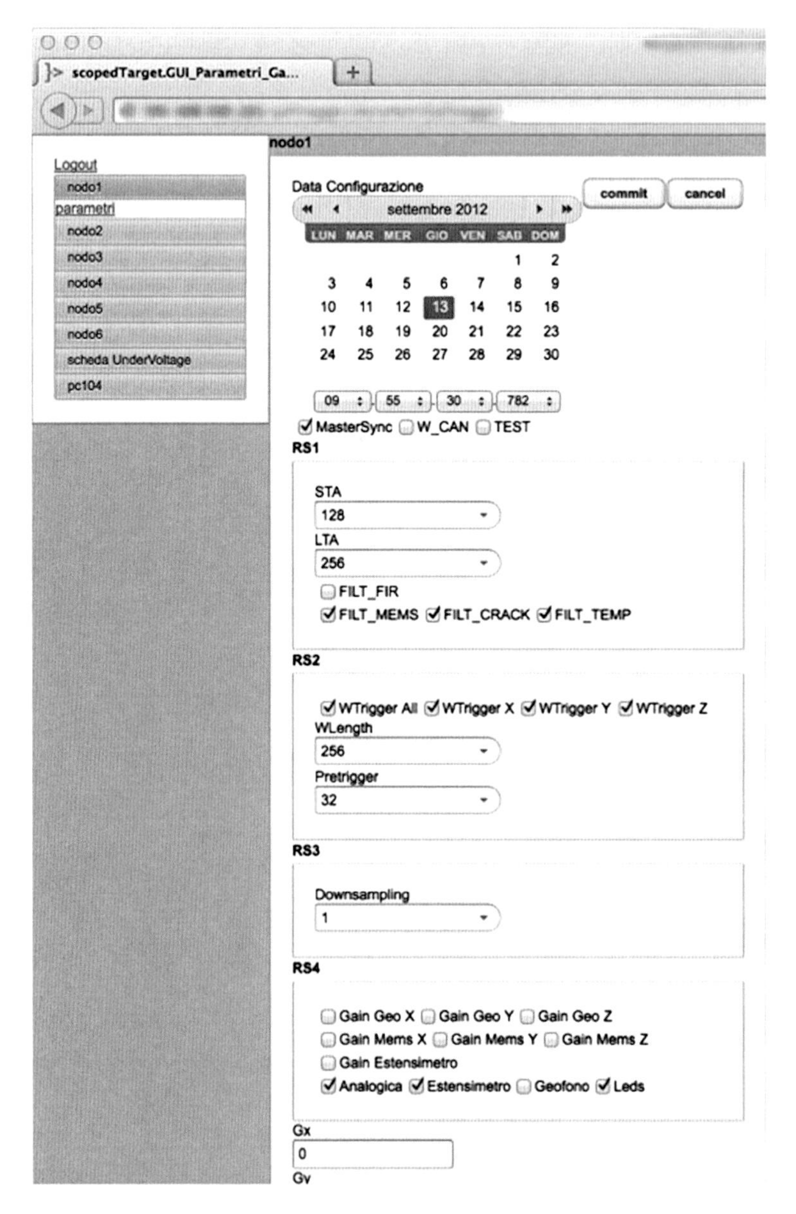

Fig. 4 Parameter gui

the more recent parameters and offers them to the client as a binary stream. Moreover, the new parameters in the database are used by the calibration routines to evaluate the data arrived after the remote update.

In the specific case of the Rialba's Tower, the monitoring system consists, beyond the data managing server, of two distinct monitoring subsystems: a

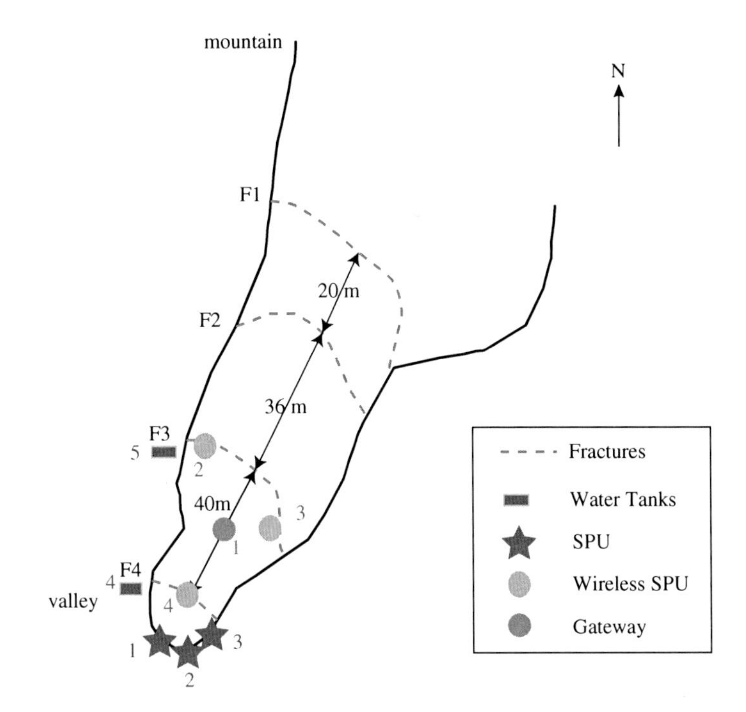

Fig. 5 Deployed subsystems

wireless sensor network on the top of the Tower and a hybrid sensor network at its basis.

A map of the deployed systems in the area is shown in Fig. 5. On this map the positions of the SPUs belonging to the hybrid (square and star markers) and wireless (circle markers) networks are depicted. The two networks communicate with the monitoring system thanks to the http protocol over the GPRS/UMTS cellular network.

3.1 Hybrid Monitoring Subsystem

As already noted, the presence and propagation of fractures within the rock at the basis of the Rialba's Tower would affect the stability of the structure and consequently induce a subsequent fall. To observe the generation and propagation of possible micro-acoustic bursts associated with the enlargement of an existing internal crack or the formation of a new one of millimetric size (sometimes they are also called microseismic bursts for the evident affinity with earthquakes) a very high sampling rate (up to 2 kHz) and a strict synchronism between sensing units (as low as 1 ms) are required. To be able to relate the crack signals to frost weathering phenomena, in addition to each micro electro-mechanical system (mems) accelerometer employed to transduce the micro-acoustic signals, temperature sensors have also been deployed.

It is worth noticing how, while the temperature signal has a low dynamic, thus ensuring that a regular slow sampling of its values is sufficient to obtain a faithful estimation of its behavior, the micro-acoustic signals have a very fast dynamic and are, by their very nature, asynchronous signals requiring a continuous and very fast sampling. Such data acquisition must be followed by a local decision whether what has been recorded could be bound to a crack event.

Since the operating conditions can change over the long time span covering the monitoring system operating life, it is important that the system operators are able to remotely modify not just the sampling and connection periods, but also the criteria with which the micro-acoustic events are evaluated. Another consequence of the long life-span of the monitoring system is that the system must be energetically totally autonomous.

To satisfy such requirements a hybrid wired-wireless architecture has been adopted for the monitoring subsystem at the lower end of the Rialba's Tower. Such choice allows to reach a trade-off between the demanding synchrony, the need to perform continuous sampling and signal processing at a high rate (2 kHz) with the consequent high consumption while at the same time to be energetically autonomous. Indeed centralizing the energy harvesting mechanisms and relying on the wired transmission medium allows to guarantee high transmission bandwidth and synchronism among the units (wired connections overcome wireless transmission in terms of both bandwidth and synchronization ability), while at the same time allow the deployment of the sensor network gateway element in an optimal place for energy harvesting.

A simple schema showing the components of the hybrid subsystem is shown in Fig. 6. What follows is a brief description of the subsystem elements, a more detailed discussion can be found in [7].

3.1.1 Base Station (Gateway/Data Collector)

The base station has a role of gateway and data collector. It acquires from the SPUs, connected to it through the CAN bus, the monitored data and relays them to the monitoring main system through a wireless (UMTS or long distance radio link) connection. It also has the task to distribute new parameters and commands received from the monitoring system to its SPUs. The base station plays a pivotal role to ensure that the monitoring system in its entirety stays responsive and adaptable to the system changes.

A block diagram of the base station is shown in Fig. 7. The heart of the base station is the main board, an industrial PC104 computer, equipped with an ARM9 microprocessor running at 200 MHz with 64 MB of DRAM memory.

For better managing the available energy, it works using a duty cycle, waking up at regular times and connecting to the remote monitoring system after a

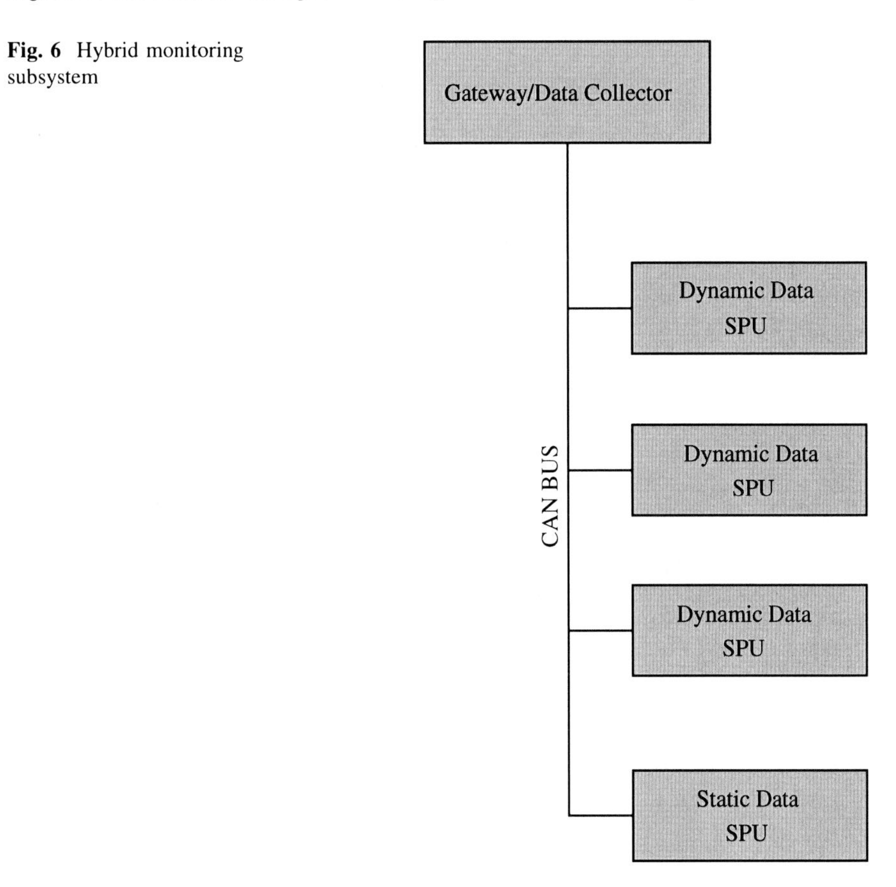

Fig. 6 Hybrid monitoring subsystem

parameterizable number of data-collecting cycles. At the start/wake-up it follows the following behavior:

1. Enquires the connected SPUs collecting the data and aggregates them on a per SPU basis.
2. On data reporting cycles, it sends the aggregated data to the monitoring system via an http connection.
3. On data reporting cycles, it collects (via http) new commands and parameters to forward them to its SPUs.
4. Updates its own parameters and if requested brings up a semi-permanent connection with the main monitoring system to allow the access to the subsystem through a remote interactive shell for maintenance and debug operations.
5. Performs a shut-down operation.

The main board also participates in the synchronization operations syncing the local time with an ntp server and seeding at regular times a global synchronization packet on the CAN bus.

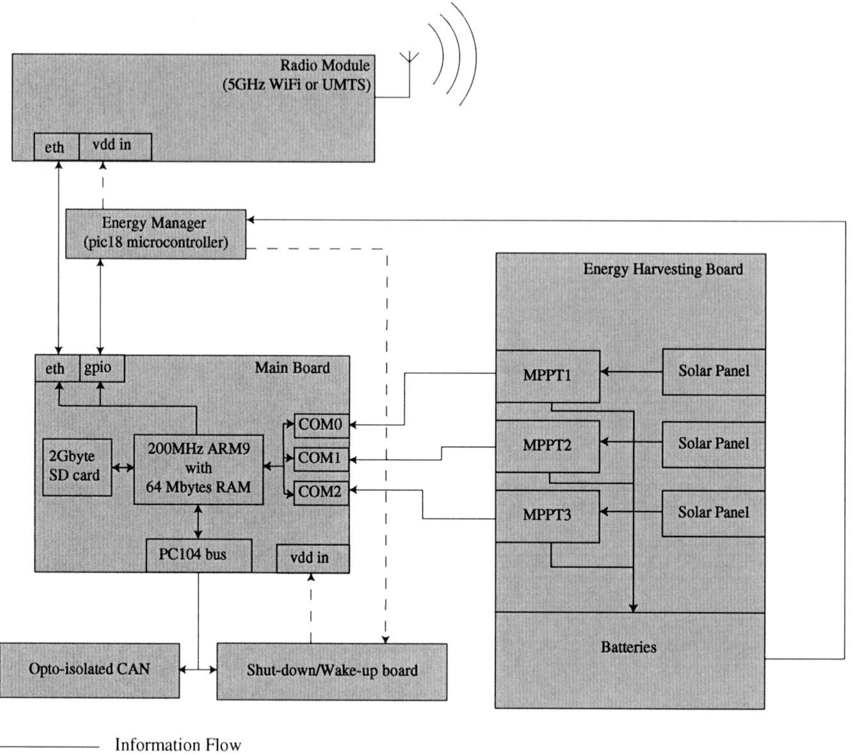

Fig. 7 Gateway/data collector

The energy harvesting board relies on solar panels with an ad hoc circuitry designed to obtain the available energy by using a maximum power point tracker (MPPT) algorithm [8]. The total number of solar panels (and of the related MPPT boards) depends on the energy needs of the subsystem. In the case of the Rialba's Tower, the base station is endowed with two 20 W solar panels and two MPPT boards. The energy harvesting board also regularly sends energy status reports to the main board, that get aggregated to the other statistics collected to keep the status of the monitoring subsystem under control.

The energy manager controls the battery status and consequently, following the state diagram in Fig. 8, enables or disables the different components of the subsystem. Such board is also equipped with an rtc, managing the main board duty cycle.

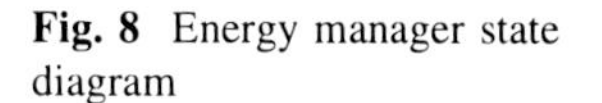

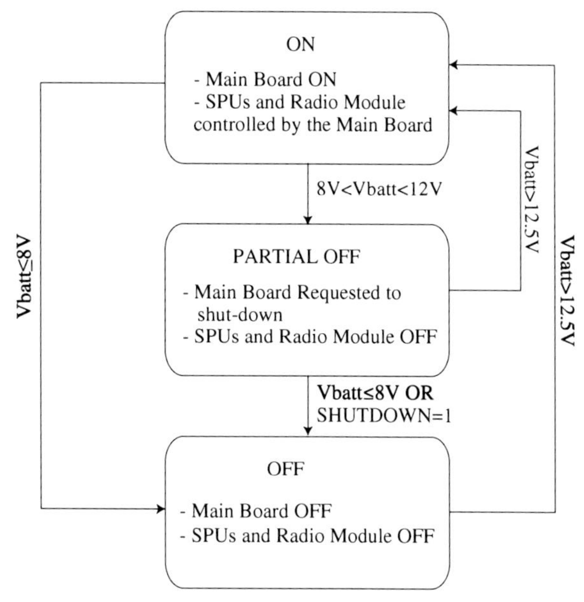

Fig. 8 Energy manager state diagram

3.1.2 Dynamic Data SPU

The dynamic data sensing and processing units (DDSPU) feature a digital signal controller (a micro-controller with dsp functionalities) able to easily filter and evaluate the acquired signals, while maintaining a low power consumption.

From a physical point of view, the DDSPU consists of two custom designed boards and a MEMS accelerometer. On the first board, the processor board shown in Fig. 9a, has a digital signal controller (a microchip dsPIC33F), a CAN bus transceiver and its relative line protection dischargers. The second board, which is the signal conditioning board also shown in Fig. 9b, contains the analog amplifiers and components used to connect the MEMS accelerometer to the ADCs present on the digital signal controller.

Since the device must operate in a mountain environment, it's enclosed in a special case specially designed for rocks, shown in Fig. 10.

The dynamic data SPU must be able to detect asynchronous micro-acoustic events. It must also be able to keep track of low bandwidth signals. Its behavior is thus regulated by the following tasks, scheduled by a rate monotonic scheduling policy:

1. A data acquisition task dedicated to continuously sample and filter the MEMS accelerometer.
2. A parameter update task dedicated to parse the received parameters and, if found correct, update them.
3. A broadcast messages manager task.
4. A CAN bus manager task.

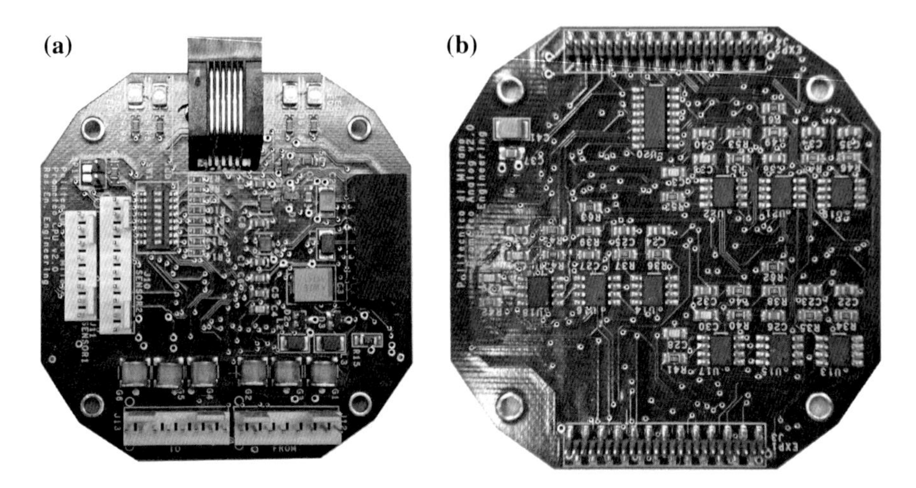

Fig. 9 Dynamic data SPU **a** processor board, **b** signal conditioning board

Fig. 10 Dynamic data SPU—deployed node

5. A message transactions manager task.
6. A protocol manager task.
7. A watchdog task.

Obeying to the parameters managed by the parameter update task, the data acquisition task filters the data coming from the adc and evaluates if a event is present or not thanks to a double moving average filter, whose parameters are also customizable as well as the gains and the paths followed by the analog signal through the signal conditioning board. The rest of the tasks pertain to the management of the communication stack and the internal diagnostics of the SPU.

Thanks to the event detection algorithm and its continuous sampling the device is able to manage the arrival of asynchronous events and save their data for the collecting requested by the base station.

3.1.3 Static Data SPU

The static data SPU (SDSPU) differs from the DDSPU with regards to the absence of dsp features, using a normal micro-controller as processor unit. The set of tasks that are executed remain the same, with the absence of the filtering and event detection operations. In the deployment scenario the SDSPU keeps track of water levels in the water tanks through sampling two piezometers.

3.1.4 CAN Bus

The SPUs rely on the CAN bus to communicate data and obtain new parameters to/from the base station. Though several transport layers for the CANbus are present in the literature, the ad hoc transport layer presented in [7] is used, since traditional approaches have been thought for scenarios where energy is not an issue and duty cycling not needed. The asynchronous events detected by the SPUs are stored in the memory of the SPU together with the data sampled from the low bandwidth signal transducers thanks to a double buffer. Such memory area is then made available to be transferred at request of the main board that acts like a master on the CAN bus, by means of a transactional communication protocol described in the previously cited article. As stated previously, the synchronization among the SPUs is a strict requirement for the evaluation of micro-acoustic bursts. To satisfy such a requirement, two levels of synchronization are adopted: global and local. The global synchronization synchronizes the SPUs with the main board clock (absolute time); the local synchronization instead relies on one SPU acting as master of the synchronization phase to keep all the slave SPUs synchronized even when the main board is switched off to reduce its energy consumption. These two levels exploit the classical synchronization algorithm presented in [9]. It has been experimentally verified that the proposed method allows the achievement of synchronization among the SPUs with a maximum variance of 500 μs.

3.2 Wireless Monitoring Subsystem

Due to the need to maintain a low power consumption and the power needs typical of wireless communications, the wireless monitoring subsystem can be properly employed just to monitor low bandwidth data.

As shown in Fig. 11, the components of the network are the coordinator node, acting as a gateway and the leaf nodes constituted by wireless SPUs. The topology

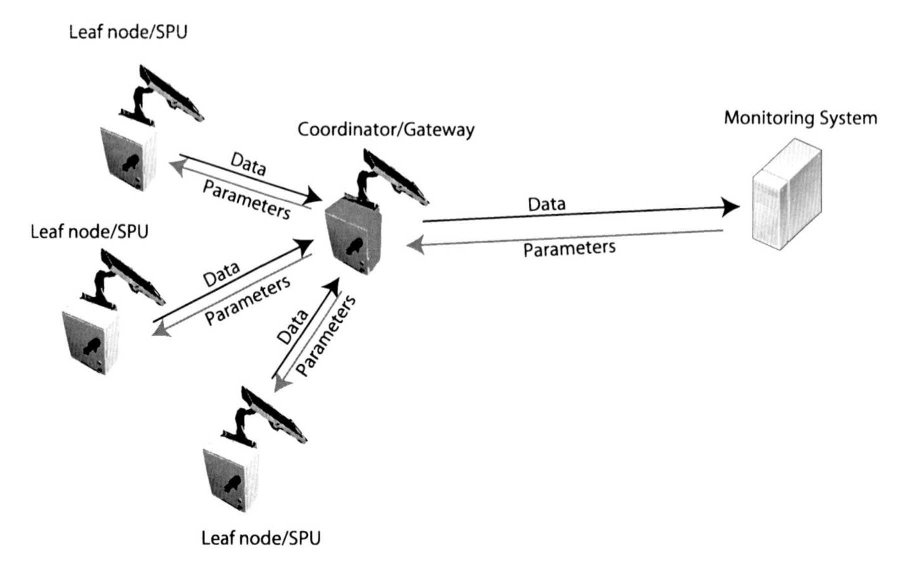

Fig. 11 Wireless monitoring subsystem

of the network is a star topology, although with just the placement of router nodes it is possible to adopt a mesh topology without any change to the software or the hardware of the nodes originally constituting the network. What follows is a brief description of the elements constituting the system. For a more detailed description the interested reader can consult [10].

3.2.1 Leaf Node/Wireless SPU

The leaf node role is that of a sensing element. A block diagram of the wireless SPU (WSPU) is shown in Fig. 12. In this case the functionalities of energy and data manager have been concentrated on the micro-controller board at the center of the diagram.

The Cortex M3 family micro-controller employed to manage the WSPU has the ability to give or negate power to all the other hardware components of the WSPU itself as well as to shut down own individual internal peripherals to shape the power consumption to fit the device needs. Even the memory and large parts of its core are involved in such powering duty cycle operations.

The energy harvesting board features a 12 V battery pack, a 10 W solar power cell and an industrial class regulator circuit designed to guarantee an efficient charge of the battery. The voltage level of the battery is acquired by the internal 12 bits ADC of the micro-controller, so that it can take the necessary decisions about the system operations.

The ZigBee [11] radio module is a Jennic JN5148 module, connected to the micro-controller through an SPI bus and an interrupt line (not shown in figure).

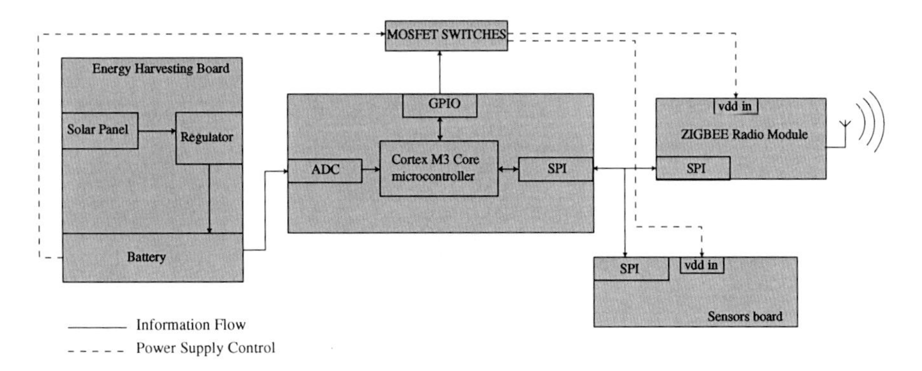

Fig. 12 Wireless SPU

The radio operates in an autonomous way from the micro-controller that manages its status through commands exchanged through the SPI bus. It is worth noticing that in order to not sacrifice network performance to power needs, the radio module status is saved on flash memory before the module gets shut down. At the module restart the first module operation is to load up its last configuration to avoid lengthy network registration operations.

The Sensor board is equipped with a set of 24 bit ADCs, able to read voltage and current measurement, and signal conditioning circuits to be able to acquire data from a wide set of transducers. In the case of the WSPUs employed in Rialba, the attached sensors comprise:

- Inclinometer
- Wired strain gauge
- Thermometer (in hole)
- Thermometer bounded to the inclinometer

The sensor board is also subject to power duty cycling, similarly to the rest of the system.

The WSPU adheres to the following behavior:

1. The micro-controller wakes up and supplies energy to the sensor board.
2. The micro-controller waits for 500 ms to give time to the peripherals to warm up and acquires data from the ADC of the sensor board.
3. The micro-controller shuts-down the sensor board and supplies energy to the radio module.
4. The micro-controller waits for the connection with the ZigBee network to be operative and once operative transmits acquired data (together with their timestamps and diagnostic data) to the collector.
5. The micro-controller enters into the receiving modality for a certain amount of time polling for new parameters or synchronization information from the collector.
6. The micro-controller shuts down the radio module and enters into the deep sleep status mode.

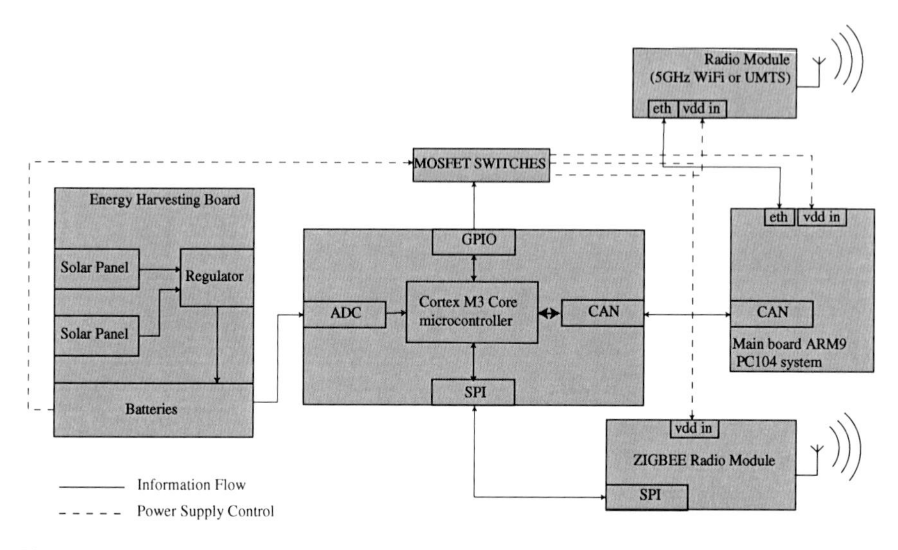

Fig. 13 Coordinator device

3.2.2 Coordinator

The coordinator is fairly more complex than the WSPU, since its role must be of network coordinator, data aggregator and gateway toward the monitoring system. A block diagram of the coordinator is shown in Fig. 13.

The micro-controller board at the basis of the WSPU is the same control board that regulates the coordinator actions. The coordinator acts like a bridge between the ZigBee sensor network and the monitoring system. Such task is performed reacting at the arrival of new data from the ZigBee module by archiving the data on an internal double buffer kept in memory, thus acting in an asynchronous way typical of the publish-subscriber paradigm to which the ZigBee protocol adheres. Such double buffer is then transferred to the main board system through a transactional protocol identical to the one employed by the hybrid subsystem. Similarly to the micro-controller board of the WSPU, it acts also like power manager of the coordinator, with the difference that it keeps the ZigBee radio module always powered and that employs its own rtc to regulate the activity periods of the main board. In addition to such tasks, it has access to a sensor board (not shown in the figure). In the case of the Rialba's Tower the micro-controller board records just the data coming from a pluviometer.

Analogously to the WSPU case, the ZigBee radio module is a Jennic JN5148 module, connected to the micro-controller through an SPI bus and an interrupt line (not shown in the figure). But differently from the other radio module, its software does not act like a bridge between the network and the micro-controller, but act also like the network coordinator, keeping track of every device registered to the wireless sensor network.

The main board module is identical to the one described in 3.1.1 with the difference that it doesn't obtain directly diagnostic data from the energy harvesting board. Moreover, the parameter data-flow coming from the monitoring system, in this case, has a single destination: the micro-controller board. It is then the micro-controller board that parses the incoming array of parameters and finally passes them to the ZigBee module that translates the logical node addresses in network addresses and offers them to the WSPUs.

4 Conclusions

The need to protect critical infrastructure systems provides the motivation to have robust monitoring systems able to promptly answer to rapidly changing conditions. In this chapter a solution is proposed that is able to combine the varying needs requested by monitoring subsystems specifically designed for different purposes, thus obtaining a coherent and holistic view of the monitored system. The proposed monitoring system has been applied to the specific case of the Rialba's Tower, where the risk of a natural disaster threatens several critical infrastructures. As already noted in Sect. 1, the monitoring system has operated without any service interruption for more than two years and it is still operating in complete energetic autonomy.

Its deployment has been performed in three distinct phases. In the first phase the subsystem has been deployed at the basis of the Tower, by installing three DDSPU to monitor the micro-fractures in the rock. Subsequently, after 6 months, the system on the top has been installed to monitor the low dynamic data. After almost a year of operation the geologists and geophysics felt the need to keep track of the water flowing down along the fractures F3 and F4. To satisfy this need, a SDSPU has been added to the CAN bus of the hybrid (lower side) system to acquire the data from the sensors in the water tanks shown in Fig. 5. This operation did not affect the operation of the rest of the system and did not require any major modification of the software or the firmware. Indeed, since both data and parameters are transferred as binary streams, such information remains transparent to the transport mechanism and requires just that the monitoring system, at the insert manager service and parameter configuration application level, is aware of the nature of the transmitted data.

In 2 years the only downtimes have been due to benchmarking/test operations and the need to stop for a few minutes the basis system to hook up the new SDSPU node. Even so, such downtimes regarded just the affected sensing subsystems, never blocking the whole monitoring system in its operation. Such high operativeness of the system is due to the choice of adopting a publish-subscriber paradigm, decoupling the monitoring system in its entirety by the single sensing subsystems. This has permitted to have independent sensor networks whose eventual failures would not affect the entire system operation, while at the same time retaining the capability to furnish a complete view of the CIS status.

Although the monitored system in such a case is a natural system, the proposed approach can be successfully applied to any kind of CIS, as well as that it may constitute a valid example of application of the design principles at the basis of intelligent CIS monitoring.

Acknowledgment This work has been partially supported by the EU INTERREG project Italy-Switzerland action 2007–2013 MIARIA (Project Id 7629775) and the European Regional Development Fund and the Republic of Cyprus through the Research Promotion Foundation, KIOS project.

References

1. Farrar, C., Park, G., Allen, D.W., Todd, M.: Sensing network paradigms for structural health monitoring. J. Struct. Control Health Monit. **13**(1), 210–225 (2006)
2. Worden, K., Dulieu-Barton, J.M.: An overview of intelligent fault detection in systems and structures. Struct. Health Monit. **3**(1), 85–98 (2004)
3. Buttyan, L., Gessner, D., Hessler, A., Langendoerfer, P.: Application of wireless sensor networks in critical infrastructure protection: challenges and design options (security and privacy in emerging wireless networks). IEEE Wirel. Commun. **17**(5), 44–49 (2010)
4. Kostopoulos, D., Leventakis, G., Tsoulkas, V., Nikitakos, N.: An intelligent fault monitoring and risk management tool for complex critical infrastructures: the SERSCIS approach in air-traffic surface control. In: UKSim 14th International Conference on Computer Modelling and Simulation (UKSim), pp. 205–210, 2012
5. Caldeira, F., Schaberreiter, T., Monteiro, E., Aubert, J., Simoes, P., Khadraoui, D.: Trust based interdependency weighting for on-line risk monitoring in interdependent critical infrastructures. In: 6th International Conference on Risk and Security of Internet and Systems (CRiSIS), pp. 1–7, 2011
6. Schreiber, F., Camplani, R., Fortunato, M., Marelli, M., Rota, G.: Perla: a language and middleware architecture for data management and integration in pervasive information systems. IEEE Trans. Softw. Eng. **38**(2), 478–496 (2012)
7. Alippi, C., Camplani, R., Galperti, C., Marullo, A., Roveri, M.: An hybrid wireless-wired monitoring system for real-time rock collapse forecasting. In: IEEE 7th International Conference on Mobile Adhoc and Sensor Systems (MASS), pp. 224–231, 2010
8. Alippi, C., Galperti, C.: An adaptive system for optimal solar energy harvesting in wireless sensor network nodes. IEEE Trans. Circuits Syst. I Regul. Pap. **55**(6), 1742–1750 (2008)
9. Lamport, L.: Time, clocks, and the ordering of events in a distributed system. Commun. ACM **21**(7), 558–565 (1978)
10. Alippi, C., Camplani, R., Roveri, M., Viscardi, G.: NetBrick: a high-performance, low-power hardware platform for wireless and hybrid sensor network. In: The 9th IEEE International Conference on Mobile Ad hoc and Sensor Systems (IEEE MASS 2012), 2012
11. Elahi, A., Gschwender, A.: ZigBee wireless sensor and control network. Prentice Hall, Upper Saddle River (2009)

Algorithms and Tools for Intelligent Control of Critical Infrastructure Systems

Mietek A. Brdys

Abstract Critical Infrastructure Systems (CIS) are spatially distributed and of a network structure. The dynamics are nonlinear, uncertain and with several time scales. There is a variety of different objectives to be reliably met under a wide range of operational conditions. The operational conditions are influenced by the disturbance inputs, operating ranges of the CIS, faults in the sensors and actuators and abnormalities occurring in functioning of the physical processes. The Chapter presents the intelligent multiagent structures and algorithms for controlling such systems. Each agent is an intelligent unit of high autonomy to perform the control functions over an allocated region of the CIS. Its mechanisms are structured in a form of a multilayer hierarchy. The regional agents are then integrated into the multiagent structure capable of meeting the operational objectives of the overall CIS. Several structures are considered starting from the completely decentralised with regard to the interactions between the local regions and ending up at the hierarchical architectures with the coordinating units, which are equipped with the instruments to coordinate activities of the agents across their functional layers. The required ability of the control system to meet the control objectives under a wide range of operating conditions is achieved by supervised reconfiguration of the agents. The recently proposed robustly feasible model predictive control technology with soft switching mechanisms between different control strategies is applied to implement the soft and robustly feasible agent reconfiguration. The generic ideas and solutions are illustrated by applications to two CIS: an integrated wastewater treatment plant and a drinking water distribution network.

M.A. Brdys (✉)
Department of Electronic, Electrical and Computer Engineering, College of Engineering and Physical Sciences, University of Birmingham, Edgbaston, Birmingham B15 2TT, UK
e-mail: mbrdys@ely.pg.gda.pl; m.brdys@bham.ac.uk

M.A. Brdys
Department of Control Systems Engineering, Gdańsk University of Technology, Ul. G. Narutowicza 11/12, 80-233 Gdańsk, Poland

E. Kyriakides and M. Polycarpou (eds.), *Intelligent Monitoring, Control, and Security of Critical Infrastructure Systems*, Studies in Computational Intelligence 565,
DOI 10.1007/978-3-662-44160-2_7

Keywords Critical infrastructure systems · Robust feasibility · Model predictive control · Intelligent agent · Operational states · Soft switching · Fault tolerant control · Decentralised control · Hierarchical control · Multilayer structures · Multiagent control structures · Wastewater treatment plants · Drinking water distribution networks

1 Introduction

Control of critical infrastructure systems (CIS) must secure their reliable and sustainable functioning in achieving the required operational objectives under a *wide range* of operational conditions, which include accidental, seriously unfavourable events such as sensor and/or actuator faults, failures of communication links, or anomalies occurring in the technological operation of the CIS physical processes. A *monitoring system* is needed in order to deliver information. The CIS are subject to deliberate attacks. The inputs resulting from these attacks can seriously deteriorate the required functioning of the CIS. The early detection of such disturbance inputs, compensation/rejection of their impact and restoration of the CIS back to its normal operation requires actions from a dedicated *security system*. The security system can be viewed as a dedicated control system to handle the special type of the disturbance inputs. This is depicted in Fig. 1.

This Chapter considers only the control component. CIS are spatially distributed and of a network structure. The dynamics are nonlinear, uncertain and with several time scales. There is a large number of variables involved in the dynamic models, which are heterogeneous. Not only the inputs but also states/outputs are constrained. The latter requires robust feasibility of the control actions generated by control units. Hence, CIS are large scale complex systems with a variety of different objectives to be met under a wide range of operational conditions (see Fig. 2).

CIS are typically distributed over a large geographical area and a centralised system for performing the monitoring, control and security functions would not be viable. Therefore, the area is decomposed into regions, as shown in Fig. 3. Performing the monitoring, control and security functions over the regions is carried out by the regional agents. These are the intelligent units of high autonomy interacting directly or/and through the CIS, as illustrated in Fig. 4.

This Chapter is organised as follows. In Sect. 2 the CIS operational states are introduced and they are mapped into the control strategies, suitable to pursue the control objectives, which can be achieved at these states. The model predictive control (MPC) technology is applied to produce the control strategies. The robustly feasible strategies (RFMPC) are outlined in Sect. 3. A reconfiguration of an agent in order to adapt the control strategy to a current operational state of the CIS is discussed in Sect. 4. A soft switching mechanism between different RFMPC strategies is proposed to produce the softly switched robustly feasible MPC

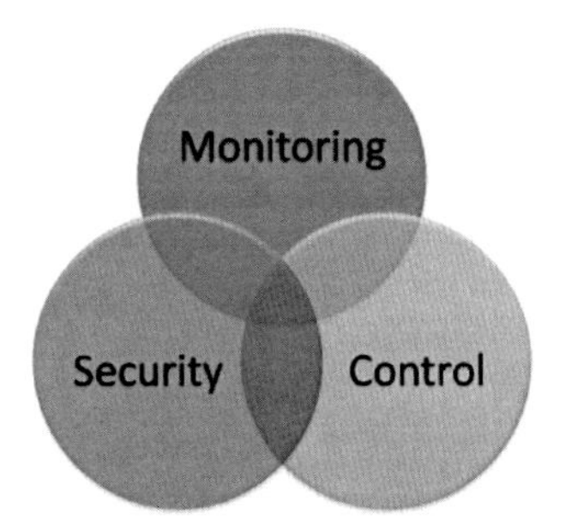

Fig. 1 Key interacting components of the system maintaining operation of CIS

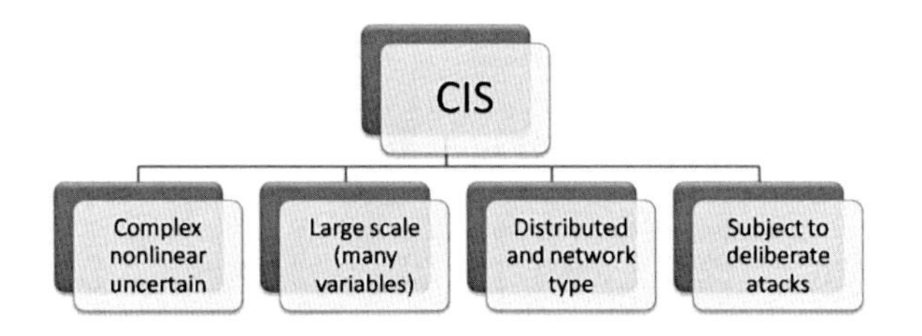

Fig. 2 CIS are large scale complex systems

Fig. 3 Spatial decomposition of CIS into regions

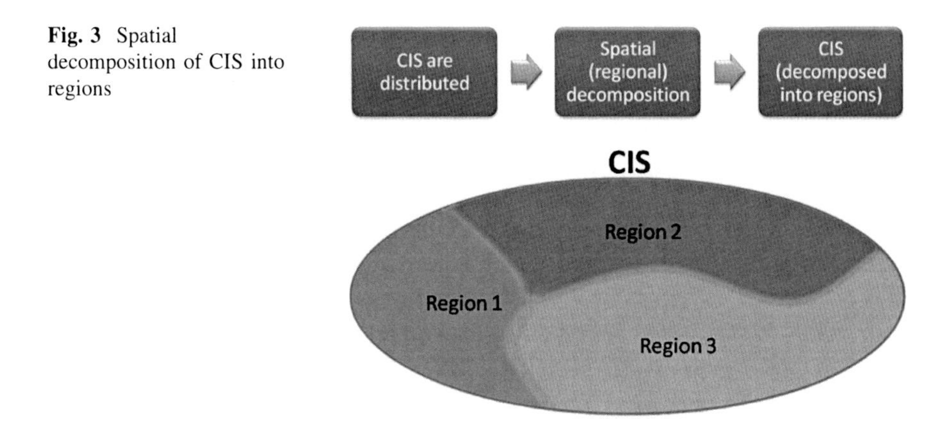

(SFRFMPC) that implements soft reconfiguration of the agent. An approach to derive a multiagent control system over an overall CIS based on the derived architecture and algorithms for a regional agent is briefly discussed in Sect. 5. An application to the integrated wastewater treatment system case study CIS is presented in Sect. 6. An application to quality control in the case study example of drinking water distribution network is described in Sects. 7 and 8. The conclusions in Sect. 9 complete the Chapter.

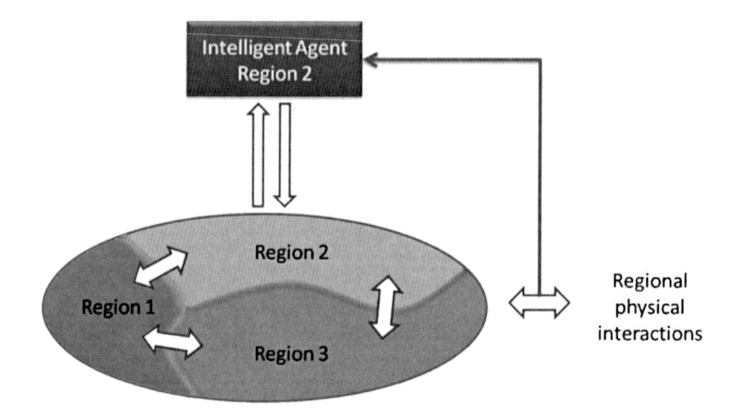

Fig. 4 Regional agents

2 Operational States and Control Strategies

A current *operational state* (OS) of a CIS is determined by the states of all the factors which influence its ability to achieve prescribed control objectives. These include: states of the CIS processes; states of the sensors, actuators and communication channels (e.g., faults), states of process anomalies, technical faults, current operating ranges of the processes, states of the disturbance inputs. The OS without security content (not security OS) are robustly estimated /predicted by using conventional state estimation algorithms (data driven, model based) and fault detection and diagnosis algorithms (FDD) [1]. The typical operational states are [2]: *normal*, *disturbed* and *emergency*. Not all control objectives can be satisfactorily achieved at a specific OS. This is identified by performing a suitable a priori analysis. Given the control objectives a control strategy suitable to achieve these objectives is designed or chosen from the set of strategies designed a priori. In this way, a mapping between the operational states and suitable control strategies to be applied at these OS is produced (see Fig. 5). It should be pointed out that there can be more than one normal, disturbed and emergency operational state and they constitute the separated clusters in the OS space equipped with the links indicating transfer between the specific operational states. In a triple of ordered and linked of the normal, perturbed and emergency operational states, a deterioration of CIS operational conditions forces the CIS to move into the perturbed operational state. The control system is expected to adapt its current control strategy to the new operational state as otherwise the CIS with no adequate control strategy can be further forced to move into the emergency operational state. Being safely in the perturbed operational state the agent senses and predicts changes in the current OS and if for example it moves back to the normal OS the intelligent agent starts adapting the control strategy back to the normal one.

The concept of operational states and the corresponding control strategies capable of meeting the control objectives associated with these states is vital for

Fig. 5 Mapping operational states into control strategies

synthesising an agent operating the CIS over a wide range of the operational conditions including the faults, threats and widely varying the disturbance inputs. Regarding the control strategies we shall consider the model predictive control (MPC) technology in order to produce the control strategies adequate to pursue the control objectives at the operational state. Indeed, the standard MPC strategy (Kerrigan and Maciejowski 2003) is uniquely determined by specifying the performance function to be optimised and the constraint functions defining a set over which the optimisation is to be performed (see Fig. 6).

In order to meet the CIS state/output constraints the model based MPC needs to be made robustly feasible. In other words, the control actions produced by the MPC controller, which satisfy state/output constraints in the model, must also satisfy these constraints when applied to the physical CIS in spite of model—reality mismatch. The Robustly Feasible MPC is presented in the following section.

3 Robustly Feasible Model Predictive Controller

There are several approaches to designing the RFMPC. A robust control invariant set can be determined for the MPC control law based on its nominal model and the uncertainty bounds so that if the initial state belongs to this set the recursive robust feasibility is guaranteed [3, 4]. Constructive algorithms were produced to determine such sets for linear dynamic systems under the additive and polytypic set bounded uncertainty models. Safe feasibility tubes in the state space were designed and utilised to synthesise RFMPC in Langson et al. [5] and Mayne et al. [6]. The reference governor approach was proposed and investigated in Bemporad and Mosca [7], Angeli et al. [8], Bemporad et al. [9] for a tracking problem, where a reference trajectory over a prediction horizon is designed with extra constraints being imposed during the reference trajectory generation. The calculated control inputs under the on—line updated reference trajectory can manoeuvre the system to the desired states without violating the state constraints under all possible uncertainty scenarios. The additional constraints on the reference are calculated based on the uncertainty bounds. In Brdys and Ulanicki [10] the hard limits on tank capacities in a drinking water distribution system were additionally reduced in the MPC optimisation task by introducing so called safety zones. This is illustrated in Fig. 7, where σ^u, σ^l denote the upper and lower safety zones modifying the original upper and lower limits y^{max} and y^{min} constraining the output to produce the modified output constraints to be used in the model based optimisation task of RFMPC (Fig. 8).

Fig. 6 Mapping operational states into MPC strategies

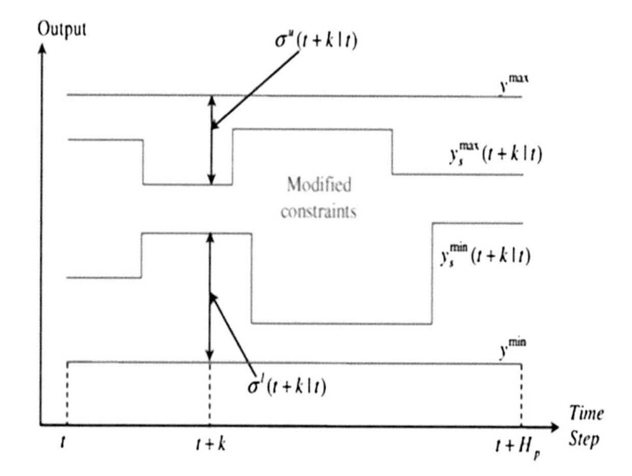

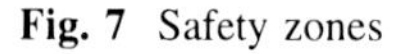
Fig. 7 Safety zones

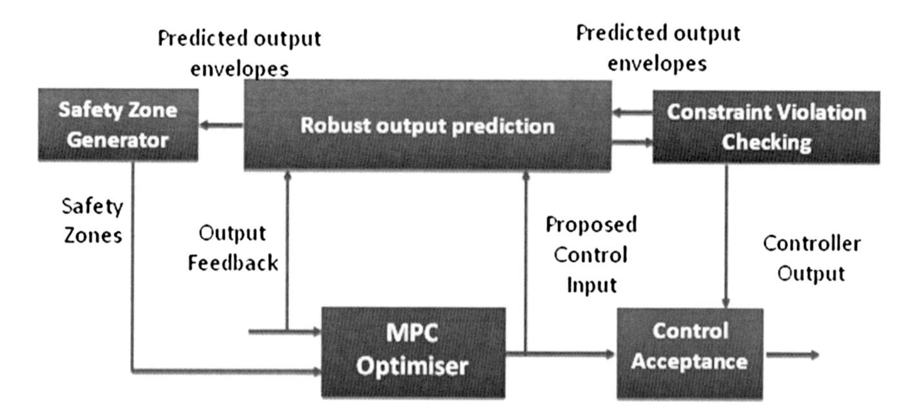

Fig. 8 Structure of RFMPC with safety zones

The safety zones were determined to be large enough in order to compensate uncertainty in the water demands so that the model based optimised control actions satisfied the original tanks constraints when applied to the physical water distribution system. Replacing the original state/output constraints in the MPC model based optimisation task with a set of more stringent constraints which preserve feasibility for any scenario of uncertainty in the system model dynamics is a key idea of this constraint restriction approach. A disturbance invariant set was designed a priori by Chisci et al. [11] to produce suitable restrictions of the constraints for linear systems. The conservatism of methods based on the invariant sets and difficulties in calculating these sets for nonlinear system dynamics impose serious limitations on the applicability of these methods. A generic approach to synthesise RFMPC that utilises the safety zones, which are iteratively updated on line based on the MPC information feedback was proposed by Brdys and Chang [12] and applied to drinking water quality and hydraulics control in Duzinkiewicz et al. [13], Tran and Brdys [14, 15] and to integrated wastewater systems in Brdys et al. [2]. The robustly feasible model predictive controller is practically applicable to nonlinear systems and the conservatism due to the uncertainty is reduced as the safety zones are updated on line over the prediction horizon H_p. We shall now briefly present this safety zone based controller.

3.1 Robustly Feasible MPC with Safety Zones

The controller structure is illustrated in Fig. 8. The control inputs are produced by solving the model based optimisation task, where the unknown disturbance inputs over the prediction horizon are represented by their updated predictions and other stationary uncertainty factors are replaced by their estimated values.

Moreover, the original state/output constraints are modified by the safety zones provided by the Safety Zones Generator. The initial state in the output prediction model is taken directly from the plant measurements or it is estimated using these measurements. The control inputs are then checked for robust feasibility. First, a robust prediction of the corresponding plant output is produced in terms of two envelopes y_p^u and y_p^l bounding a region in the output space where the plant output trajectory would lie if the inputs were applied to the plant. The robust feasibility is now verified by comparing the robustly bounding envelopes against the original output constraints (see Fig. 9).

In the upper figure the robust upper and lower envelopes determine the region in the output space that is located in a set meeting the output constraints. Hence, it can be guaranteed that the real system output is feasible for any disturbance input scenario and consequently the input is robustly feasible. This is not the case in the lower figure where it can not be guaranteed that the real system output will not violate the upper constraint. Hence, the input is not robustly feasible.

If the inputs are robustly feasible over H_p then they are applied to the plant over the following control step. Otherwise, the constraint violation is quantitatively assessed and based on this result the Safety Generator Unit updates the safety zones. The MPC optimiser is activated to produce new sequence of the control actions. Determining the robustly feasible safety zones is done iteratively and typically a simple relaxation algorithm is applied to achieve it. In order to achieve sustainable (recursive) feasibility of the RFMPC the safety zones are iteratively determined on line over the whole prediction horizon and this is still computationally demanding.

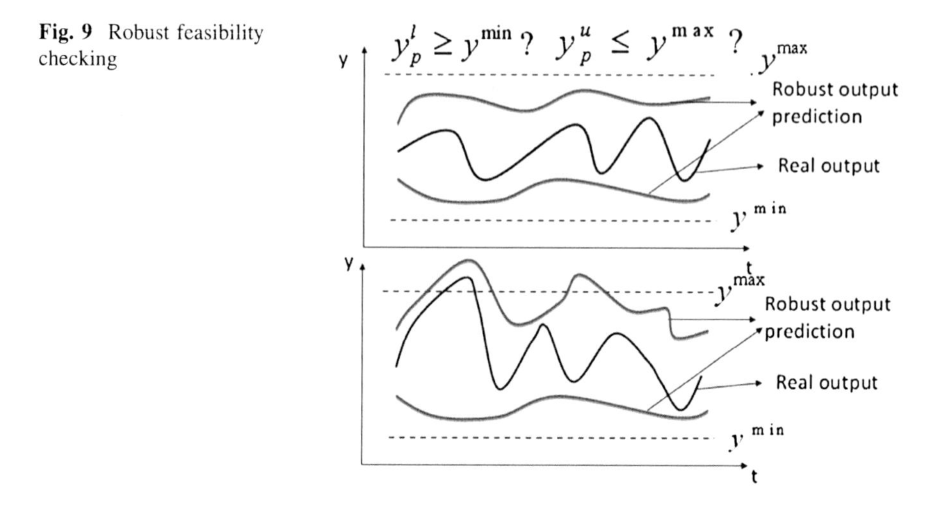

Fig. 9 Robust feasibility checking

In a recent work by Brdys et al. [16] the safety zones were used to parsimoniously parameterise recursively feasible invariant sets in the state space and a computational algorithm was derived to calculate off-line the zones and the invariant sets. An operational computation burden of the resulting RFMPC was then significantly reduced. Clearly, the safety zones and invariant sets are recalculated when a priori uncertainty bounds change.

3.2 Multilayer RFMPC with Different Time Scales

As it has been stated in Sect. 1 several time scales exist in the CIS internal dynamics. This feature of CIS can be efficiently handled by employing a temporal decomposition of the CIS dynamics into interacting and vertically ordered process dynamics operating at different time scales starting from the slowest time scale at the top and ending up with the fastest one at the bottom [17], as illustrated in Fig. 10. The control objectives can now be adequately associated with the time scales and the prediction horizons for the RFMPC controllers forcing the CIS to achieve these objectives can be determined. The horizons would be 1 month, 1 day and 1 h for the predictive controllers operating at the slow, medium and fast time scales in order to achieve the long term, medium term and short term operational objectives of CIS. Clearly, suitable lengths of the control steps are to be determined. As a result, a hierarchical multilayer architecture of the vertically nested RFMPC's, each of them operating at different time scales and hence, efficiently achieving different control objectives has now been presented. The control actions generated at an upper control layer are typically the set points to be tracked by the lower adjacent layer and/or the state/output targets to be achieved at the end of the prediction horizon at the lower adjacent layer.

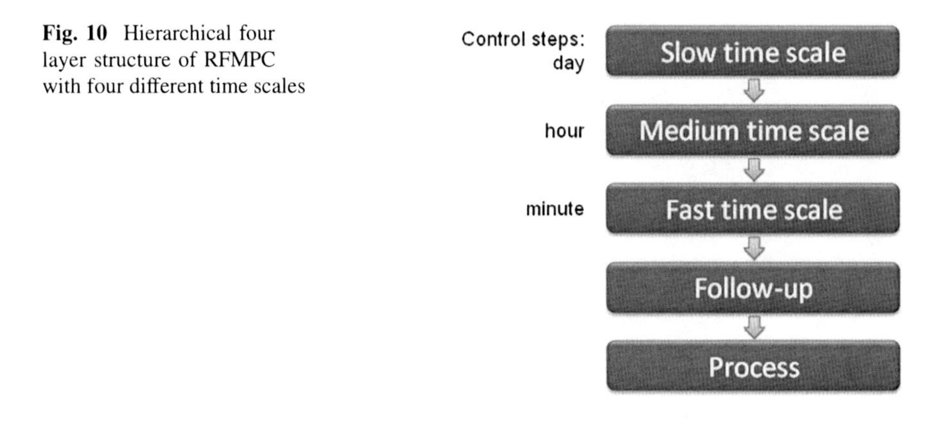

Fig. 10 Hierarchical four layer structure of RFMPC with four different time scales

The control inputs to the CIS processes are produced at the lowest follow up control layer operating in the fastest time scale. This will be applied to the wastewater treatment in Sect. 6 [2]. In the recent paper [18] a two layer predictive controller is analysed in detail and strong theoretical results are produced including the robust stability. Also, hard switching between different configurations of control handles is included into this analysis which gives rise to achieving a fault tolerant feature of the control system.

4 Agent Reconfiguration by Softly Switched RFMPC

During controlling the CIS the operational state may change and this requires adopting the control strategy to new operational conditions. This means that the current RFMPC strategy should be replaced by the new one, which is desired at the new OS.

4.1 Softly Switched RFMPC

A hard switching from the current control strategy to the new one may not be possible due to two reasons. First, the immediate replacement in the control computer of the current performance and constraint functions by those defining the new control strategy may lead to the infeasible optimisation task of the new strategy with the current initial state [19]. Secondly, very unfavourable transient processes may occur and last for certain time periods [20]. Alternatively, the switching can be distributed over time by gradually reducing the impact of the old strategy on the control inputs generated and strengthening the new strategy impact. The switching starting at $t = \bar{t}$ would complete at $t_s = \bar{t} + T_s$, where T_s denotes the duration time of the switching process. As opposed to the hard switching this is a

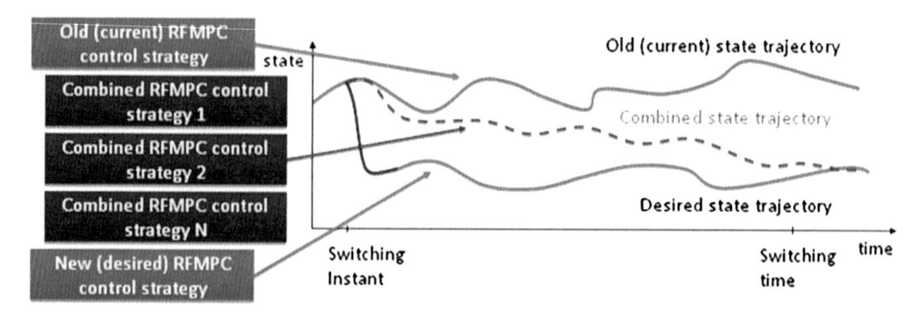

Fig. 11 Soft switching

soft switching. The soft switching was proposed and analysed by Wang et al. [19], Brdys and Wang [21]. They proposed to technically implement the soft switching by designing combined predictive control strategies in a form of a convex parameterisation of the performance and constraint functions of both strategies. Selecting on line the parameters produces a sequence of the combined strategies and the new strategy is reached in the finite time t_s. The soft switching process and its implementation is illustrated in Fig. 11.

Given the desired control strategy, the parameters of the soft switching process are: the switching time instant $\bar{t}$, the switching process duration time T_s and the switching mechanism, which is the algorithm selecting on line the combined strategies. In Wang and Brdys [22] the algorithm, which terminates the soft switching in a minimal time was proposed for linear systems and it has been recently extended to nonlinear network systems by Tran and Brdys [23]. The soft switching for hybrid RFMPC strategies was investigated in Wang and Brdys [24] producing the stability results. The soft switching was applied to optimising control of waste water treatment systems in Brdys et al. [2]. A recent research on truly Pareto multiobjective MMPC [25] has produced results indicating an enormous potential of the MMPC to develop new high dynamic performance soft switching mechanisms. There are still problems with performing on line the computing needed to produce accurate enough representation of the Pareto front. Hybrid evolutionary solvers implemented on computer grids with embedded computational intelligence mechanisms that are designed based on fuzzy-neural networks with the internal states are investigated in order to derive more efficient solvers of the multi-objective model predictive controller optimisation task.

4.2 Supervisory Control Layer

Such faults as sensor/actuator failures or packet drops in communication networks supporting operation of CIS can be handled without changing the current control strategy by activating the Fault Tolerant Control (FTC) mechanisms [26]. The new OS does not have to be defined in such cases and the switching can be avoided. It is enough to activate the FTC mechanisms. This, and serving the soft switching, needs a dedicated layer supervising an operation of the softly switched robustly feasible MPC (SSRFMPC). The supervisory control layer is responsible for the following:

- Estimation and prediction of OS
- Diagnosis if a current control strategy supported by the FTC mechanisms is powerful enough and its switching is not needed
- Selection of control strategies required by changes of OS
- Selection of soft switching mechanisms
- Starting, supervising and ending the switching process.

The layer is supported by the monitoring system and a risk based or robust Fault Detection and Diagnosis units.

4.3 Reconfigurable Multilayer Structure of Autonomous Regional Agent

The softly switched robustly feasible model predictive control layer and the supervisory control layer are the functional layers in an overall multilayer structure of the reconfigurable autonomous regional agent. It is understood that the RFMPC control strategies have multilayer structures in order to efficiently incorporate the multiple time scales in CIS dynamics as described in Sect. 3.2. The security component has not been discussed in detail as it falls into the presented methodology as a dedicated monitoring and control activity. However, it has specific features, which require dedicated solutions. Hence, it is left distinguished as strongly interacting with the other two components of the overall system. Clearly, a regional agent is truly autonomous and highly advanced. It is envisaged that there is plenty of room for dedicated agents being satellites of the main agent, which is considered in this paper. They can be particularly useful in performing distributed communication functions and wirelessly networked sensors. The regional agent overall structure is illustrated in Fig. 12.

5 Multiagent System for Control Over an Overall Distributed CIS

Integrating the regional agents into one multiagent system would produce a system capable of performing the control functions over an overall distributed CIS. Clearly, a key issue during this integration will be how to account for the interactions between the regions. Following e.g., Findeisen et al. [17] and Brdys and Tatjewski [27] a spatial decomposition would be applied with regards to the regional agent layers (see Fig. 13).

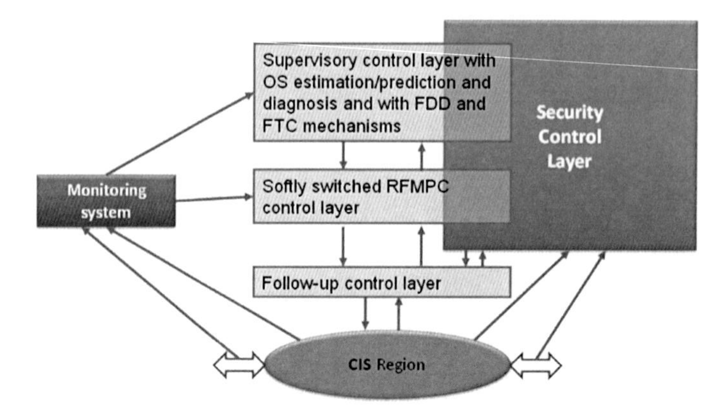

Fig. 12 Reconfigurable multilayer structure of autonomous regional agent

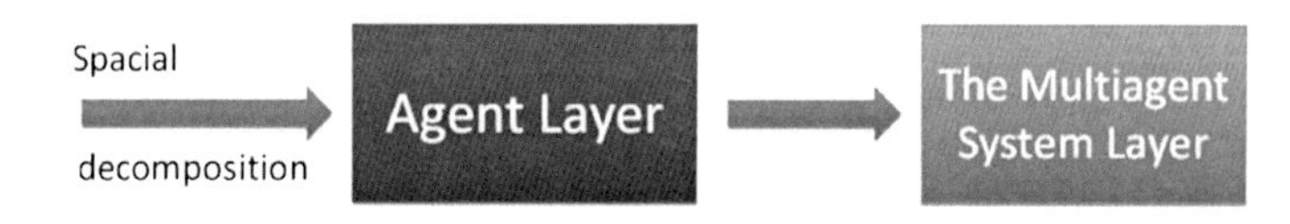

Fig. 13 Spacial decomposition of a regional agent layer

As the RFMPC is the optimisation based technology, then the well known decomposition methods of the optimisation problems can be applied to produce a hierarchical structure of the RFMPC with the regional units and a coordinator integrating the regional actions. Developing distributed MPC has attracted an immense attention of the control community during the last decade. An excellent survey can be found in Scattolini [28] and in Rawling and Mayne [29]. However, no breakthrough results have been obtained yet. The architectures are still not suitable for on-line control as a feasibility of the control actions generated is not guaranteed for MPC with uncertain models. Moreover, the horizontal information exchange between the regional agents required is immense and certainly not acceptable by real life communication networks. Finally, in order to achieve high operational performance in a cost effective manner under strong interactions between regional CIS, the distributed agents must be coordinated. The price coordination mechanism [17] is very appealing. However, it needs to be further developed so that an on-line feasibility of the actions generated by the regional agents can be guaranteed for heavily state/output constrained but not only for the control input constrained CIS.

Hierarchically structuring the soft switching mechanism is conceivable although this has not been achieved yet. Hierarchically structuring the supervisory control layer is an open problem. Although a number of applications have been reported e.g., Tichy et al. [30] the methods used are very much case study dependant and nothing generic exists. There have been significant advances made for the information systems e.g., Amigoni and Gatti [31], Shoham and Leyton-Brown [32], Keil and Goldin [33] and they would be attempted for application to the engineering systems. A good step in this direction was reported in Fregene et al. [34]. The communication protocols implementing the information exchanges between the regional agents directly or through the coordinator require security features to be embedded in these protocols and beyond with a whole information system to be applicable to the CIS. Again, although much work has been done with application to the information systems the results are not directly applicable to the engineering CIS. The decentralised and decentralised follow up control methods are directly applicable to structure the regional agent lowest layer for the multiagent purposes.

6 Application to Integrated Wastewater Treatment Case – Study System

6.1 Presentation of the Case-Study System

An integrated wastewater treatment plant at Kartuzy in Northern Poland is composed of the hydraulic storage part coupled to the sewer system and supplying the biological treatment plant with the wastewater to be cleaned before it is released to the receiving river. The system layout is illustrated in Fig. 14.

The equalization tank retention is embedded into the sewer network tank retention to produce the retention tank in Fig. 14. The treatment method is based on an activated sludge technology. The advanced biological treatment with nutrient removal is accomplished in the activated sludge reactor designed and operated according to the UCT (University of Cape Town) process.

The first zone is anaerobic where the release of phosphorus should occur. The internal recirculation of mixed liquor originates from the anoxic zone. The second zone where denitrification takes place is anoxic.

The activated sludge returned from the bottom of the clarifiers as well as the internal recirculation 1 from last aerobic zone (containing nitrates) is fed back to the anoxic zone. The last part of the biological reactor (aerobic) is aerated by a diffused aeration system. This zone is divided into three compartments of various intensities of aeration. The biologically treated wastewater and biomass (activated sludge) are separated into two parallel horizontal secondary clarifiers. The sewage is recirculated from the clarifiers to the anoxic zone. In order to ensure a high level of the phosphorus removal, iron sulphate (PIX) is added to the aerobic zone to precipitate most of the remaining soluble phosphorus. This is supported by the phosphorus precipitation in the grit chamber. The treated sewage goes to Klasztorna Struga River, which is the effluent receiver.

The plant operational objectives can be technically broken down into the following objectives:

- to stabilize the biological treatment process by properly selecting values of sludge retention time (SRT) and sludge mass (Ms),

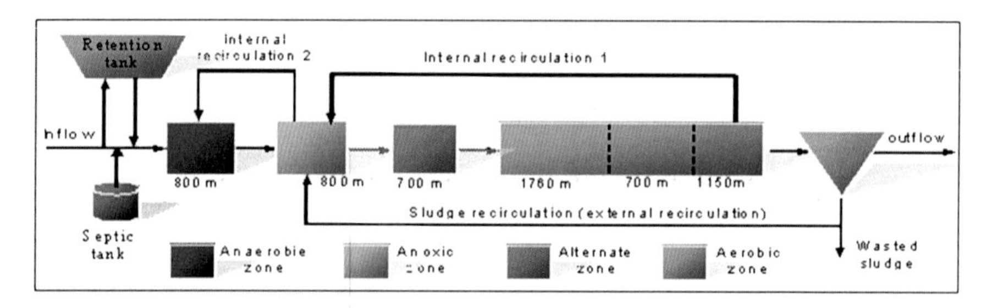

Fig. 14 Layout of Kartuzy wastewater treatment plant

- to ensure efficient hydraulic control by proper utilization of equalization and septic tanks,
- to keep to output constraints on: ammonium nitrogen (S_{NH4}), nitrate (S_{NOx}), total nitrogen (TN), total phosphorous (TP), chemical oxygen demand (COD), biological oxygen demand (BOD), soluble substrate (S_S) and total suspended solids (X_{TSS}),
- to minimize pollution load discharged to the receiver (S_{NH4}, S_{NOx}, TN, TP, COD, BOD, S_S, X_{TSS}),
- to minimize plant running costs by minimizing: air flow rate that is delivered to the aerobic zones, recirculation flow rates (internal and external), excessive sludge flow rate, equalization tank filling/emptying, PIX dosing,
- to minimize emergency overflowing.

The possible control handles (manipulated variables) are:

- air supply to aerobic zones,
- internal and recirculation flow rates,
- excessive sludge flow rate,
- the iron sulphate (PIX) dosage (or other chemicals),
- equalization tank filling/emptying, septic tank emptying.

6.2 Control Structure

Achieving the plant operational objectives requires an optimising control. A synthesis of the control structure for the optimising control of the integrated wastewater system is very complicated because of its specific following features: (i) multiple time scales in the internal dynamics of the biological processes and tanks, (ii) varying influent flow rates and its pollutant concentrations, (iii) highly non-linear and mutually interacting dynamics of large dimensions [35], (iv) the requirement of achieving biological sustainability of the system over a long time horizon while sufficiently accurate disturbance predictions over such a horizon are not available, (v) the occurrence of short term heavy rainfall disturbance events

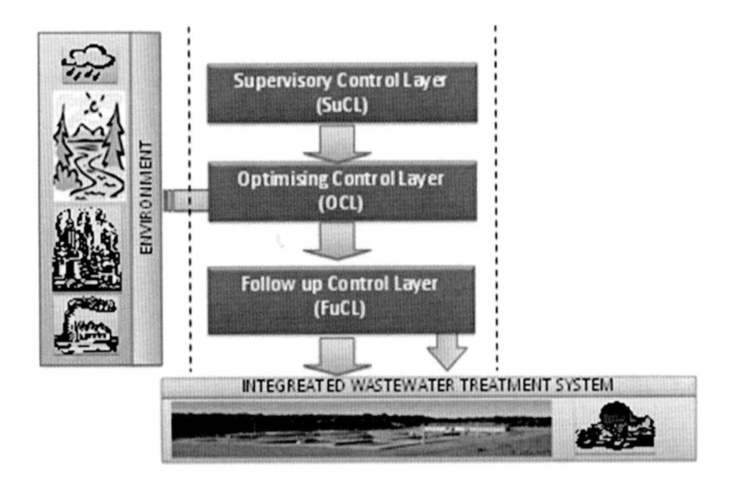

Fig. 15 Optimising control structure

hardly predictable early enough having long term consequences for the system behaviour, (vi) the lack of models that are suitable for control design and its on line implementation, (vii) the small number out of many state variables that are practically measurable. In order to efficiently handle these difficulties, the multi-layer hierarchical control structure presented in Fig. 12 is applied. The structure is illustrated in Fig. 15.

A hierarchical structuring of the control actions generation process allows for proper and thorough full utilization of all the available quantitative and qualitative information about the plant structure and dynamics, its interactions with the environment and up to date operational experience.

The Optimising Control Layer is responsible for generating the optimised control trajectories. The control objectives at the OCL can be split into the long term, medium term and short term. The different time horizons of the objectives are as a matter of fact mainly implied by the multiple time scales of the internal dynamics of the biological treatment process and the variability of the disturbance inputs. There are various ways of controlling at each time scale. It depends on the assumed/chosen control strategy, and hence it depends on the associated objective function and constraints. A core control method at OCL is the softly switched RFMPC. In order to achieve the desired objectives the OCL generates control trajectories for each of the control horizons.

It is very difficult to efficiently handle the multiple time scale dynamics in the optimising control problem by a centralized optimising controller as the required long prediction horizon and short control time steps might lead to an optimisation problem of very high dimension and under a large uncertainty radius. It obviously might cause computational problems during real time control. In order to alleviate these two fundamental difficulties a temporal decomposition of the optimising controller is applied (see Fig. 10). Three time scales in the internal dynamics of the process are distinguished: slow, medium and fast [36]. As a result, the optimising

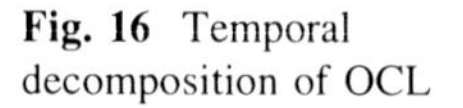

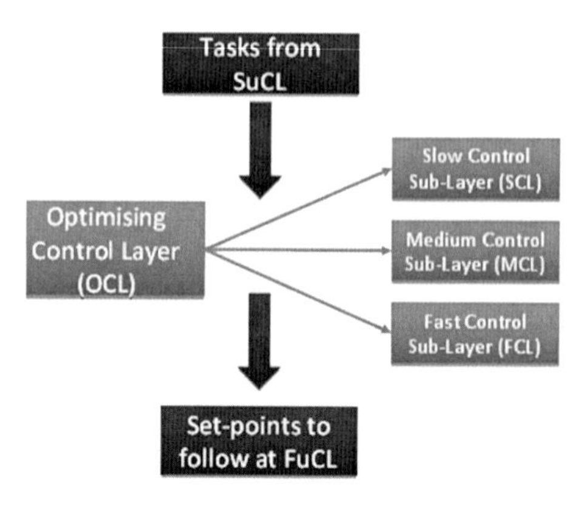

Fig. 16 Temporal decomposition of OCL

controller is decomposed and organized in the form of a three-sublayer hierarchy as shown in Fig. 16.

Each sublayer controller operates at a different time scale and handles the corresponding objectives.

The *Slow Control Sublayer (SCL)* operates at a slow time scale with a 1-day control step and handles long-term objectives over a horizon of a week up to several months. This control layer is responsible for biological sustainability of the biomass, volume control of equalization and storm water tanks and long term economical objectives under an as wide as possible range of disturbance inputs. The manipulated variables at SCL are: Sludge Retention Time (SRT), Mass Sludge (Ms) and pumping in/from equalization and septic tanks. The SRT and Ms are produced by an intelligent rule based system supported by the experienced plant engineers.

The *Medium Control Sublayer (MCL)* operates over a medium time scale with a 1-h control step and handles medium term objectives over a horizon of a day. The Medium Control Sublayer is responsible for maintaining the effluent quality within required limits and for optimising the operating cost subject to constraints prescribed by the SCL and subject to technological and actuator constraints. The manipulated variables at MCL are: dissolved oxygen concentrations at the aerobic zones, recirculation flow rates, pumping in/from equalization and septic tanks, and chemical precipitation (PIX). All of them, except for dissolved oxygen concentrations are directly transmitted to the FuCL. The SSRFMPC is applied at MCL.

The *Fast Control Sublayer (FCL)* operates in the fastest time scale with a 1-min control step and handles short term objectives over a horizon of 1 h. The Fast Control Sublayer is responsible for: effluent quality during heavy rains and of short duration time events, actuator constraints and meeting demands on desired dissolved oxygen concentrations at the aerobic zones that are prescribed by the MCL. The main functionality of the FCL is generating the set points for the FuCL so that the process is forced to follow the manipulated variable trajectories that are prescribed by MCL. The process actuators are mostly simple PLC controlled devices, except for the airflow rates that are provided by the aeration system in order to achieve required set-points for the dissolved oxygen concentrations. These airflow rates are the manipulated variables at the FCL. Due to hybrid dynamics of the aeration system, meeting the prescribed set-points with least energy cost consumed by the blowers is a complicated task [37–40]. This is done by adjusting the valve positions and the speeds of the blowers and by scheduling their *on/off* status.

6.3 Control Strategies

The first control strategy, CS1, is designed to be engaged when the system is in the *normal operational state*. The main objective of this strategy is to minimize an overall operational cost while fulfilling the discharge requirements at the same time. One of the desired approaches of controlling the plant when being in the normal operational state is to do it entirely based on biological phosphorus removal. This implies excluding the PIX dosing from the decision vector content. However, it is a common practice at the Wastewater Treatment Plants to support the phosphate precipitation by limited PIX dosing. In the CS1 control strategy the PIX upper limit of dosing was set relatively low, namely $PIX_{max} \leq 70$ (kg/d). As the system is in the normal operational state the emergency overflowing and related costs are not taken into account in the performance index.

The second control strategy, CS2, is designed for the plant being in the *disturbed operational state*. The main target of this strategy is to minimize an inevitable, due to rain fall for example, discharged pollution load during situations with increased influent pollution load. Then, in order to meet the concentration limit in the effluent for phosphorus, PIX dosing is unavoidable. Hence, the PIX dosing upper limit is raised up above its preferable level from $PIX_{max} \leq 70$ (kg/d) to $PIX_{max} \leq 140$ (kg/d). As meeting the effluent quality requirements has priority, the operational cost in the CS2 performance index is of much smaller importance than in the CS1.

The third control strategy, CS3, is designed for the situations when the system is in the *emergency operational state* with regards to hydraulic control. Hence, the strategy is engaged when influent of very high flow rate occurs at the plant input and a risk of the emergency overflowing becomes high. A main objective of CS3 is to avoid, if possible, the overflow or at least to minimize it. The operating cost is of much smaller importance. The PIX dosing limit is increased to 200 (kg/d) in order to stronger support the biological phosphorus removal under heavy pollution load into the plant.

6.4 Simulation and Field Results

The plant input testing scenario is a mixed dry-wet influent scenario, which is shown in Figs. 17, 18, 19 and 20. The total nitrogen concentration, total phosphorous concentration and chemical oxygen demand are the parameters describing influent sewage composition. The effluent discharge requirements are $TN \leq 15$ (g/m^3),

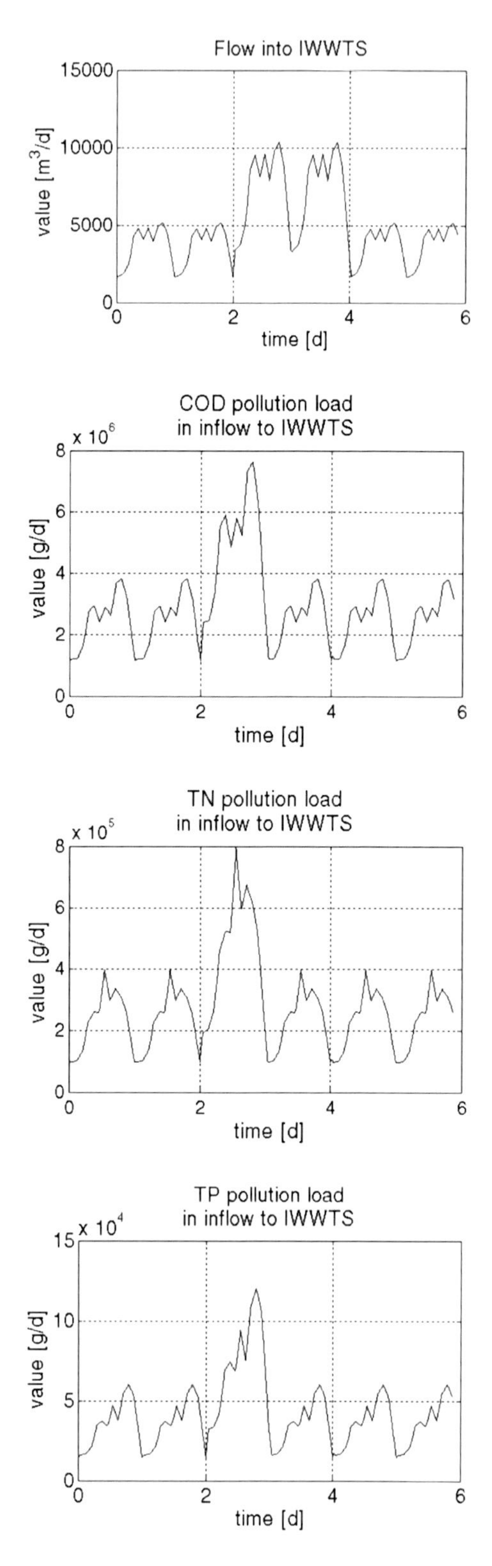

Fig. 17 Influent flow

Fig. 18 Pollution load of chemical oxygen demand in inflow into IWWTS Kartuzy

Fig. 19 Pollution load of total nitrogen in inflow to IWWTS Kartuzy

Fig. 20 Pollution load of total phosphate in inflow to IWWTS Kartuzy

TP $\leq$ 1 (g/m^3), COD $\leq$ 75 ($gO2/m^3$). The 6 days-long testing input is designed in order to capture the system dynamics. The first 2 days represent 'normal' average influent conditions. The next 2 days represent storm weather conditions and the subsequent days are 'normal days' again.

The influent flow rate and its composition correspond to real system influent and may serve as an example of rainy autumn days or heavy spring snowmelts. Usually, the influent pollution load composition has a more or less daily pattern with the exception of the heavy rains (pollution load increases at the beginning of the rain). The influent pollution concentrations from the sewer network increase greatly during the heavy rain period (day 3), as in the real system. During day 4 (the second day of rain) the sewer network is "cleaned up" by the rain, causing the decrease in the pollution concentration.

Two control strategies (CS1 and CS3) on their own as well as soft switching between them are investigated. The results are illustrated in Figs. 21, 22, 23 and 24. It follows from Fig. 21 that the CS1 is not able to avoid emergency overflowing during the 4th day. However, this is well possible by applying strategy CS3. Moreover, certain free capacity is left in the equalization tank and it might be significant in compensating the influent flow rate prediction error. Clearly, there is a price to be paid for better hydraulic control and it is an increased cost of operating the plant under the CS3 strategy. Therefore, the CS3 strategy is employed only when it is necessary.

As with the CS1 control strategy during wet weather conditions an overflow is unavoidable, the soft switching from CS1 to CS3 is activated at $t_1 = 78$ h and it is completed at $t_2 = 85$ h. It can be seen in Fig. 21 that an excellent utilization of the equalization tanks capacity is achieved. The TN trajectories are shown in Fig. 22. Again, the CS1 is not able to meet the TN effluent limit while the softly switched controller meets the requirements. The soft switching processes from the CS1 to CS3 can be clearly seen in Figs. 21, 22. It can be seen in Fig. 23 that the phosphate concentration remains with the allowed discharge limits even during heavy rain. It is because the CS3 supports stronger than the CS1 a usage of the PIX control handles (see Fig. 24).

7 Application of Two Time Scale Hierarchical Centralised MPC to Integrated Optimising Control of Quality and Quantity in Drinking Water Distribution Systems

7.1 Introduction and Problem Statement

Drinking water distribution system (DWDS) delivers water to domestic users. Hence, the main objective is to meet a water demand of required quality to every consumer [10]. For a safe and efficient process operation, monitoring and control systems are needed. In this Chapter the monitoring system is assumed in place and the control system for DWDS is pursued. There are two major aspects in control of drinking

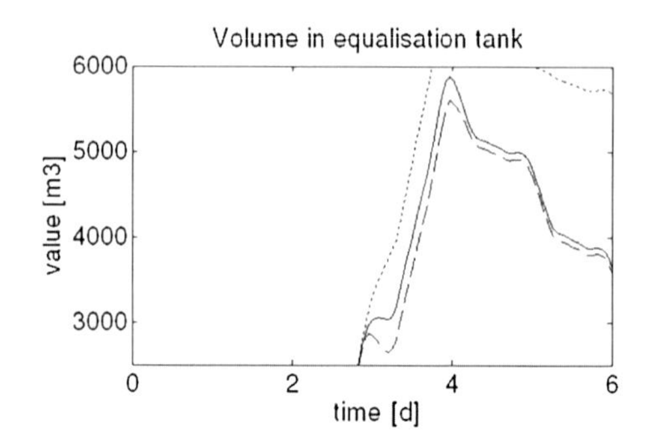

Fig. 21 Volume level in equalisation tank as an example of prescribed by SCL: CS1 (*dotted*), CS3 (*dashed*) and softly switched (*solid*)

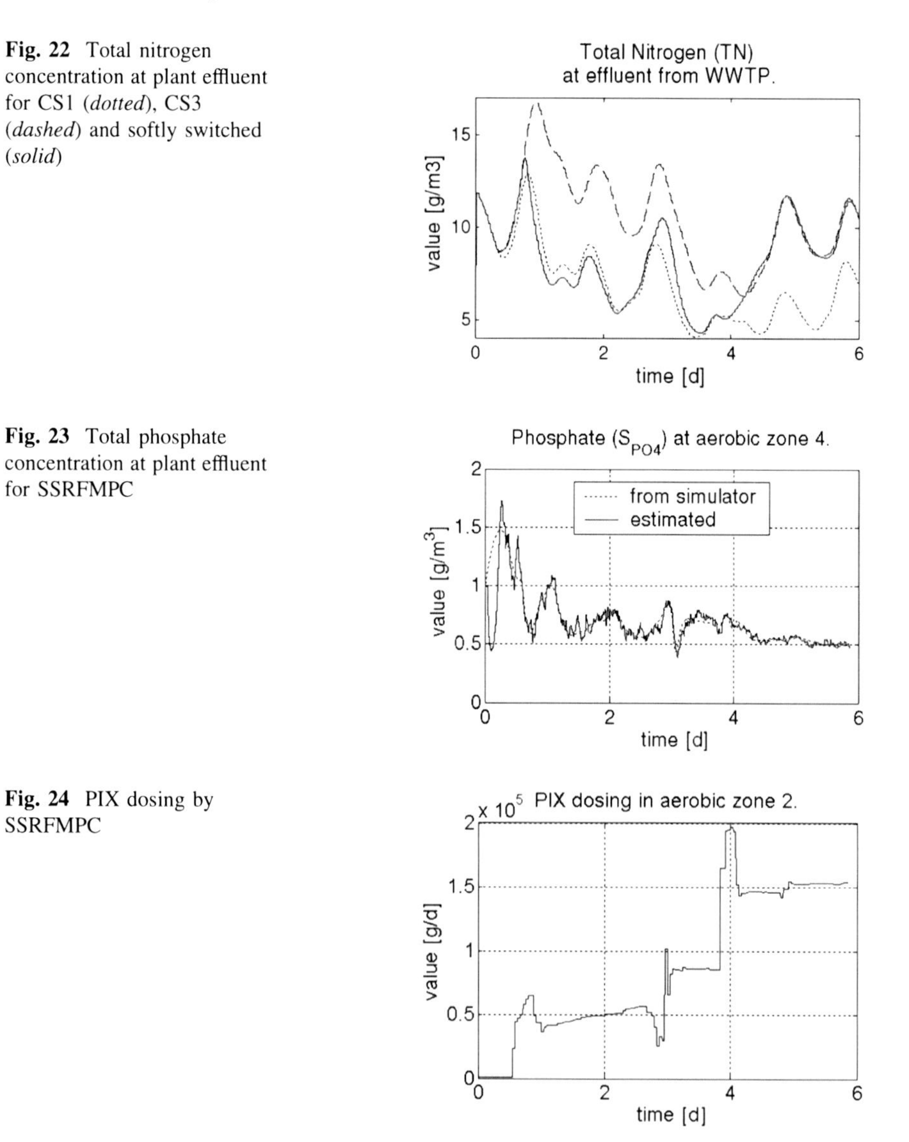

Fig. 22 Total nitrogen concentration at plant effluent for CS1 (*dotted*), CS3 (*dashed*) and softly switched (*solid*)

Fig. 23 Total phosphate concentration at plant effluent for SSRFMPC

Fig. 24 PIX dosing by SSRFMPC

water distribution systems (DWDS): quantity and quality. The quantity control deals with the pipe flows and pressures at the water network nodes producing optimised pump and valve control schedules so that water demand at the consumption nodes is met and the associated electrical energy cost due to the pumping is minimised [10, 41]. Maintaining concentrations of water quality parameters within prescribed limits throughout the network is a major objective of the quality control. In this Chapter, only one quality parameter is considered that is chlorine. It is the most common disinfectant used in the DWDS. It is not expensive and effectively controls a number of disease-causing organisms. As the chlorine reactions with certain organic compounds produce disinfectant by-products that are health dangerous [42] the allowed chlorine residuals are bounded. Hence, the objective of maintaining the desired water quality is expressed by certain lower and upper limits on the chlorine residuals at the consumption nodes. The chlorine residuals are directly controlled within the treatment plants so that the water entering the DWDS has the required prescribed residual values. However, when travelling throughout the network the disinfectant reacts and consequently its major decay may occur so that a bacteriological safety of water may not be guaranteed particularly at remote consumption nodes. Therefore, post chlorination by means of using booster stations located at certain intermediate nodes is needed. The booster station allocation problem was presented in [43, 44] and the solution methods based on multiobjective optimisation were provided. The chlorine residuals at the nodes representing outputs from the treatment plant and at the booster station nodes are the direct control variables for the quality control. Electricity charges due to pumping constitute the main component of the operational cost to be minimised. As an interaction exists between the quality and quantity control problems due to the transportation delays when transferring the chlorine throughout the network a proposal to integrate these two control issues into one integrated optimising control problem was made in [45] and a receding horizon model predictive control technique (MPC) was applied that assumed periodically varying and very similar demands over a number of subsequent days. Also, the quality and quantity modelling errors were not addressed by means of feedback. These assumptions allowed for a simplified implementation of the MPC were the feedback was taken once per day. Several solvers of the MPC optimisation task were proposed applying the genetic search [46], mixed integer linear (MIL) algorithm [45], sequential quantity-quality hybrid search and genetic-MIL approach [47] and nonlinear programming approach [48].

7.2 A Single Agent Centralised Two Level Hierarchical Controller

Due to different time scales in the hydraulic variations (slow) and internal chlorine decay dynamics (fast) the integrated optimisation task complexity did not allow applying the integrated control to many realistic size DWDS. With the hydraulic time step typically 1 h, quality time step for example 5 min, and the prediction horizon due to tank capacities typically 24 h, the problem dimension largely increases even for small size systems. A suboptimal two layer hierarchical control structure was proposed in [49] and further developed in [50] and [51]. This two level controller structure is illustrated in Fig. 25.

The *optimising controller* at the *upper control level* (UCL) operates at the hydraulic slow time scale according to a simplified receding horizon strategy. At the beginning of a 24-h time period the DWDS quantity and quality states are measured or estimated and sent to the integrated quantity and quality optimiser. The consumer demand prediction is also sent to the optimiser. The quality model assumes the same time step as the quantity dynamics model. Hence, the problem dimension is vastly reduced but the quality modelling error is significantly increased. Hence, solving the integrated quantity–quality optimisation problem produces the optimised chlorine injection schedules at the booster and treatment plant output nodes of poor quality and good suboptimal optimised pump and valve schedules over the next 24 h. The pump and valve schedules are applied to the DWDS and maintained during a so called control time horizon e.g., 2 h. The quality controls need to be improved and this is performed at the *lower correction level* (LCL) by the fast *feedback quality controller* operating at the quality fast time scale. It samples the chlorine residuals concentrations as it is required by its decay dynamics e.g., with a 1-min sampling interval. In order to take advantage of the allowed quality bounds, the robustly feasible MPC with safety zones (RFMPC) is applied (Fig. 25).

A suboptimal approach is to specify a trajectory lying within the bounds and apply an adaptive indirect model reference controller to track this reference trajectory [52], (Brdys and Chang 2001). In this Chapter the optimising MPC (RFMPC) with full information feedback is applied to synthesise the upper and lower level controllers.

7.3 Application to Gdynia DWDS Case-Study

An application to the case-study DWDS is presented in this section. A skeleton of the DWDS at Gdynia is illustrated in Fig. 26 and its data are as follows:

- 3 underground water sources
- 4 tanks and 3 reservoirs
- 10 variable speed pumps
- 4 control valves, 148 pipes, 134 pipe junction nodes, 87 demand nodes
- 5 booster stations allocated at the quality control nodes
- 129 quality monitoring nodes
- Accuracy of provided on-line demand prediction over a 24-h period was 5 % for the first 10 h and 10 for the remaining time slot of the prediction horizon
- The electricity tariff during 6–12 am and 3–9 pm was $\eta = 0.12$ (\$/kWh) and $\eta =$ (0.06 \$/kWh) during 10–5 pm.

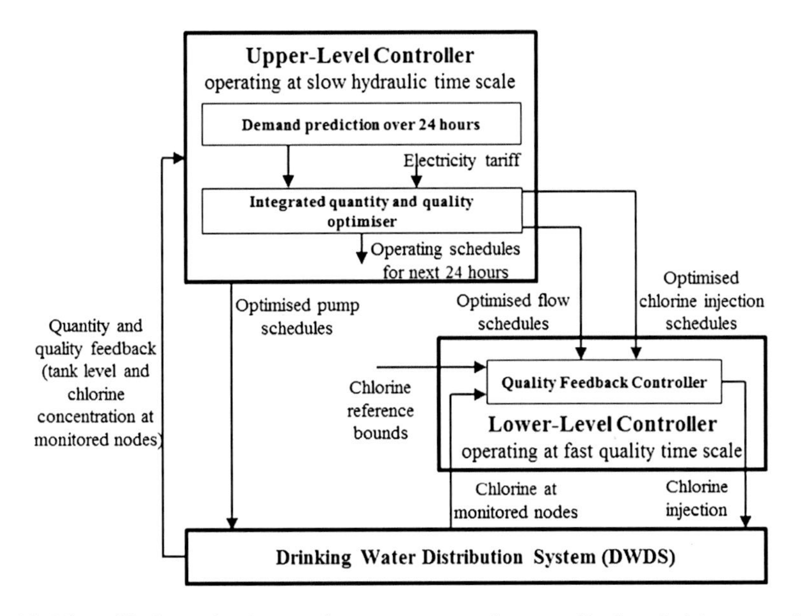

Fig. 25 Hierarchical two-level control agent structure for centralised optimising control of integrated quantity and quality

The centralised MPC controller was applied with the 2-h hydraulic time step and 10-min quality time step. A genetic solver was applied to solve on-line the MPC integrated quality and quantity optimisation tasks at the Upper Control Level and also to solve the MPC quality optimisation task at the Lower Control Level with the fast quality feedback. The results are illustrated in Fig. 27 (controlled pump speeds), Fig. 28 (resulting quality) and Fig. 29 (resulting quantity). Comparison in Fig. 27 of the quantity control actions, which are currently used at the site with the MPC actions, shows a very conservative operation of the current system. Such operation leads to high operational cost due to the electricity charges. It is implied by unavoidable difficulties in meeting the inequality constraints in this strongly interconnected system. The MPC employs an integrated approach to handling all the constraints and, in addition, a dedicated mechanism to handle the uncertainty impact. This is even more clearly seen in Fig. 29, where a trajectory of a selected tank in Witomino is illustrated. The MPC utilises the available tank capacity much better than the current operational strategy. The excellent quality control results are illustrated in Fig. 28. The chlorine concentration in a junction node lies within the prescribed limits and it gets close to the lower limit, hence assuring a limited production of harmful components due to the reactions of the organic matter with free chlorine.

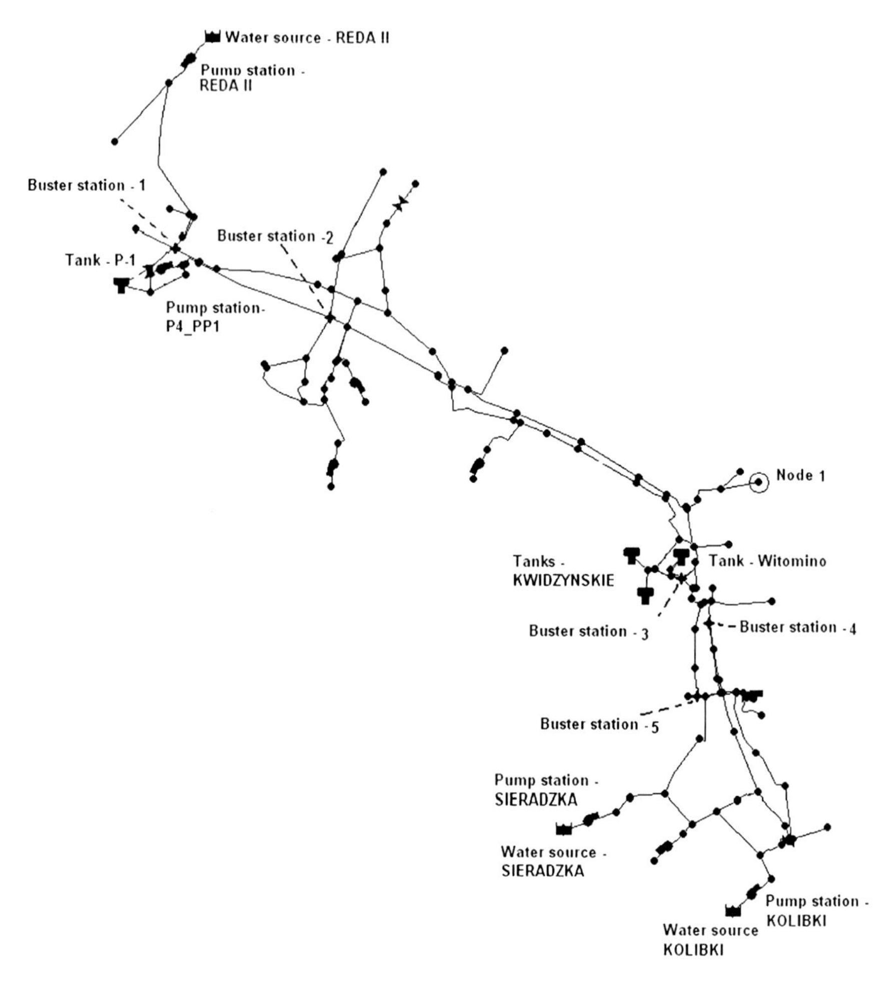

Fig. 26 A skeleton of the DWDS at Gdynia

8 Application of RFMPC to Synthesis of Fast Feedback Quality Controller in DWDS

8.1 Introduction

The RFMPC will be applied in this section to derive the lower level controller with fast feedback from the quality measurements for the control architecture presented in Fig. 25. Both the centralised and decentralised with information exchange between the control agents structures will be derived. The presentation will utilise the DWDS benchmark.

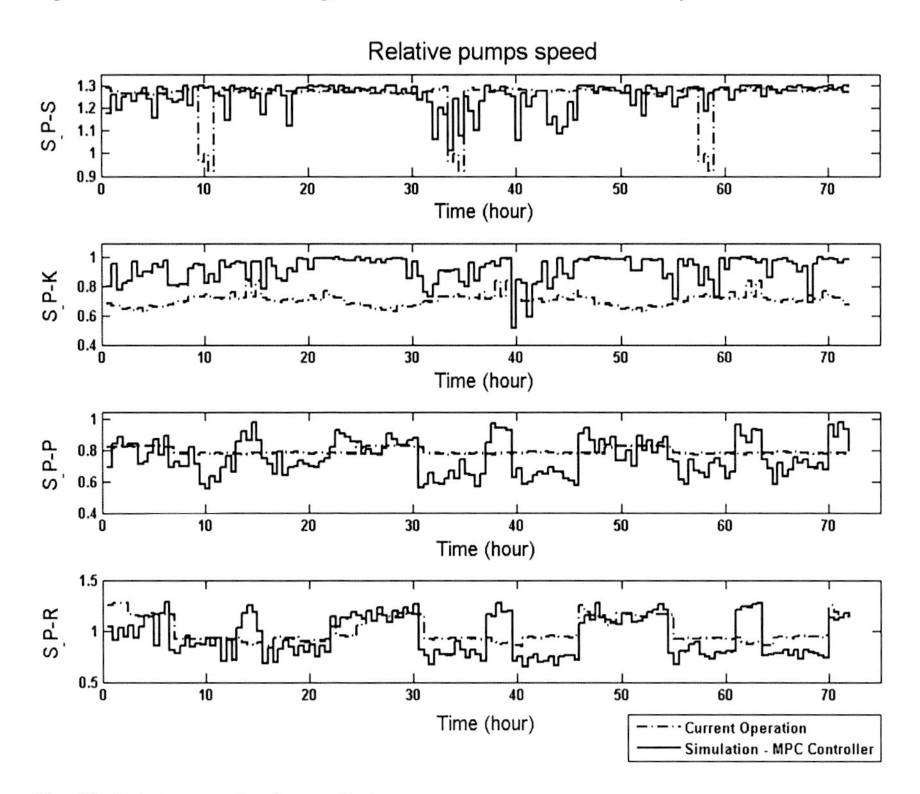

Fig. 27 Relative speeds of controlled pump

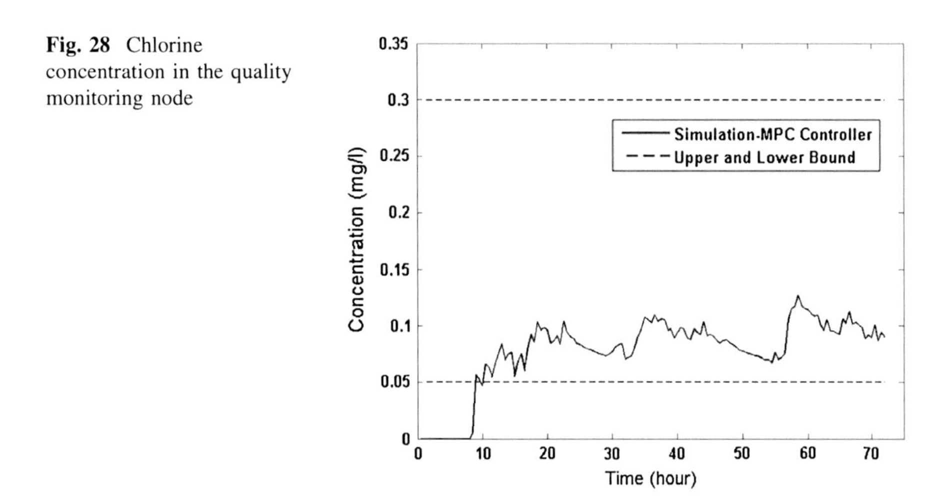

Fig. 28 Chlorine concentration in the quality monitoring node

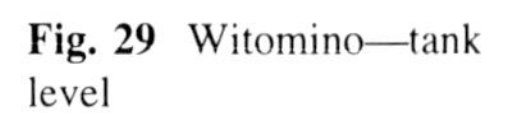

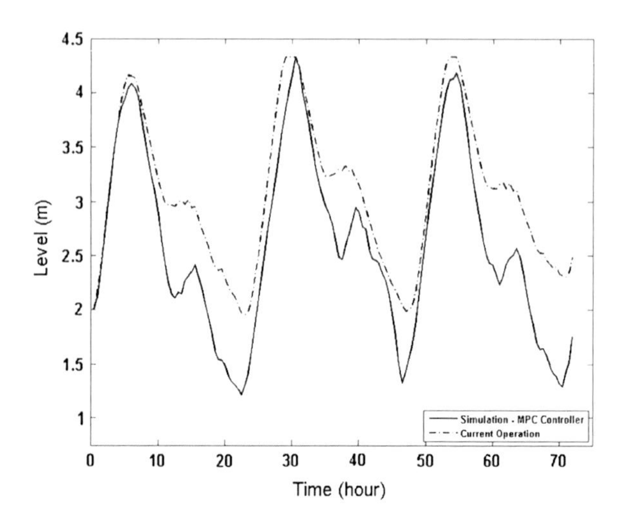

Fig. 29 Witomino—tank level

8.2 Presentation of DWDS Benchmark

The benchmark structure is illustrated in Fig. 30. There are 16 network nodes, 27 pipes and 3 storage tanks in the system. All tanks are the switching tanks and they can only be operated in repeated sequential filling and draining cycles. The water is pumped from the source (node 100 and node 200) by two pumps (pump 201 and pump 101) and is also supplied by the tanks (node 17, 18, 19). Node 16 and node 8 are selected as monitored nodes as they are the most remote nodes from the source. Hence, if the chlorine concentrations at these two notes meet the quality requirements, then these requirements are also met at all other nodes over the DWDS. The chlorine concentrations at these nodes are the two plant outputs $y_1(t)$ and $y_2(t)$, respectively. Initially, node 1 and node 5 are selected as the *quality control nodes*, where the chlorine is injected in order to control the chlorine concentration at the monitored nodes. The booster stations are to be installed at these nodes to produce the chlorine concentrations $u_1(t)$ at node 5 and $u_2(t)$ at node 1. These are the quality control inputs and the controlled output in this DWDS benchmark. The fast feedback quality controller operates under the pump control inputs determined by the upper level controller as it is shown in Fig. 25. Hence, the flows are determined. We shall now validate the quality input–output interactions and also the selection of the quality control nodes. The quantity operation status of tank 19 is shown in Fig. 31 where the flow from and into the tanks is illustrated. It can be seen that the filling periods of the tank 19 are [0, 6] [hour], [12, 19] [hour] and [20, 21] [hour], while the draining periods are [6, 12] [hour], [19, 20] [hour] and [21, 24] [hour].

The RFMPC output prediction and control horizons is 24 h while the quality control step is 5 min. Thus, the 24-h control horizon is converted into 288 discrete

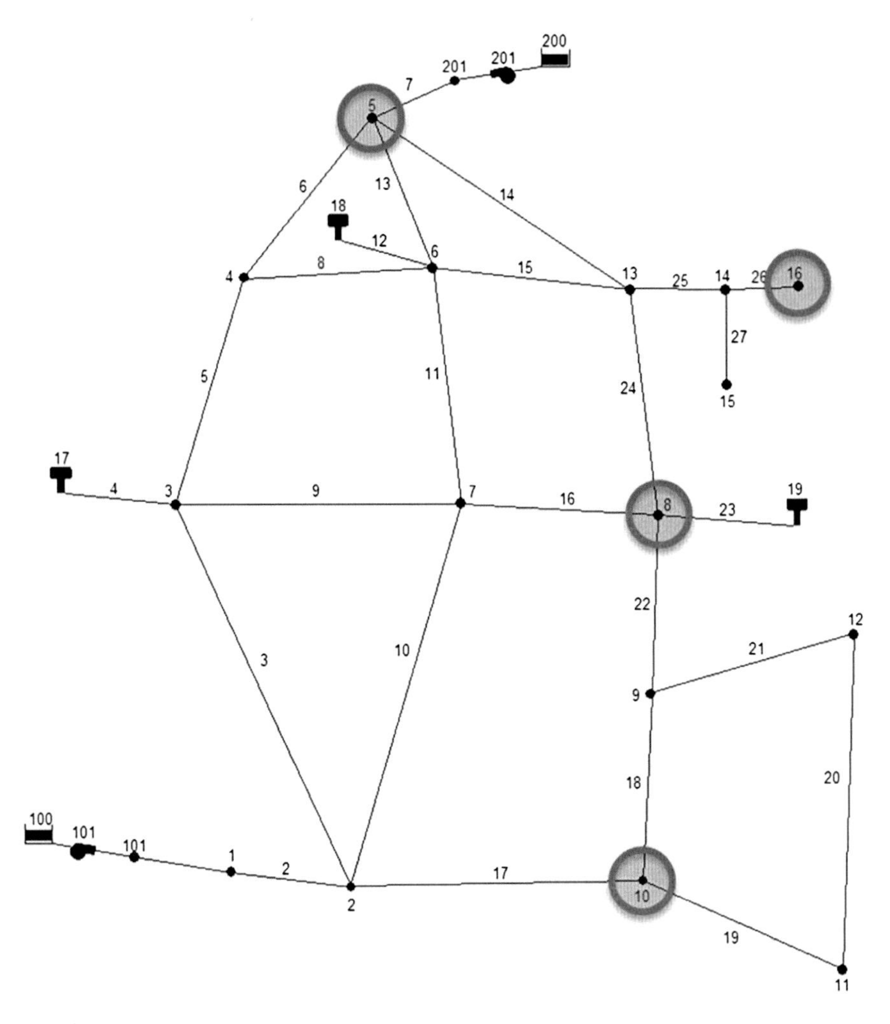

Fig. 30 Benchmark DWDS

time steps. The DWDS structure implies the following control input—controlled output pairing: (5, 16) and (10, 8). It means that the chlorine concentration at the output node 16 is mainly controlled by the injection at the node 5 and similarly the chlorine concentration at the output node 8 is mainly controlled by the injection at node 10. The interactions between these two input–output pairs exist through the flow in pipe 24, which is illustrated in Fig. 32 where the positive flow direction is defined from node 13 to node 8.

It can be seen that during most of the time over the 24-h horizon, the chlorine injection at node 5 acts on chlorine concentration at node 8 through flow 24. Over the period although the flow direction in pipe 24 changes, there is no flow from node 8 into the output node 16 because before the flow from 8 breaches the node

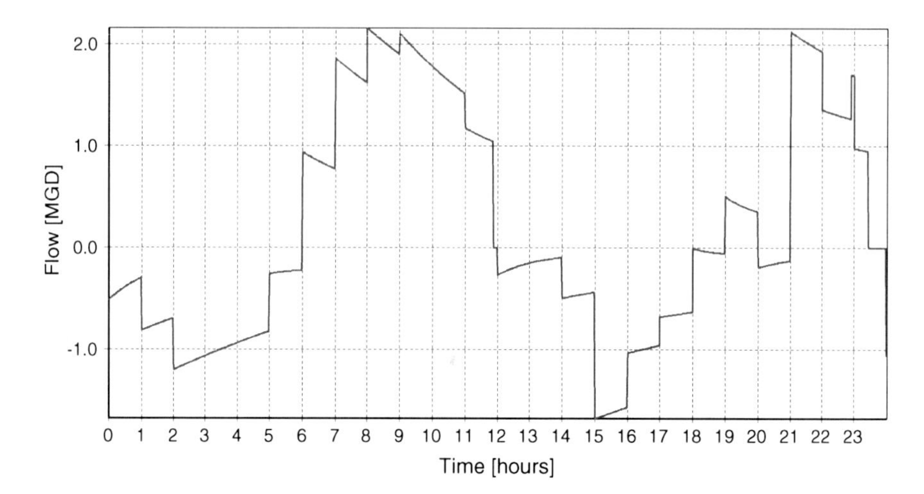

Fig. 31 Flow in pipe 23

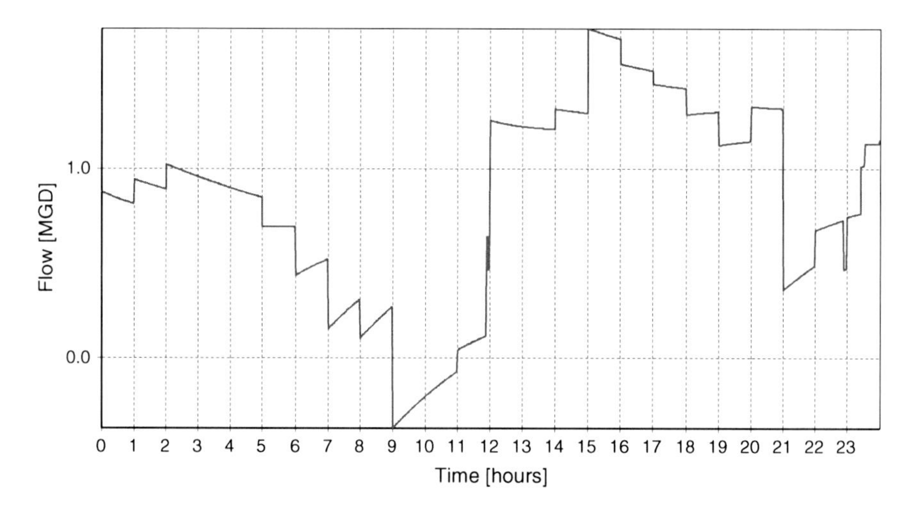

Fig. 32 Flow in pipe 24

13 the flow direction recovers. Hence, there is no interaction from the node pair 1–9 into the node pair 5–16. Moreover, hydraulic data show that water flow from tank 17 or 18 does not enter node 16 or 8 at all. This means that only one tank, tank 19, participates in producing dynamics in the outputs. During the draining period of tank 19, water demand at node 8 is mainly fulfilled by flow from the tank. Therefore, the controllability at node 8 would be weak. With the injection node 10 a strong controllability at node 8 is achieved. Hence, node 10 is finally chosen as the quality control node replacing node 5. The above considerations are

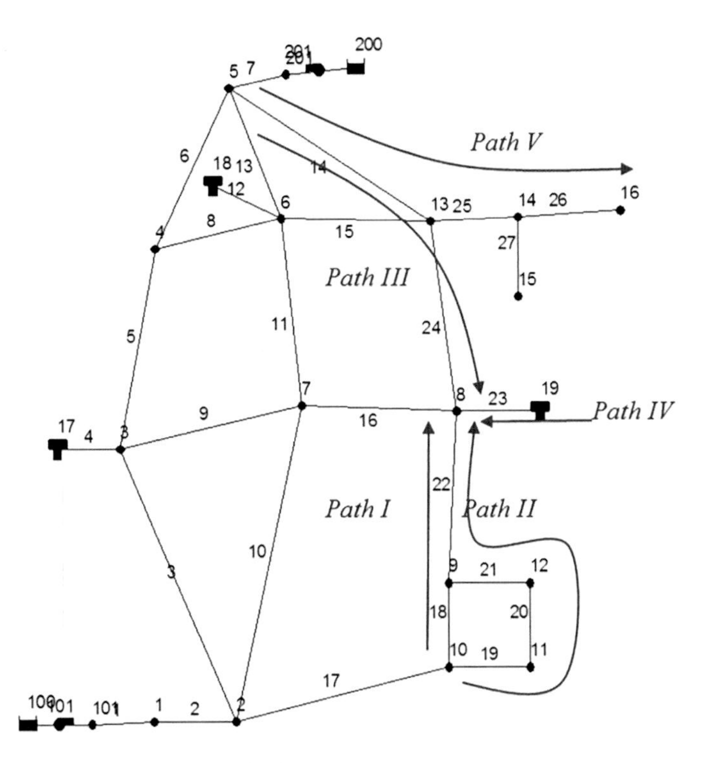

Fig. 33 Paths from the injection nodes to the monitored nodes

fundamental not only for the proper placement of the chlorine injection facilities over the DWDS, but they are also important for designing the structure of the input–output model and consequently its parsimonious parameterisation.

8.3 *Input Output Modelling for the Benchmark DWDS*

Running the path analysis [53]; Wang et al. 2001; Brdys and Chang 2001; [12] five chlorine transportation paths from the inputs to the outputs over the 24-h period are obtained, which are illustrated in Fig. 33. The minimum chlorine transportation delay in the paths is 30 min and the maximum delay is 120 min. A continuous range of a delay along each of the active paths is discretised to produce the set of all delays, which are active at a certain time instant. The model structure estimation algorithm (Brdys and Chang 2001) is applied and the set of parameters, which are associated with the active delays and constant over thirteen time-slots of the entire control horizon is determined. The resulting input–output model is point parametric, hence least conservative with respect to time varying parameter and structure uncertainty [54]. This is a piece-wise constant parameter bounded model, whose input–output equations are as follows:

$$y_1(t) = b_{1,1}^{J(t)}(t)y_1(t-1) + \sum_{j=1}^{2} \sum_{i \in I_{1,j}^{J(t)}} a_{1,j,i}^{J(t)}(t)u_j(t-i) + \varepsilon_1^{J(t)}(t) + \varepsilon_s(t)$$

$$y_2(t) = \sum_{i \in I_{2,2}^{J(t)}} a_{2,2,i}^{J(t)}(t)u_2(t-i) + \varepsilon_2^{J(t)}(t) + \varepsilon_s(t)$$

where t denotes discrete time instants, i denotes discretised delay number, $I_{1,j}^{J(t)}$ is a set of delays associated with all active paths over time slot $J(t)$, $a_{n,j,i}$ describes an impact of the injection input u_j that is delayed by i steps on the output y_n (impact coefficient), where $n = 1, 2$, and $\varepsilon_2^{J(t)}(t)$ and $\varepsilon_s(t)$ denote the delay discretisation error and model structure error, respectively. As there is no interaction from u_1 to y_2 this control input does not appear in the second equation. Figures 34 and 35 illustrate the piecewise constant envelopes robustly bounding the parameters $a_{1,1,7}$ and $a_{1,2,18}$ due to the active delay numbers 7 and 18 over a whole horizon of 288 steps, respectively. It can be clearly seen from these figures that the parameterisation of the time varying parameters into the robust piecewise constant parameters is parsimonious, indeed. The centres of these envelopes are taken as the parameter values in the nominal model, which is used by RFMPC to predict the outputs when solving its model based optimisation task.

8.4 Centralised Robustly Feasible Model Predictive Controller

The RFMPC (see Fig. 4) will be applied in this section to control the DWDS benchmark. The constraints are on the outputs in a form of upper and lower bounds on the chlorine concentrations at the monitoring outputs. A standard quadratic function is applied to design the performance function. The robustly feasible safety zones generator in Fig. 4 is designed as a simple relaxation algorithm in order to reduce a number of iterations, which require feedback information from the CIS. The model derived in the previous subsection is applied to implement the robustly feasible output prediction and consequently to design the constraint violation checking unit in Fig. 4. The MPC based on the nominal output model produces the controls, which violate the output constraints and is not applicable. A performance of the RFMPC is illustrated in Figs. 36, 37. The controlled outputs are strongly time varying between the prescribed limits and in order to maintain these limits the injections at the control nodes are also strongly varying in time to produce highly nontrivial trajectories under time varying and unknown water demand allocated at the demand nodes. The importance of the robustly feasible safety zone mechanism

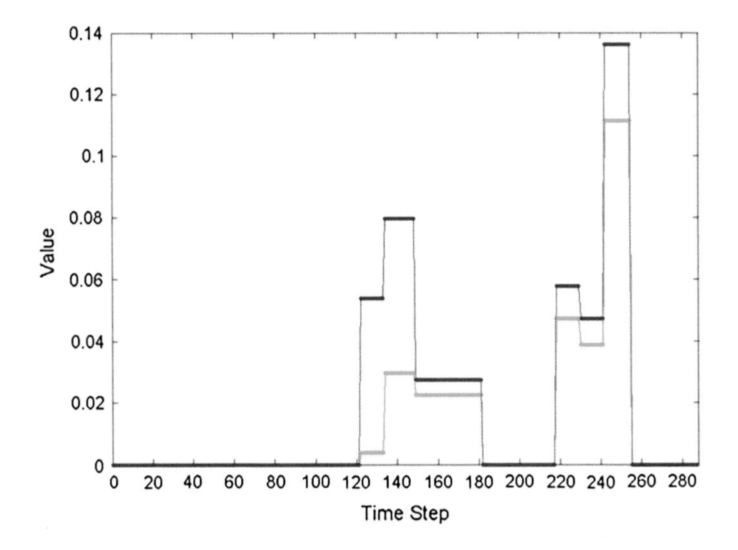

Fig. 34 Piecewise constant bounds on parameter $a_{1,1,7}$

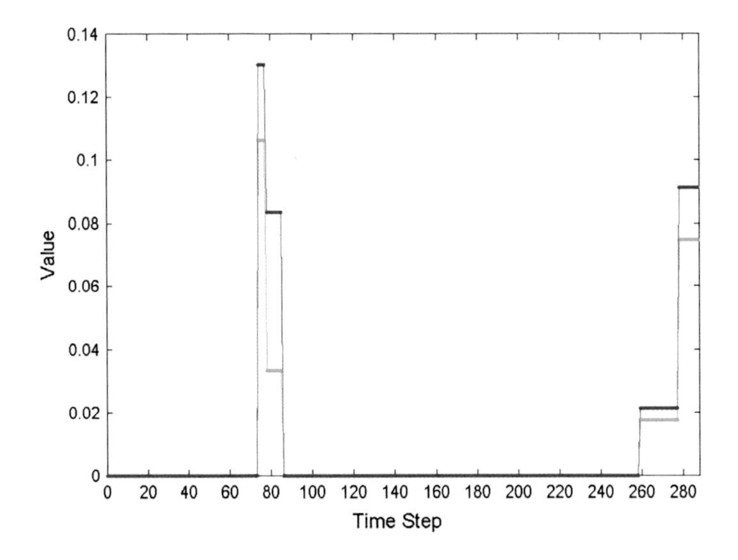

Fig. 35 Piecewise constant bounds on parameter $a_{1,2,18}$

is illustrated in Fig. 38. It can be seen that an upper bound on the output at node 16, which is used in formulation of a model based optimisation task of the RFMPC is modified by introducing the safety zones over 11–36 time steps. Clearly the solution of this task is feasible so that the corresponding output produced by the optimiser satisfies the modified output constraints. However, the output forced in the DWDS by the optimal control sequence may not be feasible due to an

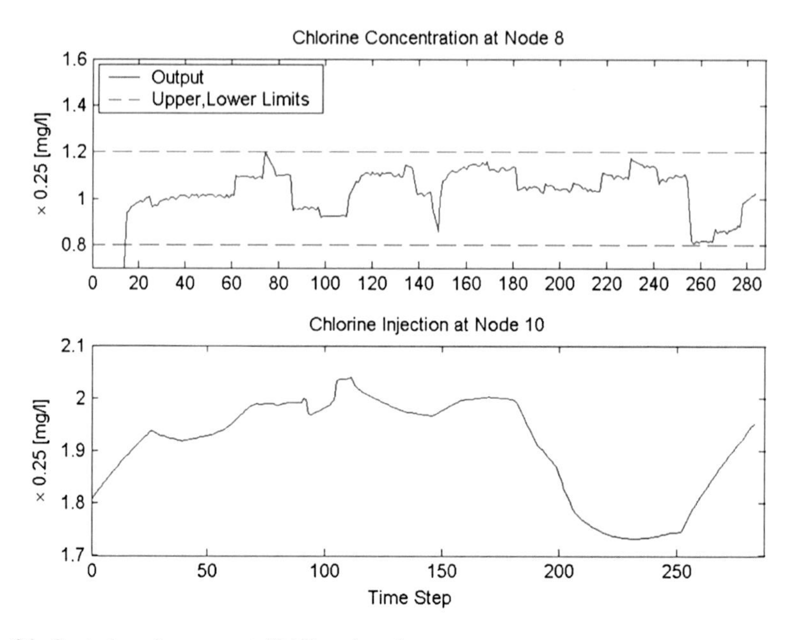

Fig. 36 Control performance at (8,10) node pair

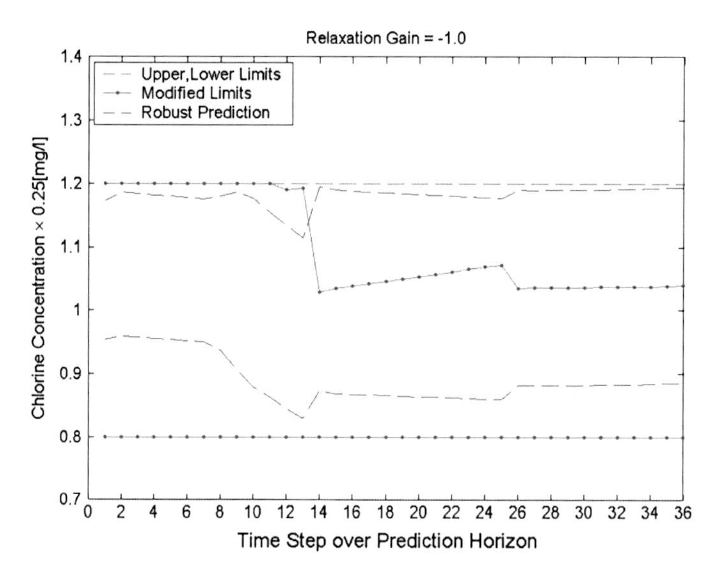

Fig. 37 On line design of Robustly Feasible Safety Zone for output node 16

uncertainty in the model used by the RFMPC to predict the outputs when solving the model based optimisation task. Its feasibility is checked by the Robust Output Prediction Unit (see Fig. 8). The envelopes robustly bounding a region where the DWDS output trajectory is contained are shown in Fig. 37 and it can be seen that

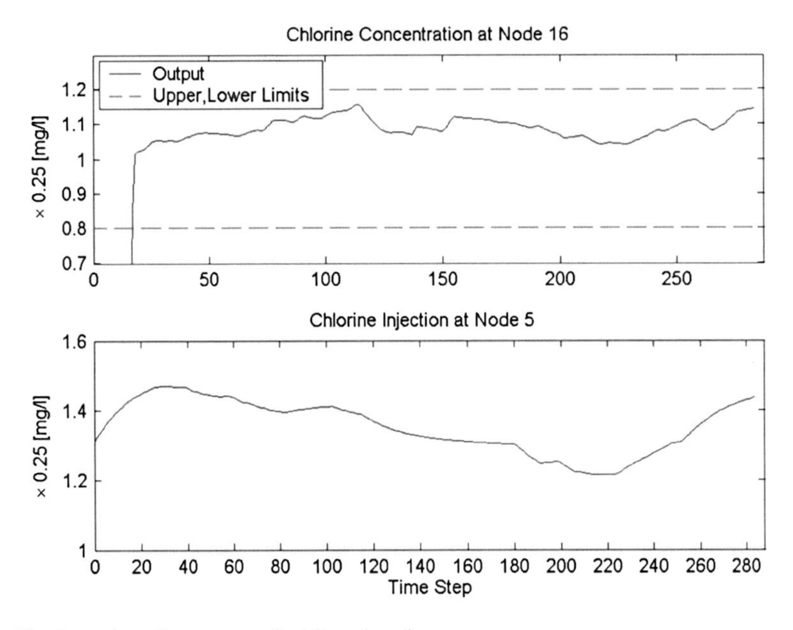

Fig. 38 Control performance at (5, 16) node pair

the upper bounding envelope violates the modified upper bounds on the output over 13–36 time steps. However, the envelope does not exceed the original upper bound on the output. It means that the output forced at the node 16 of the real DWDS is feasible although its feasibility in the DWDS model with the constraints modified by robustly feasible safety zones is not guaranteed.

The safety zones are determined on-line based on the measurements of the chlorine concentrations at the monitored nodes. These measurements are used to evaluate a robust prediction of the output constraints violation over the prediction horizon. The instantaneous violations are integrated over the horizon to produce key feedback information for the relaxation algorithm, which helps to reduce conservatism in the robustly feasible safety zones iteratively determined by the safety zone generator in Fig. 8.

8.5 Multi-Agent Robustly Feasible Model Predictive Controller with Information Exchange

8.5.1 Introduction

Decentralization of the quality feedback controller is desired due to practical reasons, e.g., computation complexity, distributed actuators and sensors and subsystem dependence on the central controller reliability. A decentralized structure is proposed in this section for robust model predictive control of chlorine residuals in drinking water distribution systems. The water network is divided into a number of interacting zones. Each zone is directly controlled by a local RMPC. However, the input–output dynamics show strong interactions. One control input may influence a number of controlled outputs if the paths exist from these inputs to various outputs. Vast majority of the control literature on decentralized regulatory control considers situations where the interactions are weak. This is not so however, in our case. In order to achieve complete decomposition of the controller into a number of independent controllers that utilize only local information, it is proposed to use the injection quality values produced at the local controllers as the real interaction estimates assuming communication between the local controllers is available. Then the error may be considered small enough to classify the interactions as weak. The error can then be rejected due to the robustness of the local controller.

In this section a decentralized controller structure with information exchange is proposed. The interaction prediction is employed in order to compensate the impact of the interactions. Robustness of the controller can still be maintained by using the safety-zone mechanism assuming bounds on the interaction prediction error are known. It is also assumed that the local controllers can exchange information about their control actions. The communication is asynchronous. This information is used to predict the zone interactions.

8.5.2 Problem Statement

The control objective is to maintain chlorine residual at monitored nodes within upper-lower limits:

$$y^{\min} \leq y_i \leq y^{\max}, i = 1, \ldots, n$$

by injecting chlorine at source nodes:

$$u^{\min} \leq u_i \leq u^{\max}, \; |\Delta u_i| \leq \Delta u^{\max}, \; i = 1, \ldots, n$$

where $y^{\min}$ and $y^{\max}$ are the limits on system output y, $u^{\min}$, $u^{\max}$ and $\Delta u^{\max}$ are actuator (booster station) constraints on system input u, given as the prescribed bounds.

The input–output model reads:

$$y_1(t) = \sum_{i=1}^{n_T} b_{1,i}(t)y_1(t-i) + \sum_{j=1}^{n} \sum_{i \in I_{1,j}} a_{1,j,i}(t)u_j(t-i) + \varepsilon_1(t)$$

$$\vdots$$

$$y_n(t) = \sum_{i=1}^{n_T} b_{n,i}(t)y_n(t-i) + \sum_{j=1}^{n} \sum_{i \in I_{n,j}} a_{n,j,i}(t)u_j(t-i) + \varepsilon_n(t)$$

where n_T denotes the number of tanks, all the model parameters $b_{n,i}$, $a_{n,j,i}$ and $\varepsilon_i(t)$ are piecewise constantly bounded as described in Sect. 8.3.

The decentralized RMPC (DRMPC) is to be developed, which is composed of n local RMPCs in which only one input–output pair is involved assuming communication among the n local RMPCs is available.

8.5.3 Structure of DRMPC

Without loss of generality, the interaction prediction can be illustrated by taking the first input–output pair as an example. The first model equation of model (3) can be rewritten as:

$$y_1(t) = \sum_{i=1}^{n_T} b_{1,i}(t)y_1(t-i) + \sum_{i\in I_{1,1}} a_{1,1,i}(t)u_1(t-i) + EL_1(t) + \varepsilon_1(t) \quad (1)$$

$$EL_1(t) = \sum_{j=2}^{n} \sum_{i\in I_{1,j}} a_{1,j,i}(t)u_j(t-i)$$

where $EL_1(t)$ is the chlorine contribution to $y_1(t)$ from external loops.

A single-input single-output (SISO) RMPC can be designed based on model (1) assuming $EL_1(t)$ is an external input. As communication between local control loops is available, the control inputs of the local loops over the prediction horizon can be exchanged among the local controllers. For the local controller 1 the following information can be obtained at time t from controllers $2, 3, \ldots, n$:

- $U_2(t-1) = [u_2(t-1|t-1) \;\; u_2(t-1+1|t-1) \;\; \ldots \;\; u_2(t-1+H_p-1|t-1)]^T$

$$\vdots$$

- $U_n(t-1) = [u_n(t-1|t-1) \;\; u_n(t-1+1|t-1) \;\; \ldots \;\; u_n(t-1+H_p-1|t-1)]^T$

(2)

where $U_i(t-1)$ is the vector of the control inputs over prediction horizon H_p, proposed at time $t-1$ by the ith RMPC. At time instant t, by using the above information, the control inputs to be generated by local controllers $2, 3, \ldots, n$ at t over the prediction horizon can be predicted as:

$$\tilde{U}_2(t) = [u_2(t-1+1|t-1) \;\; u_2(t-1+2|t-1) \;\; \ldots \;\; u_2(t-1+H_p-1|t-1) \;\; u_2(t-1+H_p-1|t-1)]^T + \varepsilon I_2(t)$$

$$\vdots$$

$$\tilde{U}_n(t) = [u_n(t-1+1|t-1) \;\; u_n(t-1+2|t-1) \;\; \ldots \;\; u_n(t-1+H_p-1|t-1) \;\; u_n(t-1+H_p-1|t-1)]^T + \varepsilon I_n(t) \quad (3)$$

where $\varepsilon I_i(t)$ is the vector of the control variable prediction error. Let us notice that if there is no uncertainty in the model then the first H_p-1 control inputs are

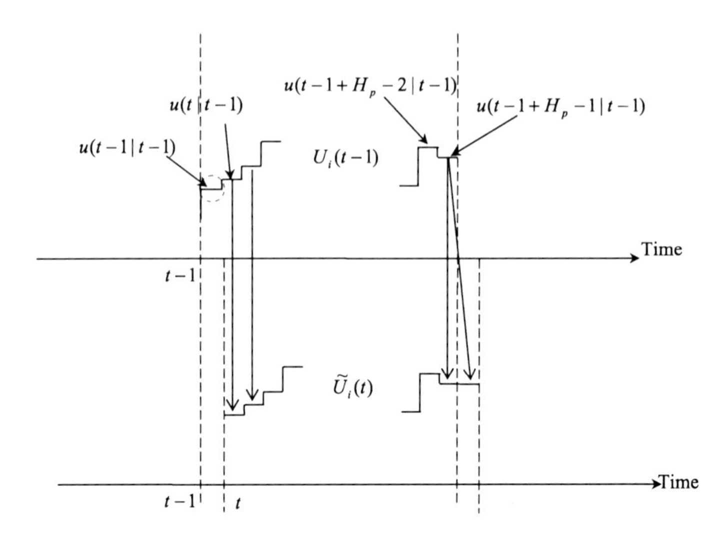

Fig. 39 Obtaining $\tilde{U}_i(t)$ from $U_i(t-1)$

predicted with zero errors. Generating $\tilde{U}_i(t)$ from $U_i(t-1)$ is illustrated in Fig. 39. It is difficult to give a least conservative estimated bound on the control variable prediction error. However, it is understood that with large enough prediction horizon H_p the difference between $U_2(t)$ and predicted $\tilde{U}_2(t)$ can be bounded efficiently. On the other hand, the use of large H_p does not allow for the accurate prediction of plant outputs. Also, as these control inputs represent corrections to the control trajectory produced by the upper level of the overall control structure in Fig. 25, the interaction prediction error is supposed not to be very excessive. In practice this value can be determined by using these error bounds as a tuning knob of the controller until satisfied performance is obtained.

Substituting (3) into (1) yields:

$$y_1(t) = \sum_{i=1}^{n_T} b_{1,i}(t) y_1(t-i) + \sum_{i \in I_{1,1}} a_{1,1,i}(t) u_1(t-i) + \widetilde{EL}_1(t) + \varepsilon_1(t) \tag{4}$$

$$\widetilde{EL}_1(t) = \sum_{j=2}^{n} \sum_{i \in I_{1,j}} a_{1,j,i}(t)[u_j(t-i|t-1) + \varepsilon I_i(t-i)]$$

Based on model (4) the SISO RMPC can be designed. The interaction prediction error enters into the modelling error, hence, it can be handled by introducing safety-zones in RMPC as well. The structure of the decentralized RMPC with information exchange is illustrated in Fig. 40.

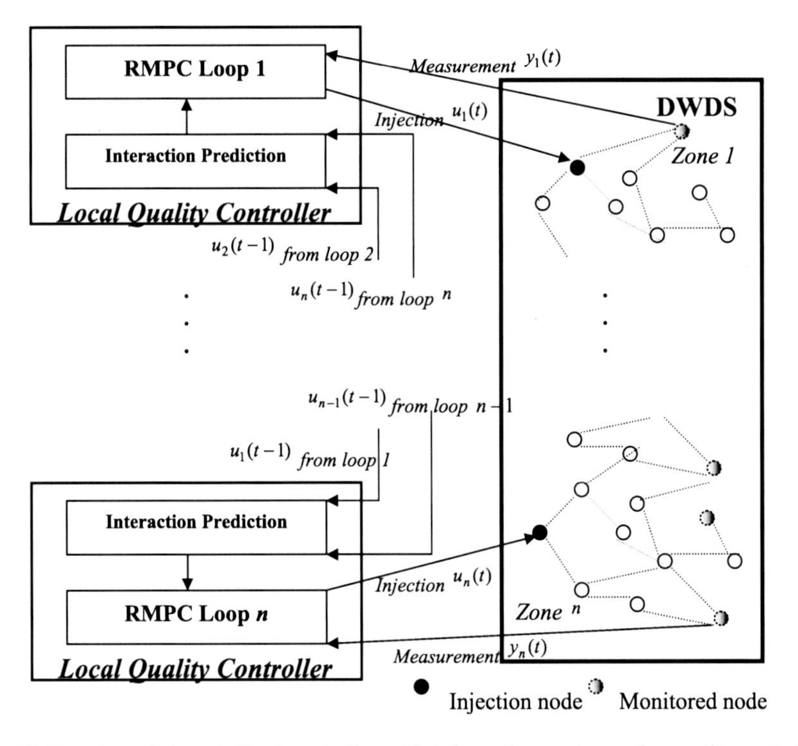

Fig. 40 Structure of decentralized controller with information exchange for quality control in DWDS

8.5.4 Application to Benchmark DWDS

A regional decomposition of the DWDS is illustrated in Fig. 41, where the interacting zone is depicted. The output reference at the end of the prediction horizon is set as:

$$y_{1r} = 1.0 \text{ and } y_{2r} = 1.0$$

The performance of the decentralised controller composed of two regional agents, exchanging information which is used to predict the interaction, is illustrated in Figs. 42, 43.

All the input and output constraints are satisfied. Comparing these results with those obtained by the centralised controller and illustrated in Figs. 36, 38, especially during the time period from step 200 to step 288, it can be found that the control inputs are quite different. The injection at node 10 of the decentralized controller is higher than in the centralized case. In the centralized controller, the control loop of node pair 10–8 obtains more chlorine contribution from loop of node pair 5–16. Coordination between the loops is weaker due to the decentralization, which leads to the corresponding compensation by the local controller.

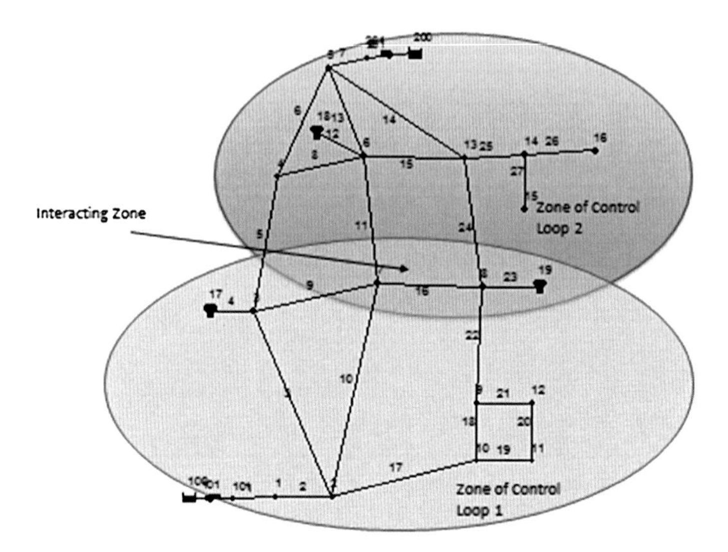

Fig. 41 Regional decomposition of DWDS

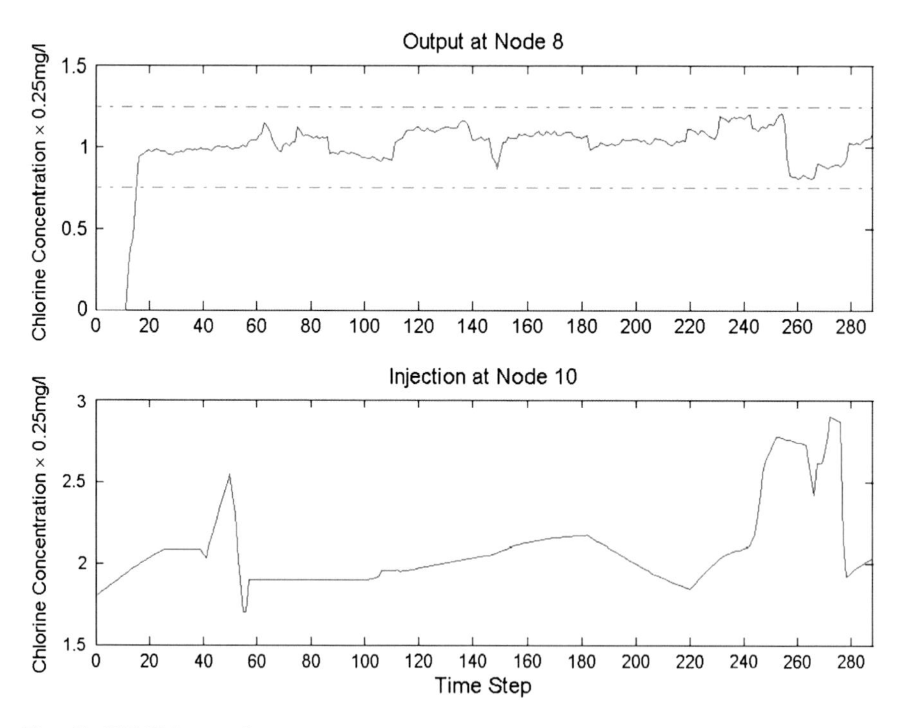

Fig. 42 DRMPC: y_1 and u_1

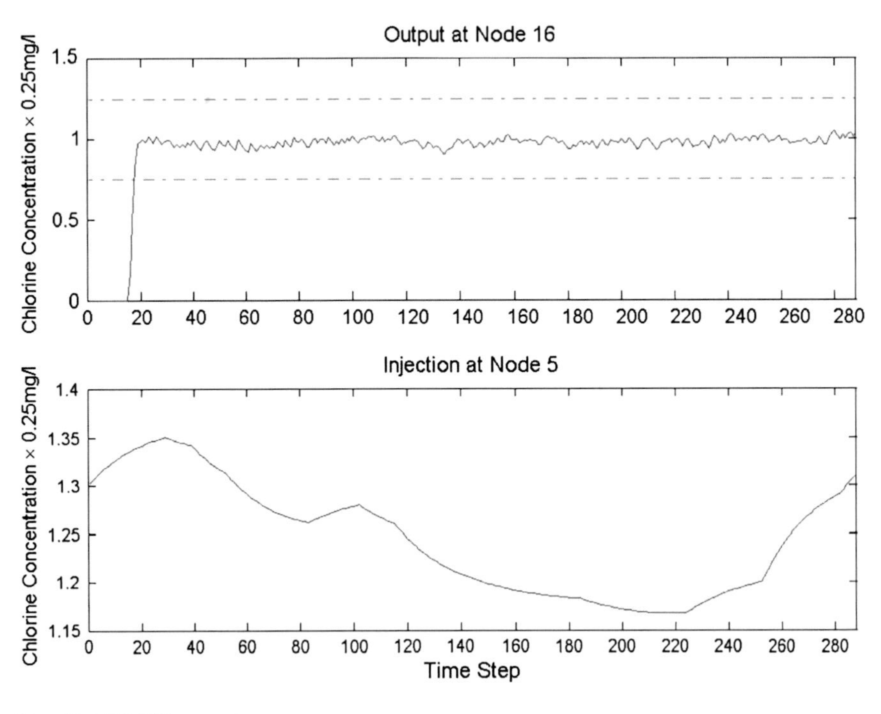

Fig. 43 DRMPC: y_1 and u_1

In this DWDS benchmark case study, such ability to compensate a weakening of the coordination between local controllers is still within the capacity of the local control loop. Hence, the output constraints are still kept within prescribed limits.

9 Summary and Conclusions

A decentralized robustly feasible model predictive control structure was presented for chlorine residual control in drinking water distribution systems. The overall water network was divided into a number of zones. Within each zone a SISO robust model predictive controller is operated. Interactions between the control zones can be predicted assuming that communications among control zones are available. The prediction errors can be handled by using a bounding method, which can then be handled by using the same RMPC algorithm developed before. Hence, the output constraints satisfaction can be maintained under uncertainties in the system. Computer simulation results show that the algorithm is applicable for practical utilization.

Acknowledgments This work was supported by the European Commission under COST Action IC0806 IntelliCIS and by Polish Ministry of Science and Higher Education under grant number 638?N – COST/09/2010 (InSIK). The author wishes to express thanks for the support.

References

1. Gertler, J.J.: Fault Detection and Diagnosis in Engineering Systems. Marcel Dekker, New York (1998)
2. Brdys, M.A., Grochowski, M., Gminski, T., Konarczak, K., Drewa, M.: Hierarchical predictive control of integrated wastewater treatment systems. Control Engineering Practice **16**(6), 751–767 (2008)
3. Grieder, P., Parrilo, P.A., Morari, M. Robust receding horizon control—analysis and synthesis. In: Proceedings of the 42th IEEE Conference on Decision and Control, Hawaii (2003)
4. Kerrigan, E.C., Maciejowski, J.M.: Robust feasibility in model predictive control: necessary and sufficient conditions. In: Proceedings of the 40th IEEE Conference on Decision and Control, Florida (2001)
5. Langson, W., Chryssochoos, I., Rakovic, S.V., Mayne, D.Q.: Robust model predictive control using tubes. Automatica **40**(1), 125–133 (2004)
6. Mayne, D.Q., Seron, M.M., Rakovic, S.V.: Robust model predictive control of constrained linear systems with bounded disturbances. Automatica **41**(2), 219–224 (2005)
7. Bemporad, A., Mosca, E.: Fulfilling hard constraints in uncertain linear systems by reference managing. Automatica **34**(4), 451–461 (1998)
8. Angeli, D., Casavola, A., Mosca, E.: On feasible set-membership state estimators in constrained command governor control. Automatica **37**(1), 151–156 (2001)
9. Bemporad, A., Casavola, A., Mosca, E.: A predictive reference governor for constrained control systems. Comput. Ind. **36**, 55–64 (1998)
10. Brdys, M.A., Ulanicki, B.: Operational Control of Water Systems: Structures, Algorithms and Applications. Prentice Hall Int, New York, London, Toronto, Sydney, Tokyo (1995)
11. Chisci, L., Rossiter, J.A., Zappa, G.: Systems with persistent disturbances: predictive control with restricted constraints. Automatica **37**(7), 1019–1028 (2001)
12. Brdys, M.A., Chang, T.: Robust model predictive control under output constraints. In: Proceedings of the 15th Triennial IFAC World Congress, Barcelona (2002)
13. Brdys, M.A., Chang, T.: Robust model predictive control of chlorine residuals in water systems based on a state space modeling. In: Ulanicki, B., Coulbeck, B., Rance, J. (eds.) Water Software Systems: Theory and Applications, vol. 1, pp. 231–245. Research Studies Press Ltd., Baldock, Hertfordshire, England (2002)
14. Tran, V.N., Brdys, M.A.: Robustly feasible optimizing control of network systems under uncertainty and application to drinking Water Distribution Systems. In: Proceedings of the 12th IFAC Symposium on Large Scale Complex Systems: Theory and Applications, Lille, 2010
15. Tran, V.N., Brdys, M.A. Optimizing control by robustly feasible model predictive control and application to drinking water distribution systems. J. Artif. Intell. Soft Comput. Res. **1**(1) (2011)
16. Brdys, M. A., Tran, V. N., Kurek, W. Safety zones based robustly feasible model predictive control for nonlinear network systems. In: 18th IFAC World Congress on Invited session on Advances in Intelligent Monitoring, Control and Security of Critical Infrastructure Systems, Milano, Italy, 28 Aug–2 Sept 2011
17. Findeisen, W., Bailey, F.N., Brdys, M.A., Malinowski, K., Tatjewski, P., Wozniak, A.: Control and Coordination in Hierarchical Systems. Wiley, London-Chichester-New York (1980)

18. Picasso, B., De Vito, D., Scattolini, R., Colaneri, P.: An MPC approach to the design of two-layer hierarchical control systems. Automatica **46**(5), 823–831 (2010)
19. Wang, J., Grochowski, M., Brdys, M.A.: Analysis and design of softly switched model predictive control. In: Proceedings of the 16th World IFAC Congress, Prague, 4–8 July 2005
20. Liberzon, D.: Switching in Systems and Control. Birkhauser, Boston (2003)
21. Brdys, M.A., Wang, J.: Invariant set-based robust softly switched model predictive control. In: Proceedings of the 16th Triennial IFAC World Congress, Prague (2005)
22. Wang, J., Brdys, M.A.: Supervised robustly feasible soft switching model predictive control with bounded disturbances. In: Proceedings of the 6th IEEE Biennial World Congress on Intelligent Control and Automation (WCICA'06), Dalian, China (2006a)
23. Tran, V.N., Brdys, M.A.: Softly switched robustly feasible model predictive control for nonlinear network systems. In: 13th IFAC Symposium on Large Scale Complex Systems: Theory and Applications, Shanghai, 7–10 July 2013
24. Wang, J., Brdys, M.A.: Softly switched hybrid predictive control. In: Proceedings of the IFAC International Workshop on Applications of Large Scale Industrial Systems (ALSIS'06), Helsinki (2006b)
25. Kurek, W., Brdys, M.A.: Adaptive multi objective model predictive control with application to DWDS. In: Proceedings of 12th IFAC Symposium on Large Scale Complex Systems: Theory and Applications, Lille, France, 11–14 July 2010
26. Ewald, G., Brdys, M.A.: Model predictive controller for networked control systems. In: Proceedings of 12th IFAC Symposium on Large Scale Complex Systems: Theory and Applications, Lille, France, 11–14 July 2010
27. Brdys, M. A., Tatjewski, P.: Iterative Algorithms for Multilayer Optimizing Control. World Scientific Publishing Co., Pte. Ltd., Main Street, River Edge, New York, USA; Imperial College Press, London, Singapore (2005)
28. Scattolini, R.: Architectures for distributed and hierarchical model predictive control—a review. J. Process Control **19**, 723–731 (2009)
29. Rawlings, J.B., Mayne, D.Q.: Model Predictive Control: Theory and Design. Nob Hill Publishing, LLC, Madison (2009)
30. Tichy, P., Slechta, P., Staron, R.J., Maturana, F.P., Hall, K.H.: Multiagent technology for fault tolerance and flexible control. IEEE Trans. Syst. Man Cybern. Part C Appl. Rev. **36**(5), 700–705 (2006)
31. Amigoni, F., Gatti, N.: A formal framework for connective stability of highly decentralized cooperative negotiations. Auton. Agents Multi-Agent Syst. (Springer), **15**(3), 253–279 (2007)
32. Shoham, Y., Leyton-Brown, K.: Multiagent Systems: Algorithmic, Game-Theoretic, and Logical Foundations. Cambridge University Press, Cambridge (2008)
33. Keil, D., Goldin, D. Indirect Interaction in Environments for Multiagent Systems. In: Weyns, D., Parunak, V., Michel, F. (eds.) Environments for multiagent systems II. LNCS 3830, Springer, Berlin (2006)
34. Fregene, K., Kennedy, D,C., Wang, D.W.: Toward a system—and control—oriented agent framework. IEEE Trans. Syst. Man Cybern. Part B Cybern. **36**(5), 999–1012 (2005)
35. Gernaey, K.V., Jorgensen, S.B.: Benchmarking combined biological phosphorus and nitrogen removal wastewater treatment processes. Control Eng. Pract. **12**(3), 357–373 (2004)
36. Weijers, S.: Modelling, Identification and control of Activated Sludge Plants for Nitrogen Removal. Veldhove, The Netherlands (2000)
37. Duzinkiewicz, K., Brdys, M., Kurek, W., Piotrowski, R.: Genetic hybrid predictive controller for optimized dissolved oxygen tracking at lower control level. IEEE Trans. Control Syst. Tech. **17**(5), 1183–1192 (2009)
38. Piotrowski, R., Brdys, M.A., Konarczak, K., Duzinkiewicz, K., Chotkowski, W.: Hierarchical dissolved oxygen control for activated sludge processes. Control Eng. Pract. **16**(1), 114–131 (2008)
39. Zubowicz, T., Brdys, Mietek A.: Decentralized oxygen control in multi-zone aerobic bioreactor at wastewater treatment plant. In: Proceedings of 12th IFAC Symposium on Large Scale Complex Systems: Theory and Applications, Lille, France, 11–14 July (2010a)

40. Zubowicz, T., Brdys, Mietek A., Piotrowski, R.: Intelligent PI controller and its application to dissolved oxygen tracking problem. J. Autom. Mob. Robot. Intell. Syst. **4**(3) (2010b)
41. Boulos, P.F., Lansey, K.E., Karney, B.W.: Comprehensive Water Distribution Systems Analysis Handbook. MWH Soft, Inc., Pasadena (2004)
42. Arminski, K., Brdys, M.: Robust monitoring of water quality in drinking water distribution systems. In: 13th IFAC Symposium on Large Scale Complex Systems: Theory and Applications, Shanghai, China, 7–10 July 2013
43. Ewald, G., Kurek, W., Brdys, M.A.: Grid implementation of parallel multi-objective genetic algorithm for optimized allocation of chlorination stations in drinking water distribution systems: Chojnice case study. IEEE Trans. Syst. Man Cybern. Part C Appl. Rev. **38**(4), 497–509 (2008)
44. Prasad, T.D., Walters, G.A, Savic, D.A.: Booster disinfection of water supply networks: multiobjective approach. J. Water Resour. Plan. Manage. **130**(5), 367–376 (2004)
45. Brdys, M. A., Puta, H., Arnold, E., Chen, K., Hopfgarten, S.: Operational control of integrated quality and quantity in water systems. In: Proceedings of the IFAC/IFORS/IMACS Symposium LSS, London (1995)
46. Ostfeld, A., Salomons, E., Shamir, U.: Optimal operation of water distribution systems under water quality unsteady conditions. In: 1st Annual Environmental & Water Resources Systems Analysis (EWRSA) Symposium, A.S.C.E. Environmental and Water Resources Institute (EWRI) Annual Conference, Roanoke (2002)
47. Trawicki, D., Duzinkiewicz, K., Brdys, M.A.: Hybrid GA-MIL algorithm for optimisation of integrated quality and quantity in water distribution systems. In: Proceedings of the World Water and Environmental Resources Congress—EWRI2003, Philadelphia, 22–26 June 2003
48. Sakarya, A., Mays, L.W.: Optimal operation of water distribution system pumps with water quality considerations. ASCE J. Water Resour. Plan. Manage. **126**(4), 210–220 (2000)
49. Brdys, M.A., Chang, T., Duzinkiewicz, K., Chotkowski, W.: Hierarchical control of integrated quality and quantity in water distribution systems. In: Proceedings of the A.S.C.E. 2000 Joint Conference on Water Resources Engineering and Water Resources Planning and Management, Minneapolis, Minnesota 30 July–2 August (2000)
50. Duzinkiewicz, K., Brdys, M.A., Chang, T.: Hierarchical model predictive control of integrated quality and quantity in drinking water distribution systems. Urban Water J. **2**(2), 125–137 (2005)
51. Brdys, M.A., Huang, X., Lei, Y.: Two times-scale control of integrated quantity and quality in drinking water distribution systems. In: 13th IFAC Symposium on Large Scale Complex Systems: Theory and Applications, Shanghai, China, 7–10 July 2013
52. Polycarpou, M., Uber, J., Wang, Z., Shang, F., Brdys, M.A.: Feedback control of water quality. IEEE Control Syst. Mag. **22**, 68–87 (2001)
53. Zierolf, M.L., Polycarpou, M.M., Uber, J.G.: Development and auto-calibration of an input–output model of chlorine transport in drinking water distribution systems. IEEE Trans. Control Syst. Technol. **6**(4), 543–553 (1998)
54. Chang, T., Duzinkiewicz, K., Brdyś, M.A.: Bounding approach to parameter estimation without priori knowledge on model structure error. In: Proceedings of the IFAC 10th Symposium Large Scale Systems: Theory and Applications, Osaka, Japan, 26–28 July 2004

Algorithms and Tools for Risk/Impact Evaluation in Critical Infrastructures

Chiara Foglietta, Stefano Panzieri and Federica Pascucci

Abstract Critical Infrastructures (CIs) are complex system of systems due to the existence of interdependencies that are not readily visible but very often play a central role. Hence, modelling and simulating networks of critical infrastructures is a crucial point to support operators and service providers in mitigating risks. In this chapter we describe how a general approach, called Mixed Holistic Reductionist, is able to join a holistic impact evaluation with a reductionist modelling of cascade effects in order to integrate situation assessment and interdependency evaluation. This approach can be realized using two different symbiotic techniques: a vertical simulation of a specific behaviour of an infrastructure and an interdependency agent-based simulator able to take into account cascading effects along several interdependency links.

Keywords Risk assessment · Critical infrastructures modelling · Interdependencies analysis

1 Simulation of Interdependent Infrastructures

In performing a risk assessment, interdependencies are a key point that must be considered in order to help operators taking right decisions. To this aim, interdependencies must be included in modelling and simulating the behaviour of facilities. Infrastructure operators are expert in responding to or mitigating outages or minor disruptions originally inside the infrastructure itself. Nevertheless, they may not be able to mitigate risks generated from other infrastructures simply due to a lack of correct information and this can be the cause of severe disruptions.

C. Foglietta · S. Panzieri (✉) · F. Pascucci
Dipartimento Di Ingegneria, Università Degli Studi "Roma Tre", Via Della Vasca Navale, 79, 00146 Rome, Italy
e-mail: panzieri@uniroma3.it

E. Kyriakides and M. Polycarpou (eds.), *Intelligent Monitoring, Control, and Security of Critical Infrastructure Systems*, Studies in Computational Intelligence 565,
DOI 10.1007/978-3-662-44160-2_8

Conceptual simulation models in this field can be divided into two categories:

1. Holistic simulation models, that are related to a single infrastructure and are able to model the evolution of an event within the CI itself, disregarding events related to other CIs. Often such models are the product of sector's experts and try to catch complex dynamics related to the CI (e.g., the transient behaviour of an electric grid when opening a switch) [21, 22].
2. Reductionist simulation models, that are more related to cascading effects and are able to simulate interdependencies among several infrastructures [14, 19, 26, 36], eventually decomposed into smaller elements. For such simulators it is important that the different CI elements are simulated into a unifying software framework, where functional dependencies take more importance than the physical/mechanical simulation. Here the service availability concept is more important than the effective service level, and fault dynamics are propagated from element to element according to several concepts of neighbourhood to exploit all the possible modelled dependencies.

Both holistic and reductionist approaches are considered as complementary approaches capable of supporting situation assessment and evaluating consequences due to interdependencies, hence able to assess an interdependent risk. In order to realize a complete simulator, the two methods must be considered at the same time but at different levels of abstraction. We could say that the holistic simulators are the ones that must take into account, and recover, all the non-emerging behaviours that the reductionist decomposition is not able to produce but are of some interest in the impact assessment. Hence, the integration that is needed between several simulators of specific infrastructures is not a federation of several simulators [15]: is not necessary to manage the contemporary evolution of different intercommunicating simulators because they can give only a misleading output that barely represents a possible evolution of the complex system. The integration must be realized starting from the reductionist level using the holistic models to evaluate single specific behaviours that can impact the reductionist decomposition.

The Mixed Holistic Reductionist approach will be defined as a general guideline to describe large complex systems like CIs, their interdependencies, and the impact of faults on these systems. This approach is generic and it can include several techniques and methods, to assess consequences and effects of faults.

The Mixed Holistic Reductionist approach (MHR in Fig. 1) [7] consists of two layers. The first one (upper) is obtained considering each infrastructure as a whole and evaluating the impact of faults or services using domain simulators. We called this layer "holistic situation assessment". The second (lower) can be considered a reductionist impact assessment layer that is built of experts' reviews and tries to assess interdependencies and how faults and their consequences are reflected on other facilities.

This approach tries to guide experts in order to consider several events and not only the simple mechanical faults. In fact, in large plants, the simple physical security is not enough to protect CIs. The physical security must be assessed

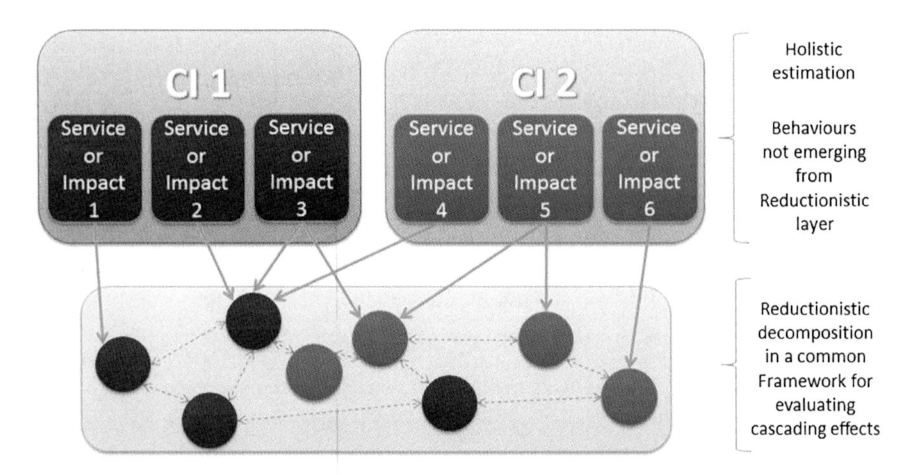

Fig. 1 MHR modelling

together with cyber vulnerabilities [9, 13] and threats and integrating outputs of firewalls, Intrusion Detection Systems (IDS) [2, 37] and also information coming from national agencies and international institutions, in order to provide business continuity and the best Quality of Service (QoS) towards customers.

The aim of the approach is to consider infrastructures not as stand-alone elements, but as complex systems that also need to consider the supply chain of services and goods, in order to guarantee a sufficiently QoS. Business Continuity standards are perfectly centred on this crucial point, considering especially operator safety. The standards BS-25999 and, recently, ISO-22301 define "Business Continuity" as "Strategic and tactical capability of the organization to plan for and respond to incidents and business disruptions in order to continue business operations at an acceptable predefined level" [31].

In the next sections, the approach is detailed in a more accurate way.

2 Holistic Cyber Impact Evaluation

The definition of the term holistic is characterized by the comprehension that the parts of something are intimately interconnected and explicable only by reference to the whole. In the field of CI protection, evaluating faults inside the facilities where they are born using CI-related tools is a holistic point of view.

Using such a point of view, a situation assessment can be realized considering several technological and organizational aspects and using techniques and methods specific to each facility. In the following Sections, some particular events/infrastructures will be described/analysed to better understand the holistic impact evaluation:

- Management of alarms, usually collected using SCADA (Supervisory Control And Data Acquisition) software and then shown to operators in order to support decisions;
- Management of physical security information, for detecting unauthorized accesses to specific areas, also using data mining algorithms;
- Evaluation of the Quality of Services (QoS) toward customers of the infrastructure, using simulators for analysing transients and outages after the fault [4, 5], such as load-flow simulation for power grids or NS-2 [5], OMNET++ [27] and Netbed [29] for telecommunication networks;
- Detection/spreading of cyber-attacks [1, 3, 6, 18, 23, 25, 30], especially worms and viruses spreading, such as Red-Code [32], Stuxnet [10] or Duqu [28] worm, through mathematical representation, such as the two-factor models;
- Use of information coming from international and national agencies, such as CERTs and other institutions, to integrate cyber-related data coming out from other infrastructures.

The aim of this analysis is the evaluation of the availability of both single elements of an infrastructure and the infrastructure as a whole, and to provide goods and services to other elements/customers at an acceptable predefined level, in the presence of faults, failures or any kind of anomaly situations.

3 Reductionist Cascading Impact with CISIA

At a lower level, analysing a domino effect in CIs means evaluating interdependencies and how these interdependencies will be used to propagate faults. Often interdependencies are not clear and well defined but they are made clear during particular situations, usually critical events. Such connection among facilities could be better detected considering simple equipment and their functionality, in terms of resources and propagated failures. In fact, at low level of abstraction, it is simple analysing the input and the output resources of each device for each infrastructure. Hence it will be easy to find not immediately evident interdependencies and connections.

After that, supposing that the system of systems has been implemented in a unifying framework, impact assessment will consist of the evaluation process of faults and failures propagation. This process can be realized at each level of abstraction, but it has an intrinsic meaning considering the equipment and devices of each infrastructure, such as router or gateway, switch or line, building or hospital. This level is called reductionist because, even if the decomposition is only functional, it is a small grain decomposition of the whole system of systems able to catch domino phenomena. Several simulators can be used at this level and in the following we analyse some of them.

CISIA (Critical Infrastructure Simulation by Interdependent Agents) [7, 8] is an agent-based simulator for modelling CI interdependencies. It was born with the aim to analyse failure propagation and performance degradation in systems composed of different, heterogeneous and interdependent infrastructures. CISIA can be applied in order to evaluate the interdependencies and, especially, the cascading and domino effects on heterogeneous equipment.

Other methodologies can be included aiming to model the interconnection among several infrastructures, such as I2Sim [35]. I2Sim is a simulation environment used to study interdependency problems in CI Protection and it allows to model physical and geographical interdependencies. The key element of I2Sim is the production cell, a functional unit that relates a set of resource levels as input to a particular resource level as output. Only one kind of resource can be associated to a single production cell. By considering a proper set of production cells it is possible to model a scenario consisting of different infrastructures and build loose or light coupling relations to model interdependencies.

CISIA has, as a key feature, the ability to deal with fuzzy numbers, in particular triangular fuzzy numbers. In general, working with infrastructures, especially considering interdependencies, is characterized by uncertainty. So, all quantities described inside CISIA are all triangular fuzzy numbers, in order to consider doubt, imprecision and uncertainty. This characteristic is more useful in context where inputs are uncertain. In fact, output data of impact evaluation and awareness contains uncertainty and it usually is described using intervals of probability. The triangular fuzzy numbers are an abstraction, helping to manage uncertain quantities.

In CISIA, each facility is modelled with macro-components at a high level of abstraction. Each macro-component is defined as an agent. Each agent has the same structure based on few common quantities, representing the state or memory of the agent:

- Operative Level (OL): the ability of the agent to perform its required job. It is an internal measure of the potential production/service; if the operative level is 100 % it does not mean that it is providing the maximum value but that it could, if necessary.
- Requirements (R): what the node needs to reach the maximum operative level.
- Faults (F): the level of failure that affects the agent, for each type of fault.

Agent inputs and outputs are necessary in order to perform interactions among agents. There are three kinds of inputs and, similarly, three kinds of outputs:

1. Induced/propagated faults: faults propagated to the considered agent from its neighbourhoods and from the considered agent to its neighbourhood, described in terms of type and magnitude.
2. Input/output requirements: amount of resources requested by/to other objects.
3. Input/output operative levels: the operative level of those objects whose resources are used in it, and the operative level of the object itself.

In CISIA, the agent dynamic is described as an input/output model among the previously listed quantities. This description of agent's behaviour is highly abstracted but it is rich enough to leave the experts to represent the model dynamics in the most appropriate way.

The relations among agents are based on their interdependencies, and incidence matrices describe them. In fact each matrix is able to spread a different type of interdependency, following Rinaldi characterization [24] among physical, geographical and cyber connections.

4 Situation Assessment Using CISIA

CISIA has been validated inside the FP7 MICIE project [12, 34] and has been adopted in the FP7 CockpitCI project [33]. Using CISIA, the consortium realized a distributed risk prediction tool [20], able to monitor and evaluate cascading and domino effects among two interconnected infrastructures, a power grid and a telecommunication network.

The simulator has been successfully applied during the implementation of the MICIE test-bed. First of all, infrastructures have been studied as separated from each other and components have been modelled as services and equipment. Then, considering interdependencies between such elements, a system of systems has been built and simulated using CISIA, as shown in Fig. 2. This software tool can evaluate consequences and effects of several kinds of faults spreading along the interconnected systems. In MICIE test-bed, failures in equipment, geographically, physically, and logically connected, have been considered and their propagation through this network studied.

CISIA is able to cope with both the reductionist and holistic evaluation, and because the main focus of MICIE project was on physical damages and mechanical faults on real equipment, all the holistic work (mainly related to the evaluation of the availability of some telecommunication connections with Remote Terminal Units—RTUs) has been performed inside the simulator. In order to share information among different CIs, the actual and estimated QoS of a CI are transmitted to another CISIA tools placed in a different control room.

In the CockpitCI project, the aim is to include also cyber interdependencies. For cyber-attacks, we consider the spreading of malicious intrusions and attacks on telecommunication networks [16, 17] in order to analyse possible impacts on real devices. As an example, a DoS (Denial of Service) attack on specific telecommunication equipment is included in CISIA, as unavailability of the equipment itself, with a well-defined timed behaviour. A DoS attack also affects the links directly connected to the equipment, generating packet congestions. As a second example, worms or a viruses are also simulated in CISIA, using the SIR model. All agents have three possible behaviour: susceptible to the worm (S), infected (I) and those who have recovered (R) and are immune. Such approach has been implemented by a particular kind of fault and a specific spreading algorithm.

The CISIA entity is also able to handle the outputs of other modules, providing some awareness coming from an external processing, as depicted in Fig. 3. For example, in CockpitCI, a Cyber-Physical Awareness is pre-computed, combining both cyber and physical threats and evaluating possible impacts.

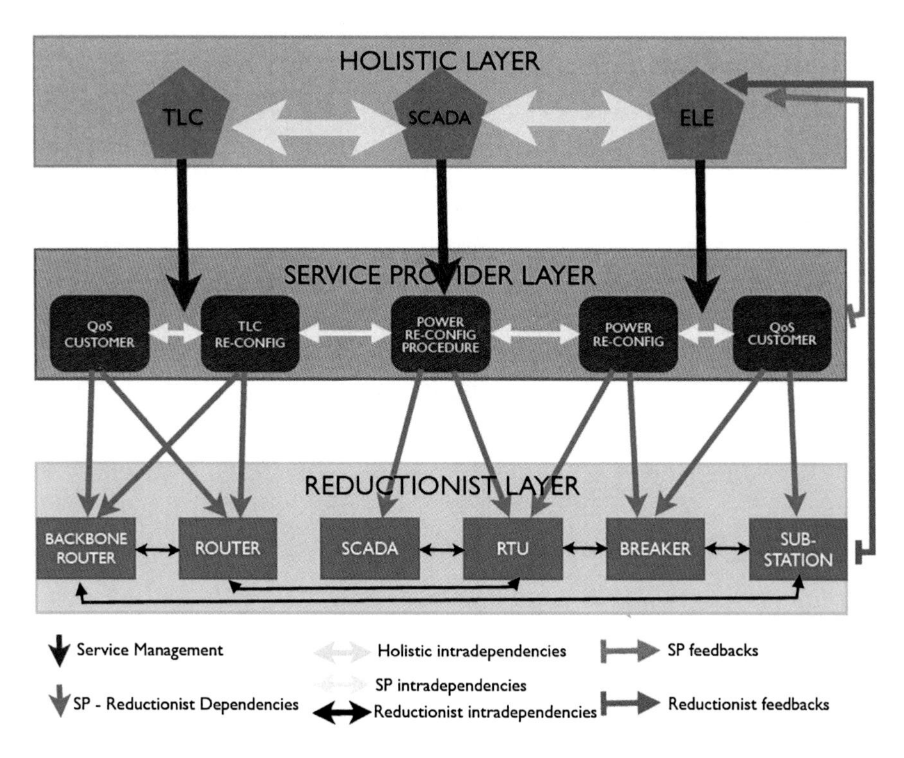

Fig. 2 CISIA three levels representation

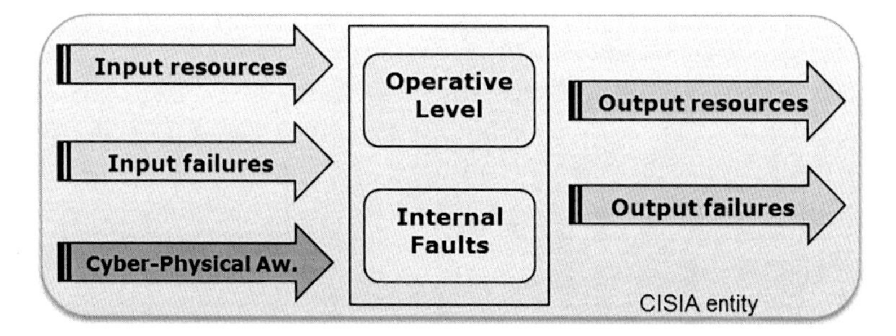

Fig. 3 CISIA with cyber-physical awareness inputs coming from external module

The holistic impact is sometimes described with some uncertainty. This is the case, for example, of a cyber-attack causing some probable unavailability. CISIA is able to manage this uncertainty thanks to triangular fuzzy numbers. In Fig. 4, a scenario is described where the output of the cascading effects evaluation is used by different people for several purposes. The cascading effects have been defined using CISIA.

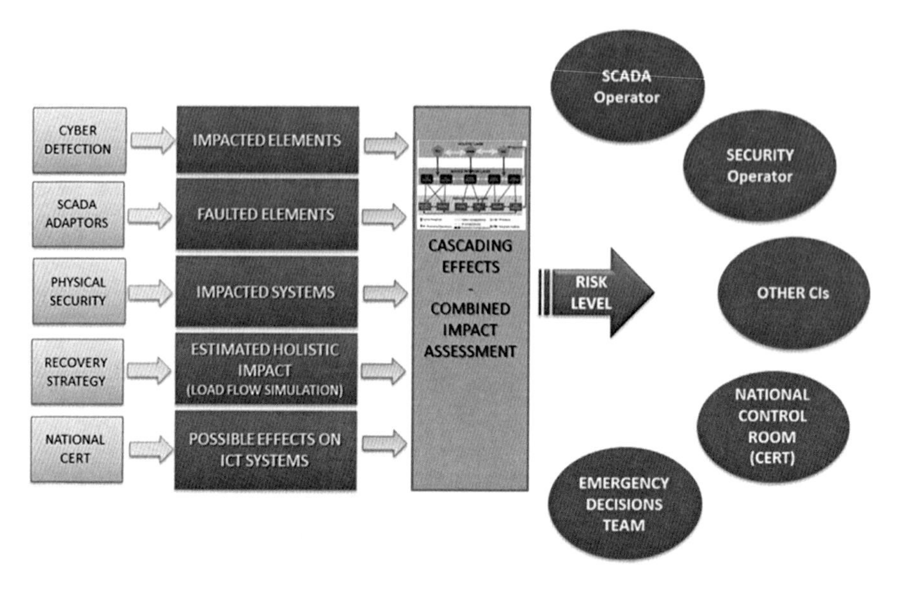

Fig. 4 Impact evaluation using the CISIA tool

5 Risk Assessment for Critical Infrastructures

Risk is traditionally tied to an outcome, such as the loss of productivity, the financial impact as gain or loss, or the time spent to restore the system, in order to provide a pre-defined quality of services towards customers.

In CI context, the risk is also related to the consequences of an adverse event. For consequences we mean not just the financial ramifications of an incident but also the impact evaluation of the threats that would result in economic instability, damage and also death.

CI systems have a great number of factors that can be impacted by system failure, some of which include catastrophic damage, significant revenue loss, environmental impact, and national economies. For this reason, the typical equation for risk definition is traditionally written as:

$$Risk = Threat \times Vulnerability \times Consequence$$

where the risk is the Cartesian product of all important threats, weaknesses and consequences. In CI, threats are not just related to the physical system, but also to the telecommunication component, the so-called cyber threats. Unlike hazards, threats are generally not predictable, even when some historical information is known about them. Threats to complex systems can come from multiple sources, such as natural disasters, environmental conditions, mechanical failures, and inadvertent actions of an authorized user, but also from hostile governments, terroristic groups, disgruntled employees, and malicious intruders.

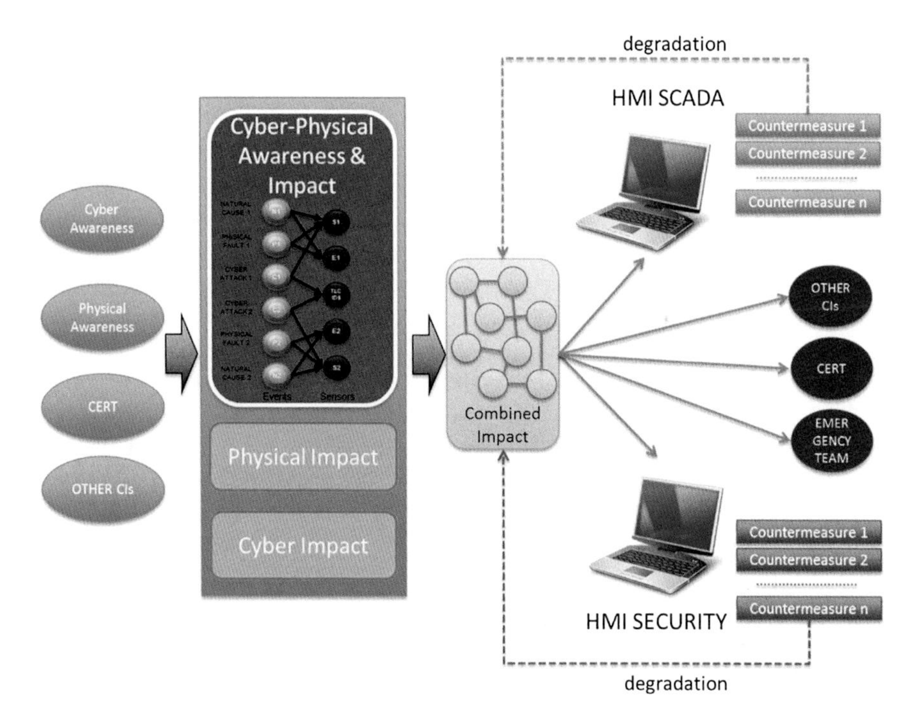

Fig. 5 process starting from impact evaluation ending to countermeasures selection

Vulnerabilities are weaknesses or inadequacies in a system that, if exploited by an attacker, could cause harm or damage to the system. The ability to detect vulnerabilities is a mandatory step in order to build strategies and create defensive activities to protect future state architecture. Vulnerabilities are usually classified into three main categories:

1. Vulnerabilities inherent in the products, such as SCADA system software;
2. Vulnerabilities caused during the installation, configuration and maintenance of the system;
3. The lack of adequate protection because of poor network design or configuration.

As a first step, see [38], we can describe vulnerabilities as a mapping between the set of possible threating events and the set of outcomes. In this view, each statement of vulnerability must be in reference to some degrees of loss or adverse outcomes. Since vulnerability was defined to be a mapping from cause to consequences, it is thus important to construct scenarios that permit meaningful statements of vulnerability.

6 Countermeasures Selection

The Risk Prediction Tool aims to show the impact assessment on equipment and, also, on services. The impact evaluation is depicted in the left side of Fig. 5, and it is a process which has as input information and data coming from the cyber and physical awareness modules, but also from other interconnected CIs and from national or international agencies, such as national CERTs. The impact assessment is a combination of several consequences: effects on physical equipment, possible outcomes on ICT infrastructure, and also by-products of physical faults on ICT and of cyber-attacks on real instruments.

The output of this module is the combined impacts on all the considered agents, as equipment and service in order to understand early if some services or hardware could be affected by failures. The right side of the picture shows how the choice of countermeasures in each control room can easily change the combined impacts of ongoing failures. Therefore, the countermeasures, applied for example by power grid operators, generate changes in the interconnected infrastructures, allowing possible unstable loops because these interconnected facilities could be applying other countermeasures generating dangerous situations. Further information on applied countermeasures or on combined impacts on equipment and services that are not controlled by operator must be exchanged increasing the complexity of the algorithm. In fact, each Risk Prediction Tool must include this new information on its process.

In Fig. 5, the module called Cyber-Physical Awareness is depicted. This module is able to fuse data and information coming from heterogeneous fields [11], in order to obtain by-products on equipment, as possible performance degradation on tool, but also possible causes of the considered failures. In this module, the impact of these failures must not be considered; otherwise the domino and cascading effects generate errors.

7 Conclusion

The Mixed Holistic Reductionist Approach is a simple guideline to model and simulate CIs behaviour, especially in huge-impact events. In fact, each facility is not able to predict the behaviour of the infrastructure itself, without considering cascading and domino effects.

This approach is described as a two-phase guide: first, the holistic view of each single infrastructure, and then, analysing the interconnections at the reductionist level of abstraction. The aim is to maintain the advantages of each of these two approaches, reducing the possible disadvantages.

MHR is a possible guide to model infrastructures, as a whole system. The application of different simulators can reduce or increase the level of detail and the capability to work in real time contexts, usually directly connected to control rooms. This approach was applied inside the FP7 MICIE project, both using NS2 and CISIA tools, in order to model interdependencies and services to customers. In fact, the main aim of the project is the effects of critical events on QoS to customer services, due to existences of interdependencies, focusing on mechanical faults.

Possible future updates regard the integration of explicit countermeasures and reactions inside the MHR approach in order to mitigate risk and outages. In general, the output of these modelling techniques is another operator interface able to show other details thanks to interconnection to similar objects, like the same tool but placed in other control rooms, or national and international agencies. The following idea is to place all together in single operator interface, where each reacting strategy is shown only if the supply chain (i.e., service or providers of goods, national governments, etc) can guarantee at least a predefined quality level.

References

1. Chabukswar, R., Sinopoli, B., Karsai, G., Giani, A., Neema, H., Davis, A.: Simulation of network attacks on SCADA systems, In: First Workshop on Secure Control Systems, 2010
2. Chappell, L., Combs, G.: Wireshark network analysis: the official Wireshark certified network analyst study guide, Chappell University 2010
3. Chunlei, W., Lan, F., Yiqi, D.: A simulation environment for SCADA security analysis and assessment. In: Proceedings of the International Conference on Measuring Technology and Mechatronics Automation, vol. 1, pp. 342–347 (2010)
4. Ciancamerla, E., Minichino, M., Rosato, V., Vicoli, G., SCADA systems within CI interdependency analysis: cyber attacks, resilience and quality of service. In: Workshop on Experimental Platforms for Interoperable Public Safety Communications—Joint Research Centre (JRC), Ispra, Italy. 10, 11 October 2011
5. Ciancamerla, E., Foglietta, C., Lefevre, D., Minichino, M., Lev, L., Shneck, Y.: Discrete event simulation of QoS of a SCADA system interconnecting a Power grid and a Telco network. In: 1st IFIP TC11 International Conference on Critical Information Infrastructure Protection 2010 World Computer Congress 2010 Proceedings, Springer, Brisbane (2010) ISSN 1868-4238
6. Davis, C.M., Tate, J.E., Okhravi, H., Grier, C., Overbye, T.J. Nicol, D.: SCADA cyber security testbed development. In: Proceedings of the 38th North American Power Symposium, pp. 483–488 (2006)
7. De Porcellinis, S., Panzieri, S., Setola, R.: Modelling critical infrastructure via a mixed holistic reductionistic approach. Int. J. Crit. Infrastruct. **5**(1/2), 86–99 (2009)
8. De Porcellinis, S., Panzieri, S., Setola, R., Ulivi, G.: Simulation of heterogeneous and interdependent critical infrastructures. Int. J. Crit. Infrastruct. 4(1/2), 110–128 (2008)
9. European Commission: Achievements and next steps: towards global cyber-security', communication from the commission to the European Parliament, the council, the European economic and social committee and the committee of the regions on critical information infrastructure protection (2011)
10. Falliere, N., O'Murchu, L., Chien, E.: W32.Stuxnet Dossier, Symantec, Mountain View, California. www.symantec.com/content/en/us/enterprise/media/securityresponse/whitepapers /w32stuxnetdossier.pdf (2011)
11. Foglietta, C., Gasparri, A., Panzieri, S.: Networked evidence theory framework for critical infrastructure modelling. In: Proceedings of Sixth Annual IFIP WG 11.10 International Conference on Critical Infrastructure Protection (2012)
12. FP7 MICIE project. http://www.micie.eu

13. Genge, B., Nai Fovino, I., Siaterlis, C., Masera, M.: Analyzing Cyber-Physical Attacks on Networked Industrial Control Systems. Crit. Infrastruct. Protect. **367**, 167–183 (2011)
14. Haimes, Y., Jiang, P.: Leontief-based model of risk in complex interconnected infrastructures. J. Infrastruct. Syst. **1**, 1–12 (2001)
15. Hemingway, G., Neema, H., Nine, H., Sztipanovits, J., Karsai, G.: Rapid synthesis of high-level architecture-based heterogeneous simulation: a model-based integration approach In: Proceedings of simulation, p 0037549711401950, March 17, 2011
16. Huitsing, P., Chandia, R., Papa, M., Shenoi, S.: Attack taxonomies for the Modbus protocols. Int. J. Crit. Infrastruct. Prot. **1**, 37–44 (2008)
17. Modicon, I.: Modicon modbus protocol reference guide. http://modbus.org/docs/PI_MBUS_300.pdf (1996)
18. Nai Fovino, I., Carcano, A., Masera, M., Trombetta, A.: An experimental investigation of malware attacks on SCADA systems. Int. J. Crit. Infrastruct. Protect. **2**(4) 139–145 (2009)
19. Oliva, G., Panzieri, S., Setola, R.: Fuzzy dynamic input-output inoperability model. Int. J. Crit. Infrastruct. Protect. **4**(3–4), pp. 165–175 (2011)
20. Oliva, G., Panzieri, S., Setola, R.: Online distributed interdependency estimation for critical infrastructures. In: 50th IEEE Conference on Decision and Control, 2011
21. OPNET: OPNET network simulation tools. http://www.opnet.com (2012)
22. PowerWorld: PowerWorld simulator. http://www.powerworld.com (2012)
23. Queiroz, C., Mahmood, A., Hu, J., Tari, Z., Yu, X.: Building a SCADA security testbed. In: Proceedings of the Third International Conference on Network and System Security, pp. 357–364 (2009)
24. Rinaldi, S.M.: Modeling and simulating critical infrastructures and their interdependencies. In: Proceedings of the 37th Annual Hawaii International Conference on System Sciences (HICSS'04)—Track 2, vol. 2 (HICSS '04), Vol. 2. IEEE Computer Society, Washington, DC, USA, 20054.1 (2004)
25. Rios, B., McCorkle, T.: 100 Bugs in 100 days: an analysis of ICS (SCADA) Software. DerbyCon 2011, Session (2011)
26. Setola, R., De Porcellinis, S., Sforna, M.: Critical infrastructure dependency assessment using input-output inoperability model, Int. J. Crit. Infrastruct. **2**, 170–178 2009
27. Varga, A.: The OMNeT++ discrete event simulation system. In: Proceedings of the European simulation multiconference (ESM'2001) (2001)
28. W32.Duqu; The precursor to the next Stuxnet, Symantec. White Paper, (2011)
29. White, B., Lepreau, J., Stoller, L., Ricci, R., Guruprasad, S., Newbold, M., Hibler, M., Barb, V., Joglekar, A.: An integrated experimental environment for distributed systems and networks. In: Proceedings of the Fifth Symposium on Operating Systems Design and Implementation, pp. 255–270 (2002)
30. Zhu, B., Joseph, A., Sastry, S. (2011) Taxonomy of Cyber Attacks on SCADA Systems. In: Proceedings of CPSCom 2011: The 4th IEEE International Conference on Cyber, Physical and Social Computing, Dalian, China, October 19-22, 2011
31. http://www.bsigroup.com/
32. Zou, C.C., Gong, W., Towsley, D.: Code red worm propagation modeling and analysis. In: Vijay, A., (ed.) Proceedings of the 9th ACM Conference on Computer and communications Security (CCS '02), 138-147. ACM, New York, NY, USA, (2002)
33. FP7 CockpitCI project, Deliverable D5.1, "system requirements"
34. FP7 MICIE project, Deliverable 4.1.1 "MICIE ICT system requirements"
35. Rahman, M., Armstrong, D.M., Marti, J.: I2Sim: A matrix-partition based framework for critical infrastructure interdependencies simulation. In: Electric Power Conference (EPEC), Vancouver (2008)
36. Leontief, W.: Input-Output Economics. Oxford University Press, New York (1966)
37. Ettercap suite for man-in-the-middle attacks on LANs. http://ettercap.sourceforge.net/
38. Kaplan, S., Garrisck, B.J.: On the quantitative definition of risk. Risk Analysis **1**(1), 11–27 (1981)

Infrastructure Interdependencies: Modeling and Analysis

Gabriele Oliva and Roberto Setola

Abstract In this chapter some of the most well established approaches to model Critical Infrastructure Interdependencies are discussed. Specifically the holistic methods, where the interaction among infrastructures is seen from a very high level of abstraction, are compared with agent-based models, where the dependency phenomena that may arise among subsystems are considered both in terms of functional and topological relations. In order to better clarify the different approaches, the Input–Output Inoperability Model is discussed as one of the most representative Holistic methodologies; Agent-based methods are then discussed with particular reference to the Agent-Based Input–Output Inoperability Model, an extension of the Input–Output Inoperability Model, developed by the authors. The increased level of detail clashes with the lack of adequate quantitative data required to tune the models, which are typically assessed based on economic exchange among infrastructures; in order to partly overcome such an issue, an Input–Output methodology based on the theory of Fuzzy Systems is discussed. Finally, some conclusive remarks and open issues are collected.

Keywords Critical infrastructures · Interdependency modeling · Input–output models · Leontief · Agent-based systems · Fuzzy systems

1 Introduction

In order to estimate the threats and to assess the vulnerabilities to modern infrastructures, as well as to identify the possible policies and countermeasures to failures and malicious attacks, it is becoming more and more mandatory to assume

G. Oliva · R. Setola (✉)
University Campus Bio-Medico of Rome, Via A. del Portillo 21, 00128 Rome, Italy
e-mail: r.setola@unicampus.it

G. Oliva
e-mail: g.oliva@unicampus.it

E. Kyriakides and M. Polycarpou (eds.), *Intelligent Monitoring, Control, and Security of Critical Infrastructure Systems*, Studies in Computational Intelligence 565,
DOI 10.1007/978-3-662-44160-2_9

a System of Systems perspective and to take into account the *interdependency* that may arise among different infrastructures and their subsystems. The above tasks and actions are part of what is called *Critical Infrastructure Protection* (CIP) [1], while *Critical Infrastructures* are defined as those systems and assets such that a disruption or failure affecting them "would have a serious impact on the health, safety, security or economic well-being of Citizens or the effective functioning of governments" [2].

The study of modern infrastructures starts from the consideration that, nowadays, a disruption, a failure or a malicious attack affecting a small portion or subsystem of an infrastructure may have severe effects also on very distant subsystems, as well as on other infrastructures, due to interdependency and domino effects.

One of the main reasons of such an increasing coupling among once isolated systems is the diffusion of telecommunication technologies and in particular, the pervasiveness of the cyberspace, that potentially couples each and every system connected to the world wide web.

In 2010, the discovery of the Stuxnet worm [3] became a concrete proof that cyber attacks on industrial control systems and SCADA systems were possible. Stuxnet was able to infect Windows computers used to supervise industrial control systems, and to control them. Since 2010, the amount of SCADA vulnerability disclosures and exploits exploded. Terry McCorkle and Billy Rios found 100 SCADA bugs in 100 days, thanks to free software available on-line [4].

Given the dimension of the problem, the methodologies available to perform impact assessment on such complex systems are very immature, mainly due to the difficulty to retrieve adequate quantitative data to properly set up the models, as well as to the scale and the geographical dispersion of systems and subsystems.

Although it is possible to categorize interdependency modeling methodologies according to different criteria (e.g., probabilistic or deterministic, static or dynamic, etc.) one of the most widely adopted categorizations is to divide such frameworks in *holistic* and *agent based* (or *reductionistic*) approaches.

Within the holistic paradigm, the interaction among infrastructures is considered at a very high level, typically taking into account only the (inter)dependency relations that involve whole infrastructures, without providing insights on how such dependency phenomena may be explained referring to the interaction of subsystems, components and equipment. One of the most adopted holistic frameworks is the *Input–Output Interdependency Model* (IIM) [5, 6], which is based on the input–output economic tables (i.e., the economic loss for an economic sector A of a nation due to the complete absence of the resources provided by an economic sector B) provided yearly by governments and public institutions, such as BEA for the United States of America or ISTAT for Italy.

Conversely, the approach to decompose the infrastructures into their elements with a given (and often heterogeneous) degree of detail is often referred to as *reductionistic* (meaning that the overall System of Systems is described by the interaction of small and well known composing elements) or *agent based*, meaning that such subsystems are characterized in terms of their behavior (e.g., their resource production and consumption, the degradation of their performances in the event of a failure, etc.).

However, as noted in [7, 8], the economic aspects are just one component of a complex six-dimensional phenomenon that characterizes the dependencies/interdependencies among sectors and infrastructures. To overcome such a limit in [9] it is suggested to complement economic data with operational information elicited by stakeholders and experts of the different infrastructure sectors. In this chapter the problem of modeling interdependencies among Critical Infrastructures is discussed, and the holistic and agent based methodologies are compared with particular reference to the IIM model, and to an agent-based extension of such a methodology, namely *agent-based IIM*, developed by the authors [10].

As stated above, the increased level of detail of agent-based approaches requires a non-trivial effort and very often, the lack of reliable quantitative data limits the applicability of such methodologies. Conversely, the standard IIM model is based on the economic dependencies existing among infrastructures, which are easy to obtain, but are not able to fully describe the complexity of the relation existing among coupled infrastructures, which in many cases is not completely correlated with the intensity of the economic interaction.

In order to partly overcome such an issue, in this chapter a modeling methodology developed by the authors, namely *Fuzzy IIM* [11–14], is described as a tool to assess the dependencies existing among infrastructures based on the codification of the knowledge of human experts, stakeholders and technicians in terms of *fuzzy numbers* [15]. In fact, from one side, human experts have a direct knowledge of the behavior of infrastructures and subsystems, encompassing not only economical, but also virtually every possible cause of dependency. Conversely, they typically express via qualitative and vague statements and there is a need to provide a formalism capable to handle and represent such an ambiguous information, and the theory of fuzzy sets and systems appears a reasonable choice.

The chapter is organized as follows: Sect. 2 contains a discussion on interdependency modeling and a comparison between holistic and agent-based methodologies; in Sects. 3 and 4 the IIM and the agent-based IIM frameworks are discussed, respectively; the issues related to the lack of adequate quantitative data, together with the Fuzzy IIM methodology are discussed in Sect. 5, while Sect. 6 contains some conclusive remarks.

2 Interdependency Modeling: Holistic Versus Agent-Based Approaches

In order to discuss interdependency modeling among Critical Infrastructures, let us provide some useful definitions. The *inoperability* of an infrastructure or subsystem is the inability to perform its intended function; a *failure* is a negative event which influences the inoperability of infrastructures and subsystems (i.e., a natural or an accidental event, as well as a malicious attack). An infrastructure or sub-system A is *dependent* on another infrastructure or subsystem B when a degradation of B, i.e., an increment of its level of inoperability, induces a degradation into A [16]. Obviously, A and B are *interdependent* if they are mutually dependent. These definitions are very general and, besides including evident and direct dependencies, embrace more complex behaviors, such as amplifications, domino effects and loops.

While modeling dependencies and interdependencies, there is a need to consider several, often coexisting dimensions. In [7] four not mutually exclusive classes are defined: physical, cyber, geographical and logical dependencies. Physical dependency accounts for the relations that may arise among two infrastructures A and B, when A depends on the physical outputs (i.e., goods, services, etc.) of B. Conversely, if the correct functioning of A relies on the information transmitted by B by means of the cyberspace, then A has a cyber dependency on B. A geographical dependency may arise when two elements of different infrastructures, even if not dependent from a cyber of physical point of view, are in close spatial proximity, so that particular events (e.g., a fire blast) affecting B may have direct consequences on A. A Logical dependency, finally, arises when A depends on B according to control, regulatory or other mechanisms that cannot be considered physical, geographical or cyber. The above categories have been further enriched in [17] explicitly considering also the Sociological Dependency: if a "disorder" related to human activities affects B, it may also have an effect on A, i.e., due to the arise and spreading of collective behaviors that have some negative impact on the capability of the infrastructure to correctly work.

In [18] it is emphasized that, to correctly understand the behavior of interdependent infrastructures, there is a need to consider, at the same time, the set of equipment and assets that constitute the infrastructures (Physical Layer), the subsystems that control and manage the infrastructures (Cyber Layer) and the behavior of human operators, technicians, employees, etc. (Organizational Layer). A similar kind of decomposition was also used to analyze the 2003 blackout in US and Canada [19].

As discussed above, most of the existing methodologies can be referred to as *Holistic* [20], since each infrastructure is represented as a unique, monolithic entity. Among others, the *Input–Output Inoperability Model* (IIM) [5] gained large attention. Within this class of models, however, the interactions among different infrastructures are modeled with a high level of abstraction, while the behavior of subsystems is masked (Fig. 1). Such a high level of abstraction (and simplification) does not take into account the structure and the geographical extension of the infrastructures; in fact, any critical infrastructure is a complex, geographically dispersed cluster of systems.

Considering each infrastructure as an atomic entity represents a very crude simplification that does not take into account its geographical extension and its structure. There is, therefore, the need to adopt bottom–up approaches, as largely done when dealing with scarce or ill-defined macro-scale information, like in the field of bio-complexity.

Following the bottom–up philosophy, each infrastructure can be decomposed into a set of elementary interconnected components, taking into account both intra-

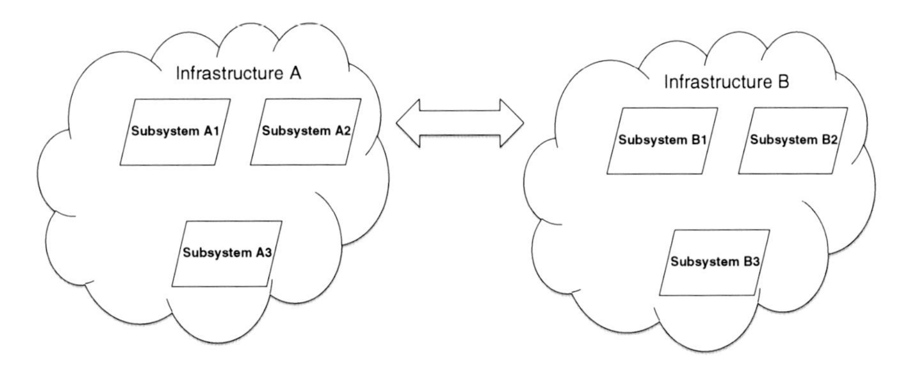

Fig. 1 Example of holistic interdependency modeling

infrastructure and cross-infrastructure dependencies and interdependencies, as shown by Fig. 2 (see also [20]).

In order to obtain more insight on the behavior of interdependent infrastructures, a first step is to represent them as complex networks composed of similar basic elements (i.e., a network of distributed and interconnected generators may represent an electrical infrastructure), inspecting emerging behaviors generated by the interconnection of such elements [21–24]. The assumption of homogeneity, however, limits the applicability of these methodologies, since in real cases infrastructures are composed of highly heterogeneous subsystems; moreover topological methods typically limit their scope to the geographical interaction of subsystems.

A step further is taken by adopting an *agent-based* perspective, focusing on a sophisticated representation of the isolated behavior of subsystems and then considering their interaction by means of simulation platforms and tools [25–27]. Specifically, as shown in Fig. 3, each subsystem can be characterized by modeling its input–output behavior, i.e., how the subsystem produces its outputs (i.e., resources but also failures, thus modeling the propagation and spread of negative events such as a fire blast) based on the received inputs (again, resources and failures). The complexity of *agent-based* models therefore lies in the network structure of the connections and relations among entities, in terms of resource exchange and diffusion of failures.

The adoption of sophisticated models, however, poses non-trivial issues related to the availability of data, which are in general difficult to obtain, mainly because such data are related to the interaction among elements belonging to different domains.

3 Input–Output Inoperability Model

The input–output inoperability model (IIM) [5, 6] was developed to express the global effects of negative events in highly interdependent infrastructures. Such model has been extended in a variety of ways; for example, the dynamic

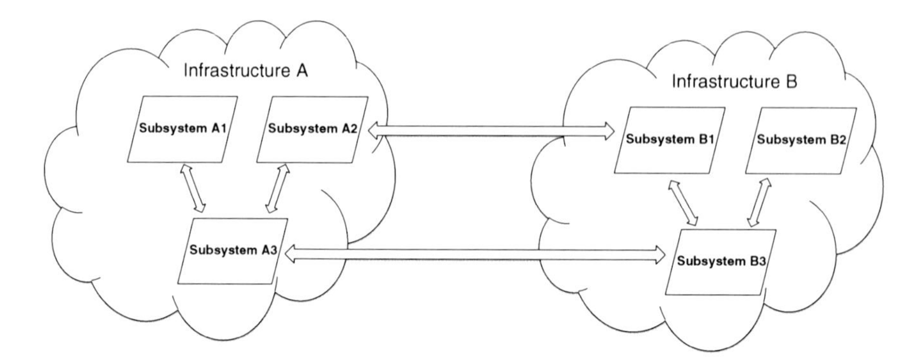

Fig. 2 Example of reductionistic interdependency modeling

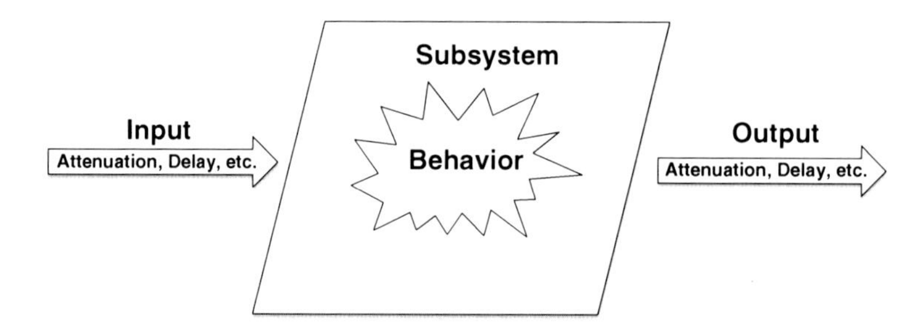

Fig. 3 Representation of an agent within agent-based models

input–output inoperability model (D-IIM) [6] considers system dynamics during crisis situations while the multi-regional input–output inoperability model (MR-IIM) [28] expresses multi-sector and multi-regional economic interdependencies.

The input–output inoperability model helps analyze how a natural outage or attack on an infrastructure may affect other infrastructures, highlighting the cascading effects and the intrinsic vulnerabilities. The main assumption is that two entities with a large amount of economic interaction also have a large amount of physical interdependency [6, 8].

Following the economic equilibrium theory of Leontief [29], a static demand-reduction model [5, 6] for n infrastructures is given by:

$$\delta x = A^{*}\delta x + \delta c^{*} \tag{1}$$

where δx is the difference between the planned production (x_0) and the degraded production (x_d) production, δc^{*} is the difference between the planned final demand (c_0) and the degraded final demand (c_d), and A^{*} is a square $n \times n$ matrix whose elements a^{*}_{ij} (Leontief technical coefficients) represent the ratio of the input from infrastructure i to infrastructure j with respect to the overall production requirements of infrastructure j.

The inoperability provides an indication of the state of each infrastructure, i.e., its inability (as a percentage) to operate properly. To this end, an $n \times n$ transformation matrix P is introduced [6]:

$$P = [diag\{x_0\}]^{-1} \tag{2}$$

The inoperability is computed by applying the following transformation to the reduction of production:

$$q = P\delta x \tag{3}$$

The input–output inoperability model has the form:

$$q = Aq + c \tag{4}$$

where $A = PA^*P^{-1}$ and $c = Pc^*$. Note that c assumes the role of an externally induced inoperability. Thus, it can be viewed as an actual perturbation generated by an adverse or malicious event. Although it is mathematically possible to have $a^*_{ij} > 1$, the a_{ij} coefficients are strictly less than one, due to the normalization procedure of Eq. (3) [8].

The inoperability q corresponding to a perturbation c is given by:

$$q = (I - A)^{-1}c = Sc \tag{5}$$

In the following, let us refer to A and $S = (I - A)^{-1}$ as the open-loop and closed-loop interdependency matrices, respectively. Matrix A models the direct effects due to first-order dependencies while matrix S also takes into account the amplifications introduced by domino effects (i.e., second-order and higher-order dependencies).

In the literature some indices to express the criticality of a scenario have been introduced [10, 16] in order to evaluate the level of dependency of an infrastructure, based on their economic counterpart [30]. The *dependency index* is defined as the sum of the coefficients of matrix A along a row:

$$\gamma_i = \sum_{j=1}^{n} a_{ij} \tag{6}$$

This index measures the robustness of the infrastructure with respect to the inoperability of other infrastructures. As a matter of fact, it represents the maximum inoperability of the i-th infrastructure when every other infrastructure is fully inoperable. The lower the value, the greater the ability of the i-th infrastructure to preserve some working capabilities (e.g., using buffers, back-up power, etc.) despite the inoperability of its supplier infrastructures.

The *influence gain*, conversely, measures the influence exerted by one infrastructure over the others. It is defined as the sum of the coefficients of matrix A along a column:

$$\theta_j = \sum_{i=1}^{n} a_{ij} \tag{7}$$

A large influence gain means that the inoperability of the j-th infrastructure induces significant degradations to the entire system. When $\theta_j > 1$, the negative effects (in terms of inoperability) induced by cascading phenomena on the other infrastructures are amplified. The opposite is the case when $\theta_j < 1$. The dependency index and influence gain allow quick global evaluations of the resilience of a given infrastructure. However, they are mainly related to the nominal, open-loop behavior.

Setola et al. [10] have introduced analogous indices for the closed-loop matrix (based on the economic indices in [31]), namely the overall dependency index ($\bar{\gamma}_i$) and the overall influence gain ($\bar{\theta}_j$). These two indices are defined as the row-wise and column-wise sums of matrix S, respectively:

$$\bar{\gamma}_i = \sum_{j=1}^{n} s_{ij} \tag{8}$$

$$\bar{\theta}_j = \sum_{i=1}^{n} s_{ij} \tag{9}$$

Such indices express the resilience or the influence of a given infrastructure considering high-order dependency phenomena. Comparing $\bar{\gamma}_i$ with γ_i ($\bar{\theta}_i$ with θ_i) provides useful information about the amplification due to second-order and higher-order cascading effects.

Note that for a positive-definite matrix A, the inverse of matrix $(I - A)$ is given by [32]:

$$(I - A)^{-1} = I + A + A^2 + A^3 + \cdots = \sum_{p=0}^{\infty} A^p \tag{10}$$

Such an equation provides an immediate understanding of the cumulative effects of high-order dependencies in matrix S. Moreover, the equation defines a procedure for the estimation of the closed-loop interdependency matrix without explicitly computing the inverse of matrix $(I - A)$.

Let us now discuss a dynamic extension of the IIM model.

3.1 Dynamic IIM

The static input–output inoperability model defined in Eq. (1) can be extended by incorporating a dynamic term [6]:

$$\delta\dot{x}(t) = K(A^* - I)\delta x(t) + K\delta c^*(t) \tag{11}$$

Also in this case, the application of the transformation matrix P to Eq. (11) yields the dynamic input–output inoperability model, which is given by:

$$\dot{q}(t) = K(A - I)q(t) + Kc(t) \tag{12}$$

Matrix K is referred to as the *industry resilience* coefficient matrix; each element k_{ii} measures the resilience of the i-th infrastructure. The k_{ii} coefficient can also be seen as the recovery rate with respect to adverse or malicious events. Matrix K assumes the role of an actual control parameter; in fact, countermeasures and risk mitigation strategies for the i-th infrastructure increase k_{ii}, minimizing economic losses and impact, with shorter recovery times.

The IIM model, in its dynamic fashion, can be adopted to represent the response of interdependent infrastructures to an induced perturbation, until the equilibrium (if any) is reached. The following results [33] provide some conditions for the stability of the system and correlate the static and dynamic IIM models:

Lemma 3.1 *If all the eigenvalues of the matrix* $(A - I)$ *have strictly negative real parts and* $||c(t)||$ *is bounded, system* (12) *is stable. Moreover, if* $c(t)$ *is stationary, then system* (12) *reaches an equilibrium that coincides with* Eq. (5).

Lemma 3.1 provides a first condition for the stability of the system, which is not dependent on the particular matrix K considered; moreover it defines the relationship between the static and dynamic models.

The following theorem [33] provides a further condition for the stability of the system.

Theorem 3.2 *If* $||c(t)||$ *is bounded, then a sufficient condition to guarantee the stability of system* (12) *is that the maximum of the dependency indices* (6) *of matrix* A *is less than one.*

Theorem 3.2 represents the condition that, if the degree of dependency is sufficiently small, then the increment of inoperability is bounded; moreover, the presence of UPS, batteries or buffers in each infrastructure assures the stability of the system, that does not depend on K.

The dynamic input–output inoperability model can be used to express the response of interdependent infrastructures to an induced perturbation until an equilibrium is reached. In many application scenarios, however, it is more useful to consider a discrete-time representation of the input–output inoperability model.

Given a sampling rate T_s, a simple approximation for the derivative $\dot{q}(t)$ is given by:

$$\dot{q}(t) \simeq \frac{q(t+T_s) - q(t)}{T_s} \tag{13}$$

Choosing the starting time $t_0 = 0$ and considering uniform sample periods yields the discrete-time input–output inoperability model, which is given by:

$$q((k+1)T_s) = q(k+1) = [T_s A - T_s + I]q(k) + T_s c \Rightarrow q(k+1) = A^d q(k) + c^d \tag{14}$$

Often, the discrete-time interdependency model is directly set up to assess the values of the elements of matrix A [10, 16].

3.2 Discussion

Although the IIM framework is very compact and elegant, and is able to model cascading effects, its high degree of abstraction does not allow to perform accurate analyses on the real nature of dependencies; in fact, such an approach considers only relations that involve whole infrastructures, while it is impossible to understand and represent the contribution of each subsystem. This latter aspect is fundamental, in order to address the huge complexity of geographically dispersed and highly coupled systems.

There is, then, the need to decompose the infrastructures into a set of more concrete systems, whose working capability is mainly characterized by the availability of resources (goods or services).

Moreover, the economic origin of the IIM model represents a structural limitation: in fact, even if use/make matrices are considered, taking into account the production and consumption of multiple commodities for each infrastructure, only the economic value of such commodities is typically available. Nevertheless, this data is related to the infrastructures seen as monolithic entities.

On the other hand, when decomposing infrastructures with a finer grain perspective, it is difficult to retrieve exact economic data for the resources involved in the exchanges among subsystems or components.

In the following section a methodology for the representation of interdependencies with a lower level of abstraction and considering the exchange of different resources among subsystems, namely Agent-Based IIM approach, will be discussed.

4 Agent-Based IIM

As highlighted in the previous section, there is a need to provide a more expressive, yet well-grounded, interdependency modeling framework; the availability of significant data, however, is one of the main issues that have to be addressed in order to adopt a finer grain perspective.

The abstract Leontief coefficients can be assessed while considering the macroeconomic evidence. When dealing with more descriptive models, however, it is not easy to obtain adequate economic data; in fact, focusing on subsystems, the assumption that interdependency is proportional to economic loss appears not realistic. Notwithstanding, the IIM model is very diffused and well established.

A feasible approach, then, would be the application of the IIM framework in the reductionistic fashion (i.e., with a lower abstraction degree). Moreover, since no economic data is currently available while adopting such an in-the-small perspective, the model has to be extended, considering the *production*, *consumption* and *transmission* of resources, in order to enable the different stakeholders to contribute to the modeling activities and encode their knowledge in a flexible and simple way. An attempt in such a direction has been done in [10], where the IIM coefficients have been assessed by means of specific questionnaires and technical interviews.

In this section a different approach will be discussed, namely *Agent-Based IIM model* (AB-IIM) [33]. One of the main characteristics of such an approach is that the reductionistic Leontief technical coefficients can be easily derived. This latter feature is very important, since the AB-IIM model can be used to derive a simple IIM model, using the knowledge of the different stakeholders in a very efficient way.

4.1 Static AB-IIM Model

The key idea of the AB-IIM model is that each element interacts via the production, exchange and consumption of *resources*, without directly exchanging their inoperability. Such an abstract quantity becomes an *internal* variable that represents the overall status of each element, driving its dynamic.

In order to highlight the role of the working condition of the infrastructures, the static IIM model can be rearranged, considering the *operativeness*, defined as:

$$op = 1_n - q \tag{15}$$

Therefore, the IIM model, in the operativeness form, becomes:

$$op = Aop + \hat{c} \tag{16}$$

where $\hat{c} = (I - A)1_n - c$. Note that $\hat{c}_i$ can be expressed as:

$$\hat{c}_i = (1 - c_i) - row_i\{A\}1_n = \hat{c}_i^* - \gamma_i \tag{17}$$

where $\hat{c}_i^*$ is the external, induced operativeness and γ_i represents the maximum operativeness that may be received by the i-th entity when each other element is fully operative.

In such a perspective, therefore, the operativeness of an element depends on the operativeness of the others, by means of the *same* coefficients of the original IIM model, considering also an induced operativeness and a negative, constant balancing term. This latter form can be very convenient when there is a need to analyze the relations between the state of the element and the production of resources.

An intuitive choice for the model is that the *produced* resources are proportional to the *operativeness* that, in turn, depends on the *received* resources (see Fig. 4). However, in order to keep the complexity down, only the produced resources are directly taken into account, while received resources are obtained as a weighted sum of resources produced by other elements, eventually considering the *attenuations* during transportation (i.e., dissipations).

Let us consider m different resources, and let r_i^j be the normalized *production* of the j-th resource from the i-th element. Such a quantity is proportional to the operativeness by means of the coefficient $\phi_i^j \in [0, 1]$:

$$r_i^j = \phi_i^j op_i \Rightarrow r = \Phi op \tag{18}$$

where Φ is an $nm \times n$ matrix. On the other hand, the operativeness of an element depends on the weighted sum of the total amount of *received* resources:

$$op_i = \sum_{j=1}^{m} \psi_i^j \bar{r}_i^j + \hat{c}_i \Rightarrow op = \Psi \bar{r} + \hat{c} \tag{19}$$

where ψ_i^j is the coefficient that represents the influence of the j-th resource received by the i-th element, $\bar{r}_i^j$ is the total amount of resource j received by i and Ψ is a $n \times nm$ matrix. Received resources, for the i-th entity, depend on the production of each element other than i, eventually considering the attenuation δ_{pi}^j during the transmission of j-th resource from element p to element i:

$$\bar{r}_i^j = \sum_{p=1; p \neq i}^{n} \delta_{pi}^j r_p^j \Rightarrow \bar{r} = \Delta r \tag{20}$$

where the *attenuation* matrix Δ is $nm \times nm$. Replacing the above equation inside Eq. (19), and shifting to the inoperability form, the *static AB-IIM model* is obtained:

$$\begin{bmatrix} q \\ r \end{bmatrix} = \begin{bmatrix} 0 & -\Psi\Delta \\ -\Phi & 0 \end{bmatrix} \begin{bmatrix} q \\ r \end{bmatrix} + \begin{bmatrix} A \\ \Phi \end{bmatrix} 1_n + \begin{bmatrix} I_n \\ 0 \end{bmatrix} c \tag{21}$$

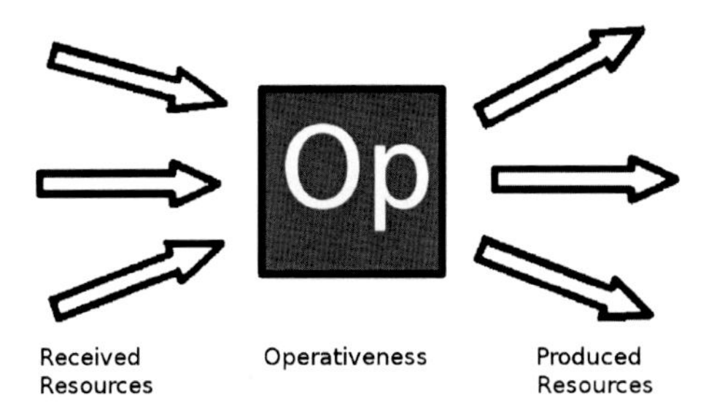

Fig. 4 Relation among received resources, operativeness and produced resources within the AB-IIM approach

The following Lemma [33] provides a useful parallelism to the standard IIM model.

Lemma 4.1 *If* $\Psi\Delta\Phi = A$*, the static AB-IIM model, described in* (21) *, coincides with the static IIM model described in* (4).

In other terms, the abstract matrix A of the classical IIM model is decomposed into the product of three matrices Φ, Δ and Ψ, with a physical meaning. Indeed, Ψ represents how the availability of the different resources influences the operativeness of a single component; Φ the capability of the element to produce its different outputs, while Δ gives a measurement of the eventual losses due to the links or other factors. Similarly to the IIM standard model, let us define some indices able to depict the level of dependency of the elements. Let $\Theta = \Psi\Delta$ be an $n \times nm$ matrix, whose elements are

$$\theta_a^b = \sum_{k=1}^{nm} \psi_{ak}\delta_{kb}; \quad a \in [1, n]; b \in [1, nm] \tag{22}$$

The *AB-IIM dependency index* $\tilde{\gamma}_i$ is defined, for the i-th element, as:

$$\tilde{\gamma}_i = -\sum_{j=1}^{nm} \theta_i^j \tag{23}$$

This index represents the cumulative effect of resources received by the i-th entity on its operativeness. The following theorem [33] correlates the AB-IIM dependency index with the IIM model.

Theorem 4.2 *If* $\Psi\Delta\Phi = A$*, then* $\tilde{\gamma}_i$*, defined in* (23) *, coincides with* γ_i*, defined in* (6) *, for each* $i = 1, \ldots, n$

4.2 Dynamic AB-IIM Model

Similarly to the dynamic IIM model (12) lets define the *dynamic* IIM model with resource exchange as:

$$\begin{bmatrix} q(t) \\ r(t) \end{bmatrix} = \begin{bmatrix} 0 & -\Psi\Delta \\ -\Phi & 0 \end{bmatrix} \begin{bmatrix} q(t) \\ r(t) \end{bmatrix} + \begin{bmatrix} A \\ \Phi \end{bmatrix} 1_n + \begin{bmatrix} I_n \\ 0 \end{bmatrix} c - \begin{bmatrix} K^{-1} \\ W^{-1} \end{bmatrix} \begin{bmatrix} \dot{q}(t) \\ \dot{r}(t) \end{bmatrix} \quad (24)$$

where K is the $n \times n$ industry resilience matrix and W is an $nm \times nm$ diagonal matrix (note that K and W are always invertible).

The above model can be rearranged as:

$$\begin{bmatrix} \dot{q}(t) \\ \dot{r}(t) \end{bmatrix} = \begin{bmatrix} -K & -K\Psi\Delta \\ -W\Phi & -W \end{bmatrix} \begin{bmatrix} q(t) \\ r(t) \end{bmatrix} + \begin{bmatrix} KA \\ W\Phi \end{bmatrix} 1_n + \begin{bmatrix} K \\ 0 \end{bmatrix} c(t) \quad (25)$$

The following results [33] provide some stability conditions.

Lemma 4.3 *If the dynamic matrix of system* (25) *has all eigenvalues with negative real part and* $||c(t)||$ *is bounded, system* (25) *is stable. Moreover, if* $A = \Psi\Delta\Phi$ *and* $c(t)$ *is stationary, the system reaches an equilibrium that coincides with* Eq. (5).

Theorem 4.4 *If* $A = \Psi\Delta\Phi$, $||c(t)||$ *is bounded and* $\gamma_i < 1$ *for* $i = 1, \ldots, n$, *system* (25) *is stable.*

A sufficient condition for the stability of the system, then, is that $\rho_i \leq 1$ for $i = 1, \ldots n(m+1)$. Since, in the second case $\rho_i = \phi_i^{j*} \leq 1$, where ϕ_i^{j*} is the sole non-zero entry of the considered row of the dynamic matrix, the condition has to be verified just for $i = 1, \ldots, n$. From Theorem (4.2), it is sufficient that $\gamma_i \leq 1$ for $i = 1, \ldots, n$.

4.3 Discussion

The AB-IIM method is, hence, a flexible, reductionistic, IIM-based modeling framework and represents a very simple example of agent-based model. Moreover, since the parameter tuning is mainly related to the production transportation and consumption of resources, it is really easy and straightforward to encode the knowledge of the different stakeholders. Finally, the possibility to derive the reductionistic Leontief coefficients, transforming the AB-IIM into a simple, although agent-based IIM model, is one of the main features of such a methodology.

5 Fuzzy Dynamic IIM

In this Section a fuzzy extension of the IIM methodology, namely the *Fuzzy Dynamic Input–output Inoperability Model* (FD-IIM) [11] will be discussed as a formalism able to handle ambiguity and linguistic expressions for the Leontief coefficients, the perturbation and the initial conditions.

Modeling real systems often represents a hard challenge and is generally subject to non trivial issues; in fact in many situations the system to be modeled is very complex, or the model may be affected by subjective choices. This is particularly true when humans are directly involved in the functioning of the system (i.e., when linguistic opinions have to be handled quantitatively). Fuzzy theory, [15, 34–36], appears a quite natural choice to address the problem of modeling complex systems affected by vagueness.

5.1 Fuzzy Systems

The key idea of fuzzy theory is to extend traditional set theory by allowing an element to partially belong to a set.

Let a fuzzy subset of $\mathbb{R}$ be defined by a *membership function* $x : \mathbb{R} \rightarrow [0, 1]$ which assigns to each point $p \in \mathbb{R}$ a grade of membership in the fuzzy set [36].

Let an α-level set $[x]^\alpha$ of a fuzzy set x be defined as

$$[x]^\alpha = \{p \in \mathbb{R} | x(p) \geq \alpha\} \tag{26}$$

where $x(p)$ is the degree of membership of $p \in \mathbb{R}$ to the set x and $\alpha \in [0, 1]$. The α-levels allow to treat a fuzzy set as a collection of nested real intervals.

According to [35], let the space $\mathbb{E}$ of Fuzzy Numbers (FN) [15, 37] be the space of fuzzy subsets x of $\mathbb{R}$ such that the membership functions are compact and the α-levels are nested, i.e., those with greater values of α are contained into those with smaller α (all the intervals are contained in the support). In other terms:

1. x maps $\mathbb{R}$ onto $[0, 1]$;
2. $[x]^0$ is a bounded subset of $\mathbb{R}$;
3. $[x]^\alpha$ is a compact subset of $\mathbb{R}$ for all $\alpha \in (0, 1]$;
4. x is *fuzzy convex*, that is:

$$x(\phi p + (1 - \phi)q) \geq \min[x(p), x(q)]; \quad \forall p, q \in \mathbb{R}.$$

Let us now discuss a particular subclass of FNs denoted as *Triangular Fuzzy Numbers* (TFN). A triangular fuzzy number $x \in \mathbb{E}$ is described by an ordered triple $\{x_l, x_c, x_r\} \in \mathbb{R}^3$ with $x_l \leq x_c \leq x_r$ and such that $[x]^0 = [x_l, x_r]$ and $[x]^1 = \{x_c\}$, while in general the α-level set is given, for any $\alpha \in [0, 1]$ by:

$$[x]^{\alpha} = [x_c - (1-\alpha)(x_c - x_l), x_c + (1-\alpha)(x_r - x_c)] \tag{27}$$

Triangular Fuzzy Numbers are widely used in many applications, due to their compact representation. In fact they can be described by the triple of the abscissae of their vertices. Moreover, a singleton is obtained for $\alpha = 1$; hence the α-level with the strongest belief collapses into a crisp point.

A *Complete Discrete-Time Fuzzy System* (C-DFS) [14] is a linear and stationary system defined as follows:

$$x(k+1) = Hx(k), \quad x(0) = x_0 \tag{28}$$

where H is an $N \times N$ fuzzy valued matrix, i.e., its entries $h_{ij} \in \mathbb{E}$ are fuzzy numbers so as initial conditions, hence $x, x_0 \in \mathbb{E}^N$.

The extension of arithmetic operations to triangular fuzzy numbers is relatively straightforward. The sum of two fuzzy numbers is a linear operation and the resulting fuzzy number is also triangular [15]. Specifically, the sum of two triangular fuzzy numbers $w = \{w_1, w_2, w_3\}$ and $z = \{z_1, z_2, z_3\}$ is given by:

$$w + z = \{w_1 + z_1, w_2 + z_2, w_3 + z_3\} \tag{29}$$

The multiplication of a fuzzy number by a real scalar value η is given by:

$$\eta \times w = \{\eta w_1, \eta w_2, \eta w_3\} \tag{30}$$

Note that, while fuzzy numbers – specifically, triangular fuzzy numbers—are closed with respect to the sum and scalar product operations, the product of two triangular fuzzy numbers is, in general, not triangular. Several triangular approximations are given in the literature for the product of triangular fuzzy numbers. The simplest definition, in the case where $w, z \geq 0$ (i.e., $w_1, z_1 \geq 0$), is given in Eq. (31).

$$w \times z = \{w_1 \times z_1, w_2 \times z_2, w_3 \times z_3\} \tag{31}$$

However, using the theory of inclusions [14, 35, 38, 39], it is possible to overcome such a problem and provide an exact characterization of the product. Let us first provide a simple example. Consider the scalar equation example provided in Eq. (32), where w and z are triangular fuzzy numbers.

$$y = w \times z \tag{32}$$

Note that, even if w and z are represented by triangular fuzzy numbers, their product is no more triangular: this renders the direct calculation of Eq. (32) not feasible in practice. A solution is to resort to the theory of fuzzy inclusions [35, 39]. Specifically, each fuzzy set is considered as a collection of nested intervals (the α-levels); the product is performed *levelwise*, i.e., each alpha level of the product is given by the product of the intervals of the factors in the multiplication.

For any $\alpha \in [0,1]$ let the interval matrix

$$[H]^{\alpha} = [\underline{H}^{\alpha}, \overline{H}^{\alpha}]$$

where $\underline{H}^{\alpha}, \overline{H}^{\alpha}$ denote the matrices whose elements are the lower and the upper end points of the interval, respectively. Moreover, let H^{α} denote the generic matrix within $[H]^{\alpha}$, i.e., such that each coefficient $h_{ij}^{\alpha} \in [\underline{h}_{ij}^{\alpha}, \overline{h}_{ij}^{\alpha}]$.

Analogously, for any $\alpha \in [0,1]$ and for any $k \in \mathbb{N}_{+}$, let

$$[x(k)]^{\alpha} = [\underline{x}^{\alpha}(k), \bar{x}^{\alpha}(k)]$$

be a vector whose components are intervals

$$[x_i(k)]^{\alpha} = [\underline{x}_i^{\alpha}(k), \bar{x}_i^{\alpha}(k)]$$

and $\underline{x}^{\alpha}(k), \bar{x}^{\alpha}(k)$ represent the vectors composed of the left and right endpoints respectively, while the generic element in the interval is denoted by $x^{\alpha}(k)$.

The linear and stationary C-DFS (28) can be rewritten as the FDI:

$$x^{\alpha}(k+1) \in [H]^{\alpha}[x(k)]^{\alpha}; \quad x^{\alpha}(0) \in [x_0]^{\alpha}; \quad 0 \leq \alpha \leq 1 \tag{33}$$

Notice that, in (33) the symbol "=" is substituted by "∈". Hence (33) provides the set of all possible values that the fuzzy variables x may assume at time $k+1$ with a membership grade greater than α, knowing the set of possible values assumed by x at time step k and the extrema of the interval matrix. Evaluating this for all $\alpha \in [0,1]$ it is possible to estimate the whole admissible set spanned by $x(k+1)$.

Corollary 5.1 *Let a Linear and Stationary FDI* (33) *and suppose that the crisp matrix* $\underline{H}^0 \geq 0$ *and that, for all* $i = 1, \ldots, N$

$$\sum_{j=1; j \neq i}^{N} (\overline{H}^0)_{ij} < 1 - (\overline{H}^0)_{ii} \tag{34}$$

then the FDI (33) *is asymptotically stable. Analogously the result holds if* $\overline{H}^0 \leq 0$ *and*

$$\sum_{j=1; j \neq i}^{N} (\underline{H}^0)_{ij} > -1 - (\underline{H}^0)_{ii} \tag{35}$$

Proof 5.2 See [39].

Theorem 5.3 *Let a linear and stationary FDI* (33) *and suppose that the crisp matrix* $\underline{H}^0 \geq 0$, *where* $\underline{H}^0$ *is the left extrema of the interval matrix* $[H]^0$; *then the following holds for each* $\alpha \in [0,1]$:

$$\underline{\mathbb{S}}_f^{\alpha}(x_0,k) = (\underline{H}^{\alpha})^k \underline{x}_{\alpha 0}; \quad \overline{\mathbb{S}}_f^{\alpha}(x_0,k) = (\overline{H}^{\alpha})^k \overline{x}_{\alpha 0} \tag{36}$$

where $\underline{x}_{\alpha 0}$ *and* $\overline{x}_{\alpha 0}$ *are the left and right bounds of* $[x_0]^{\alpha}$, *respectively and* $\underline{\mathbb{S}}_f^{\alpha}(x_0,k)$, $\overline{\mathbb{S}}_f^{\alpha}(x_0,k)$ *are the left and right bounds of the set of solutions* $[\mathbb{S}_f(x_0,k)]^{\alpha}$ *of the Inclusion* (33).

Proof 5.4 See [39].

5.2 Fuzzy Dynamic IIM

Let us define a FD-IIM model for n infrastructures as follows:

$$q(k+1) = Aq(k) + c, q(0) = q_0 \tag{37}$$

where A is an $n \times n$ matrix with triangular fuzzy entries, i.e., $a_{ij} \in \mathbb{E}$. The initial condition $q_0 \in \mathbb{E}^n$ is a triangular fuzzy number as is the external perturbation $c^d \in \mathbb{E}^n$. Moreover, in accordance with the classical input–output inoperability model, the fuzzy matrix A is assumed to be positive semi-definite (i.e., for all $\alpha \in [0,1]$ and for all crisp matrices $A_{\dagger} \in [A]^{\alpha}$, $A_{\dagger}$ is positive semi-definite). In other words, the entries of $\underline{A}_0$ satisfy the property: $\underline{a}_{0ij} \geq 0$ for all i,j.

In the fuzzy model, the inoperability of each infrastructure is described by a fuzzy variable, i.e., by a set of values with different degrees of membership (or belief). Also, the dependency coefficients are expressed as fuzzy numbers. In this way, the model defined by Eq. (37) can handle uncertain perturbations and also take into account the uncertainty and vagueness that characterize human knowledge about the dependency coefficients.

If each state variable and input is described by a triangular fuzzy number, the fuzzy evolution of the system has a clear meaning. The center of the triangle represents the trajectory associated with the maximum belief, and the width of the support provides a measure of the uncertainty of each state variable or input.

The following theorem [11] provides a stability condition for this class of fuzzy systems.

Theorem 5.5 *Consider a system that conforms with the fuzzy dynamic input–output inoperability model* (Eq. 37) , *and suppose that* $\underline{A}_0 \geq 0$ *and*

$$\sum_{j=1;j\neq i}^{n} (\bar{A}_0)_{ij} \leq 1 - (\bar{A}_0)_{ii} \forall i = 1,\ldots,n \tag{38}$$

Then, the system is stable.

As in the case of the standard input–output inoperability model, open-loop and closed-loop indices can be defined for the FD-IIM model. In particular, the open-loop dependency index and the influence gain are obtained by summing the fuzzy entries in the i-th row or column of matrix A, respectively, according to Eq. (29). However, the closed-loop dependency index and influence gain are more complex because it is nontrivial to compute the fuzzy matrix $(I - A)^{-1}$. In fact, the inverse of a fuzzy matrix can be defined in several ways [40]. In the following, let us resort to a level-wise approach known as Rohn's Scheme [41].

For each α-level, the inverse of the interval matrix $[A]^{\alpha}$ is defined as the smallest interval matrix that contains the set $\{B : \hat{A}B = I, \hat{A} \in [A]^{\alpha}\}$. This interval matrix $[B]^{\alpha}$ satisfies the equations:

$$(\underline{B}_{\alpha})_{ij} = \min\left\{(\hat{A}^{-1})_{ij} : \hat{A} \in [A]^{\alpha}\right\} \tag{39}$$

$$(\bar{B}_{\alpha})_{ij} = \max\left\{(\hat{A}^{-1})_{ij} : \hat{A} \in [A]^{\alpha}\right\} \tag{40}$$

Clearly, the inverse of a fuzzy matrix exists only if the matrix $\hat{A} \in [A]^{\alpha}$ is invertible for each α-level.

The following theorem [11] characterizes the bounds of $[(I - A)^{-1}]^{\alpha}$.

Theorem 5.6 *If the conditions of Theorem 5.5 hold, then*

$$[(I - A)^{-1}]^{\alpha} = \left[(I - \underline{A}_{\alpha})^{-1}, (I - \bar{A}_{\alpha})^{-1}\right] \tag{41}$$

According to Theorem 5.6, if the matrix is composed of triangular entries, it is sufficient to consider the bounds of the level sets for $\alpha = 0$ and $\alpha = 1$ to characterize the fuzzy inverse. In fact, in this case, the interval matrix for $\alpha = 1$ collapses into a single crisp matrix, while the level set for $\alpha = 0$ expresses the maximum level of uncertainty of the inverse.

Under such a triangularity assumption, the fuzzy matrix $S = (I - A)^{-1}$ is therefore defined by:

$$s_{ij} = \left\{\left((I - \underline{A}_0)^{-1}\right)_{ij}, \left((I - A_1)^{-1}\right)_{ij}, \left((I - \bar{A}_0)^{-1}\right)_{ij}\right\} \tag{42}$$

where $A_1 = \underline{A}_1 = \bar{A}_1$. Note that the approach suggested by Eq. (42) is simple and compact in that it only requires the inversion of three crisp matrices. Also, it yields results that are numerically identical to those obtained by using the fuzzy version of Eq. (10).

Let us now extend the closed-loop indices $\bar{\gamma}_i$ and $\bar{\theta}_i$ (see Eqs. (8) and (9), respectively) for the i-th infrastructure by summing the i-th row and column of the fuzzy matrix S, respectively.

Table 1 Sample influence estimation table

Impact	Description	Value
Nothing	Event does not induce any effect on the infrastructure	0
Negligible	Event induces some very limited and geographically bounded consequences on services that have no direct impact on infrastructure operability	0.005
Very limited	Event induces some geographically bounded consequences on services that have no direct impact on infrastructure operability	0.008
Limited	Event induces consequences only on services that have no direct impact on infrastructure operability	0.010
Some degradation	Event induces limited and geographically bounded consequences on the ability of the infrastructure to provide services	0.020
Moderate degradation	Event induces geographically bounded consequences on the ability of the infrastructure to provide services	0.030
Significant degradation	Event significantly degrades the ability of the infrastructure to provide services	0.050
Some services provided	Event causes the infrastructure to provide only some essential services nationwide	0.100
Stop	Event causes the infrastructure to provide only some essential services in some geographically area	0.300
Quit (complete stop)	Event causes the infrastructure to be unable to provide services	0.500

The coefficients for the fuzzy IIM can be obtained by focusing directly on the operative impact that the degradation or absence of the services provided by each infrastructure experts on all the other infrastructures. This can be done with the help of domain experts, who can be asked to estimate the impact produced on "their infrastructures" by the complete absence of services provided by each of the other infrastructures. In this framework, the domain experts are provided with questionnaires and are required to quantify the impact on the operational capability of infrastructures using some linguistic expressions, as those presented in Table 1. In addition, each expert is asked to estimate his/her confidence in the data provided using some linguistic expressions, as those listed in Table 2.

On the basis of the meanings of the expressions reported in Tables 1 and 2 the linguistic expressions can be converted into numerical values; which are reported in the last column of each table.

5.3 Discussion

The fuzzy IIM methodology is a valuable framework to overcome the limitation of the economic origin of the coefficients, by asking directly the experts and operators and converting the linguistic values into numerical ones and taking into account the associated ambiguity using fuzzy numbers. Note that the framework is suitable

Table 2 Sample confidence estimation scale

Confidence	Description	Value
+	Good confidence	0
++	Relative confidence	±0.005
+++	Limited confidence	±0.010
++++	Uncertain	±0.015
+++++	Strongly uncertain	±0.020

for both an holistic (i.e., high level) and reductionistic (i.e., low level) IIM model, hence it can be applied with the desired level of detail (although requiring a nontrivial effort as the level of detail is increased).

6 Conclusions

In this chapter holistic and agent based methodologies to represent Critical Infrastructure Interdependencies have been discussed with particular reference to the Input–Output Inoperability Model and some of its extensions. The chapter also describes a methodology to overcome the lack of quantitative data by resorting to fuzzy theory and by asking the experts and technicians to assess the dependencies in a linguistic way.

References

1. Suter, M., Brunner, E.: International CIIP Handbook 2008/2009. Center for Security Studies, ETH Zurich (2008)
2. E.U. Commission: Green Paper on a European Programme for Critical Infrastructure Protection COM(2005)576, Brussels (2005)
3. Falliere, N., O' Murchu, L., Chien, E.: W32. Stuxnet Dossier. White Paper, Symantec Corp., Security Response, Version 1.4 (2011)
4. Rios, B., McCorkle, T.: 100 Bugs in 100 days: an analysis of ICS (SCADA) Software. DerbyCon 2011, Session
5. Haimes, Y., Jiang, P.: Leontief-based model of risk in complex interconnected infrastructures. J. Infrastruct. Syst. **7**, 1–12 (2001)
6. Haimes, Y., Horowitz, B., Lambert, J., Santos, J., Lian, C., Crowther, K.: Inoperability input–output model for interdependent infrastructure sectors. I: theory and methodology. J. Infrastruct. Syst. **11**(2), 67–79 (2005)
7. Rinaldi, S., Peerenboom J., Kelly, T.: Identifying understanding and analyzing critical infrastructure interdependencies. IEEE Control Syst. Mag. **26**, 11–25 (2001)
8. Kujawski, E.: Multi-period model for disruptive events in interdependent systems. Int. Syst. Eng. **9**(4), 281–295 (2006)
9. Lewis T.G.: Critical Infrastructure Protection in Homeland Security: Defending a Networked Nation. Wiley, New York (2006)

10. Setola, R., De Porcellinis, S., Sforna, M.: Critical infrastructure dependency assessment using the input–output inoperability model. Int. J. Crit. Infrastruct. Protect. **2**, 170–178 (2009)
11. Oliva, G., Panzieri, S., Setola, R.: Fuzzy dynamic input–output inoperability model. Int. J. Crit. Infrastruct. Prot. **4**(3–4), 165–175 (2011). doi:10.1016/j.ijcip.2011.09.003
12. Oliva, G., Panzieri S., Setola, R.: Fuzzy input-output inoperability model. Crit. Inf. Infrastruct. Secur. Lect. Notes Comput. Sci. **6983**, 200–204 (2013)
13. Oliva, G., Panzieri, S., Setola, R.: Distributed synchronization under uncertainty: a fuzzy approach. Fuzzy Sets Syst. (Elsevier eds.) **206**, 103–120 (2012). doi:10.1016/j.fss.2012.02.003
14. Oliva, G.: Stability and level-wise representation of discrete-time fuzzy systems. Int. J. Fuzzy Syst. **14**(2), 185–192 (2012)
15. Dubois, D., Prade, H.: Possibility Theory: An Approach to the Computerized Processing of Uncertainty. Plenum, New York (1993)
16. Setola, R.: How to measure the degree of interdependencies among critical infrastructures. Int. J. Syst. Syst. Eng. **2**(1/2010), 38–59 (2010) (to appear)
17. De Porcellinis, S., Panzieri, S., Setola, R., Ulivi, G.: Simulation of heterogeneous and interdependent critical infrastructures. In: Proceedings of 50th ANIPLA Conference (2006)
18. Macdonald, R., Bologna, S.: Advanced modelling and simulation methods and tools for critical infrastructure protection. ACIP project report (2001)
19. U.S. and Canada Power System Outage Task Force: Final Report on the August 14, 2003 Blackout in the United States and Canada: Causes and Recommendations (2004)
20. De Porcellinis, S., Oliva, G., Panzieri, S., Setola, R.: A holistic-reductionistic approach for modeling interdependencies. In: Papa, M., Shenoi S. (eds.) Critical Infrastructure Protection III, vol. 311/2009, pp. 215–227. Springer, Berlin (2009)
21. Watts, D.J., Strogartz, S.H.: Collective dynamics in small–world networks. Nature **393**, 440 (1998)
22. Jeong, H., Mason, S.P., Barabasi, A.L., Oltvai, Z.N.: Lethality and centrality in protein networks. Nature **411**, 41 (2001)
23. De Porcellinis, S., Issacharoff, L., Meloni, S., Rosato, V., Setola, R., Tiriticco F.: Modelling interdependent infrastructures using interacting dynamical models. Int. J. Crit. Infrastruct. (IJCIS) **4**(1/2), 63–79 (2008)
24. Duenas-Osorio, L., Craig, J.I., Goodno, B.J., Bostrom, A.: Interdependent Response of networked systems. J. Infrastruct. Syst. **13**(3) 185–194 (2007)
25. Hopkinson, K., Birman, K., Giovanini, R., Coury, D., Wang, X., Thorp, J.: EPOCHS: integrated commercial off-the-shelf software for agent-based electric power and communication. In: Proceedings of Winter Simulation Conference, pp. 1158–1166. Luisiana, 7–10 Dec 2003
26. Pederson, P., Dudenhoeffer, D., Hartley S., Permann, M.: Critical Infrastructure Interdependency Modeling: A Survey of U.S. and International Research, Idahho National Laboratory, August 2006 (2006)
27. De Porcellinis, S., Panzieri, S., Setola, R.: Modelling critical infrastructure via a mixed holistic reductionistic approach. Int. J. Critical Infrastruct. (Inderscience eds.) **5**(1/2), 86–99, Inderscience (2009)
28. Crowther, K.G., Haimes, Y.Y.: Development of the multiregional inoperability input–output model (MRIIM) for spatial explicitness in preparedness of interdependent regions. Syst. Eng. **13**(1), 28–46 (2010)
29. Leontief, W.: The Structure of the American Economy 1919–1939. Oxford University Press, Oxford (1951)
30. Chenery, H.B., Watanabe, T.: International comparison of the structure of production. Econometrica **26**(4), 98–139 (1958)
31. Rasmussen, P.: Studies in Intersectoral Relations. North Holland, Amsterdam, The Netherlands (1956)
32. Luenberger, D.: Introduction to Dynamic Systems: Theory, Models, and Applications. Wiley Ed., New York (1979)

33. Oliva, G., Panzieri, S., Setola, R.: Agent based input–output interdependency model. Int. J. Crit. Infrastruct. Protect. (Elsevier) **3**(2), 76–82 (2010). doi:10.1016/j.ijcip.2010.05.001
34. Oliva, G., Setola, R., Panzieri, S.: Distributed consensus under ambiguous information. Int. J. Syst. Syst. Eng. **4**(1), 55–78 (2013). doi:10.1504/IJSSE.2013.053504
35. Lakshmikantham, V., Mohapatra, R.N.: Theory of Fuzzy Differential Equations and Inclusions. Taylor & Francis, London (2003)
36. Zadeh, L.A.: Fuzzy sets. Inf. Control **8**, 338–353 (1965)
37. Hanss, M.: Applied Fuzzy Arithmetic, An Introduction with Engineering Applications. Springer, Berlin. ISBN 3-540-24201-5 (2005)
38. Kellett, C.M., Teel, A.R.: On the robustness of KL-stability for difference inclusions: smooth discrete-time Lyapunov functions. SIAM J. Control Optim. **44**(3), 777–800 (2005)
39. Oliva, G., Panzieri, S., Setola, R.: Discrete-time LTI fuzzy systems: stability and representation. In: 51th Conference on Decision and Control (CDC2012) Maui USA (2012)
40. Dehghan, M., Ghatee, M., Hashemi, B.: Inverse of a fuzzy matrix of fuzzy numbers. Int. J. Comput. Math. **86**(8), 1433–1452 (2009)
41. Rohn, J.: Inverse interval matrix. SIAM J. Numer. Anal. **30**(3), 864–870 (1993)

Fault Diagnosis and Fault Tolerant Control in Critical Infrastructure Systems

Vicenç Puig, Teresa Escobet, Ramon Sarrate and Joseba Quevedo

Abstract Critical Infrastructure Systems (CIS) are complex large-scale systems which in turn require highly sophisticated supervisory-control systems to ensure that high performance can be achieved and maintained under adverse conditions. The global CIS Real-Time Control (RTC) need of operating in adverse conditions involves, with a high probability, sensor and actuator malfunctions (faults). This problem calls for the use of an on-line Fault Detection and Isolation (FDI) system able to detect such faults and correct them (if possible) by activating Fault Tolerant Control (FTC) mechanisms, as the use of soft sensors or using the embedded tolerance of the controller, that prevent the global RTC system from stopping every time a fault appears. To exemplify the FDI and FTC methodologies in CIS, the Barcelona drinking water network is used as the case study.

1 Introduction

Critical Infrastructure Systems (CIS), such as water, gas or electrical networks, are complex large-scale systems which in turn require highly sophisticated supervisory-control systems. They are geographically distributed and decentralized with a hierarchical structure. Each sub-system is composed of a large number of elements

V. Puig (✉) · T. Escobet · R. Sarrate · J. Quevedo
Automatic Control Department, Technical University of Catalonia (UPC), Rambla Sant, Nebridi, 10, 08222 Terrassa, Spain
e-mail: vicenc.puig@upc.edu

T. Escobet
e-mail: teresa.escobet@upc.edu

R. Sarrate
e-mail: ramon.sarrate@upc.edu

J. Quevedo
e-mail: joseba.quevedo@upc.edu

E. Kyriakides and M. Polycarpou (eds.), *Intelligent Monitoring, Control, and Security of Critical Infrastructure Systems*, Studies in Computational Intelligence 565,
DOI 10.1007/978-3-662-44160-2_10

with time-varying behavior, exhibiting numerous operating modes and subject to changes due to external conditions (e.g., weather) and operational constraints. But, in order to take profit of these expensive infrastructures, it is also necessary to have a highly sophisticated real-time control (RTC) scheme which ensures that high performance can be achieved and maintained under adverse conditions [27, 49]. The advantage of RTC applied to CIS has been demonstrated by an important number of researchers during the last decades. Comprehensive reviews that include a discussion of some existing implementations are given by [32, 47, 49] and cited references therein, while practical issues are discussed by [48], among other. The RTC scheme in CIS might be local or global. When local control is applied, regulation devices use only measurements taken at their specific locations. While this control structure is applicable in many simple cases, in large–scale systems with a strongly interconnected and complex infrastructure of sensors and actuators, it may not be the most efficient alternative. Conversely, a global control strategy, which computes control actions taking into account real-time measurements all through the network, is likely the best way to use the infrastructure capacity and all the available sensor information. Global RTC deals with the problem of generating control strategies for the control elements in a CIS, ahead of time, based on a predictive dynamic model of the system, and readings of the telemetry system, in order to optimize its operation [27, 32]. The multivariable and large-scale nature of CIS have lead to the use of some variants of Model Predictive Control (MPC), as global control strategy [14, 20, 31, 32, 34].

The global RTC need of operating in adverse conditions involves, with a high probability, sensor and actuator malfunctions (faults). This problem calls for the use of an on-line fault detection and isolation (FDI) system able to detect such faults, and correct them (if possible) by activating fault tolerant control (FTC) mechanisms. FTC techniques, such as the use of soft sensors or the retuning of the controller, prevent the global RTC system from stopping every time a fault appears.

The FDI process aims to carefully identify which fault (including hardware or software faults, and malicious attacks) can be hypothesized to be the cause of monitored events. In general, when addressing the FDI problem, two strategies can be found in the literature: hardware redundancy based on the use of redundancies (adding extra sensors and actuators), and software (or analytical) redundancy based on the use of software/intelligent sensors (or model) combining information provided by the sensor measurements or using other actuators to compensate the faulty actuator. In CIS, hardware redundancy is preferred. However, for large-scale systems, the use of hardware redundancy is very expensive and increases the number of maintenance and calibration operations. This is the reason why, in CIS applications, systems that allow combining both hardware and analytical redundancy [12] must be developed.

Instrumentation plays a crucial role in FDI and FTC of CIS. On the one hand, the performance of FDI depends on the set of measurements that are available in the system. Most of them are provided by sensors that are installed in the system. On the other hand, FTC not only depends on the set of sensors, but also on the set of actuators available. Therefore, for a given set of sensors and actuators already installed in the system the maximum FDI and FTC performances that can be achieved are bounded. Thus, designing an FDI module and an FTC ultimately aims at reaching those performance bounds.

To achieve a better performance the instrumentation set up should be redesigned. This is just the main goal of a sensor and actuator placement methodology, i.e. deciding which sensors and actuators would be needed to achieve certain performance specifications. In fact, these performance specifications could be diverse, but in this chapter those related to fault diagnosis and fault tolerance will be covered.

To exemplify the FDI and FTC methodologies in CIS, the Barcelona drinking water network is used as the case study.

The structure of this chapter is the following: Sect. 2 presents the modeling of CIS and its structural analysis. Sections 3–5 present FDI, FTC and the integration of both in CIS. Section 6 discusses the sensor (and the extension to actuator) placement for FDI (and FTC) in CIS. Section 7 presents a description of the Barcelona drinking water network used as a case study, and the results obtained using the proposed FDI and FTC as well as the sensor placement analysis. Finally, Sect. 8 closes the chapter drawing the conclusions.

2 Modeling of CIS for FDI and FTC

2.1 Mathematical Modeling Principles

Most of the widely used methods for FDI and FTC for large-scale CIS rely on the use of models. But, those models are required to be

- Representative of the CIS dynamic behavior.
- Simple enough to allow for a large number of evaluations in a limited period of time, imposed by real-time operation.

In general, mathematical models are primarily obtained by either theoretical/physical modeling and/or experimentally by identification methods. Typically, theoretical and experimental modeling mutually complements each other. Theoretical modeling provides the functional description between physical process data and its parameters, while experimental modeling contains parameters as numeric values whose functional relation with the physical process data remains unknown [23].

However, when building a model of a dynamic process to monitor its behavior, there is always a mismatch between the modeled and the real behavior. This is because some effects are neglected in the model, some non-linearities are linearized in order to simplify the model, some parameters have tolerance when they are compared between several units of the same component, some errors in parameters or in the structure of the model are introduced in the model estimation process, etc.

These modeling errors introduce uncertainty in the model and interfere with the fault detection. To manage this uncertainty properly several authors in the FDI [35, 37] and automatic control [10, 19] communities have suggested the use of interval models. Using this modelling approach, a nominal model plus uncertainty intervals are provided, guaranteeing that all seen data from the system in non-faulty scenarios will be included in the interval for the model prediction. Interval methods are very suitable when additionally to additive uncertainty (noise), modeling uncertainty is present. In particular, no assumption about the noise statistical distribution should be introduced. An alternative to interval based methods are stochastic models and methods [3].

In order to obtain the CIS mathematical model, the constitutive elements and their basic relationships should be considered. Then, a convenient description of the mathematical model of a CIS regarding the application of FDI and FTC is by means of the following discrete-time non-linear model:

$$\begin{aligned} x_{k+1} &= g(x_k, u_k, \theta_k) + w_k \\ 0 &= f(x_k, u_k, \theta_k) + \eta_k \\ y_k &= h(x_k, u_k, \theta_k) + v_k \end{aligned} \tag{1}$$

where: $x \in \mathbb{R}^{n_x}$ is the vector of system states, $u \in \mathbb{R}^{n_u}$ is the vector of control actions and $y \in \mathbb{R}^{n_y}$ is the vector of system outputs; $\theta_k \in \mathbb{R}^{n_\theta}$ is a vector of uncertain parameters; $w_k \in \mathbb{R}^{n_w}$ and $\eta_k \in \mathbb{R}^{n_\eta}$ are unmodelled dynamics and disturbances; $v_k \in \mathbb{R}^{n_v}$ are measurement noises; $g : \mathbb{R}^{n_x} \to \mathbb{R}^{n_x}$ and $h : \mathbb{R}^{n_x} \to \mathbb{R}^{n_y}$ are the state-space and measurement non-linear functions, respectively; and f is the non-linear static relation function.

It should also be noted that the controller could generate control actions at discrete-time points than can change the operational models of the plant (for example, by turning components *on* and *off*, changing component parameter values and the set-point of regulators [8]). These operating mode changes produce discrete changes in the dynamic models of the system behavior. Thus, multiple system models are required to analyse their behavior. Current techniques propose modeling this complex system using a hybrid system model [4, 5] which combines continuous dynamic modeling approaches with a discrete-event model given by an automaton representing the transitions between operation modes, non linearities and faulty situations. The continuous dynamics in each one of the discrete modes could be modeled by (1).

2.2 Structural Analysis

When analyzing the model of a large-scale system for FDI and FTC design, the designer is invariably confronted with a dilemma: to use a more accurate model which is harder to manage, or to work with a simpler model which is easier to manipulate but with less confidence. A hierarchy of models with increasing complexity and fidelity is often used for different purposes (control/supervision design, simulation). However, as the number of variables increases, it is worth to start analyzing the system with simple structural models that offer relatively easy ways to identify unsuitable system configurations, causes for lack of desired properties and straightforward remedies.

The structural model of a system is an abstraction of its behavioral model in the sense that only the constraints (i.e., the link between variables and parameters) are considered but not the constraints themselves. Structural analysis aims at the study of system properties using a structural graph [9, 51]). Structural properties of interest in control that can be derived from structural analysis are input/output reachability and observability/controlability.

With regards to fault diagnosis and tolerant control, structural properties that can be derived from structural analysis are: diagnosability (i.e., capability to detect and isolate one fault from the others) [30] or reconfigurability (i.e., capability to find alternative paths to the variables of interest in case of a fault in sensors or actuators) [9]. It also allows determining critical components (i.e., set of system components that are indispensable to satisfy a determined property) or redundant components (i.e., system components which are not critical for the correct functionality of the system, therefore could be subtracted from the system and the satisfaction of the objective will still be achieved) [52]. Structural analysis also allows decomposing a system into subsystems [50] and can be used to place sensors and actuators for control or supervision [44, 46].

In the structural approach, the behavioral model of a system $\mathcal{M}$ introduced in (1) can be seen as a set of n equations, which depend on a set of m variables $\mathcal{Z} = \mathcal{X} \cup \mathcal{O}$ where $\mathcal{X}$ is the set of unknown variables and $\mathcal{O}$ is the set of observed variables, $(u_i, y_i) \in \mathcal{O}$. A structural model can be formalized as a bipartite graph, $\mathcal{G} = (\mathcal{M}, \mathcal{Z}, \mathcal{A})$ where $\mathcal{M}$ and $\mathcal{Z}$ are the set of vertices and $\mathcal{A}$ is the set of edges, such that $(e_i, z_j) \in \mathcal{A}$ as long as equation $e_i \in \mathcal{M}$ depends on variable $z_j \in \mathcal{Z}$.

A structural model is usually represented by a biadjacency matrix, that relates equations as rows and variables as columns. An element b_{ij} of the biadjacency matrix is 1 as long as $(e_i, z_j) \in \mathcal{A}$.

The Dulmage-Mendelsohn (DM) decomposition [18] is a well-known theoretical tool in the structural model-based fault diagnosis community. The DM decomposition is usually applied to the structural model $\mathcal{G}_{\mathcal{X}} = (\mathcal{M}, \mathcal{X}, \mathcal{A})$, that relates equations and unknown variables. It defines a partition on the structural model. The biadjaceny matrix in Fig. 1 shows the Dulmage-Mendelsohn decomposition of $\mathcal{G}_{\mathcal{X}}$. The gray-shaded areas contain ones and zeros, while the white areas only contain zeros. Three main parts of $\mathcal{M}$ can be identified in the partition, namely, the under-determined part $\mathcal{M}^-$, the just-determined part $\mathcal{M}^0$ and the over-determined part $\mathcal{M}^+$. In the over-determined part, there are more equations than unknown variables, which implies that there exists some degree of redundancy, and this is the part of the model that is useful for process monitoring.

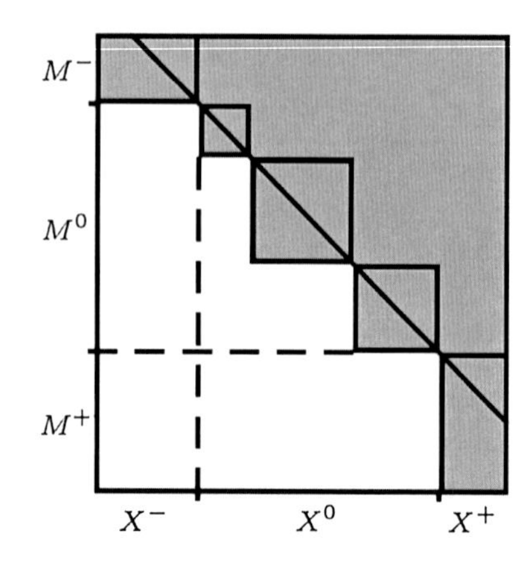

Fig. 1 Dulmage-Mendelsohn decomposition of a model $\mathcal{M}$

3 FDI in CIS

3.1 Background

The ability to detect and isolate faults is an important task in order to safeguard the integrity of critical infrastructure systems. This problem has been attacked from many angles, using very different techniques, and by many researchers, applying different schools of thought, theories and assumptions. An overview of techniques in this area is given in a series of review papers [55–57]. A historical review of these techniques has been published recently in [17].

The diagnostic process aims to carefully identify which fault (including hardware faults, software faults, and malicious attacks) can be hypothesized to be the cause of monitored events. A *fault* must be understood as an unexpected change in a component or system. Although it may not represent physical *failure* or *breakdown*, it may be due to an erroneous state of hardware or software resulting from failures of components, physical interference from the environment, operator error or even an incorrect design. An *error* is the way in which a fault manifests itself, that means the deviation of the system behavior from the required operation. And, a failure is defined as the inability of a component or system to operate according to its specifications. Notice that if a fault occurs, not always manifests as an error, except under certain conditions; and, also, if an error occurs not always results in a system failure. For example, in a water distribution system there are many redundant valves; if a hardware fault causes that a valve remains stuck in a close position, this fault may not manifest until the valve is eventually commanded to open. But, due to physical redundancy the system can operate to perform its required function.

In general, an FDI system consists of three modules. The first is *fault detection*, implemented with a set of fault detection tests based on checking errors, which allows deciding whether a fault has occurred, and its apparition time instant. The second module is *fault isolation*, which is typically achieved through algorithms

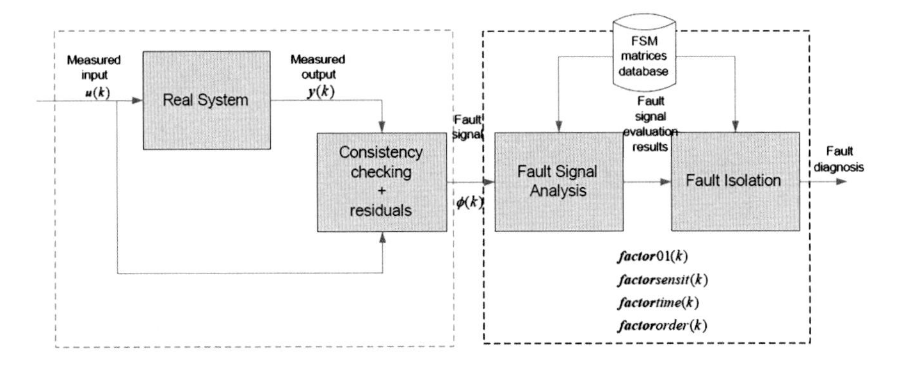

Fig. 2 Fault detection and isolation block diagram

that compute a possible faulty component(s). And the third is *fault identification and estimation*, which aims at determining the kind of fault and its severity. In general, FDI approaches are implemented using a network structure in which the lower level includes a detection module designed to detect abnormal behaviors and triggering alarms, and the higher level tries to isolate and localize the problem. A reduced number of approaches include the third one, unless it is necessary for achieving a fault tolerant control system.

Typically the interface between these fault detection and fault isolation modules is through a binary codification of the evaluation of every residual; this binary interface could lead to wrong diagnosis when the residuals present different sensitivities and order/time of activation after the fault appearance [15], and also produce undesirable decision instability (chattering) due to the effect of the noise and uncertainties. In the literature, there are different approaches to dealing with this problem. For example, Ragot et al. [40] proposed an improved fault diagnosis approach based on the fuzzy evaluation of the residuals that considers not only binary information but also signs/sensitivities as well as the persistence of residual activation. In Puig et al. [36], the use of the Kramer function [33] is proposed for evaluating the residuals gradient and to compute a fault diagnosis signal. The history of this fault diagnosis signal is compared to the stored fault patterns based on an extension of the fault signatures matrix (which includes other signal properties such as signs, occurrence order and time) and, finally, a decision logic algorithm proposes the most probable fault candidate.

The strategy proposed in this chapter for building an FDI system for CIS is shown in Fig. 2. The proposed FDI procedure checks the consistency between the observed and the normal system behavior using a set of analytical redundancy relations, which relate the values for measured variables according to a normal operation (fault-free) model of the monitored system. When some inconsistency is detected, the fault isolation mechanism is activated in order to identify the possible fault.

3.2 ARR Generation

The design of a model-based FDI system is based on utilizing the system model (1) to build a set of consistency tests that only involve observed variables, known as Analytical Redundancy Relations (ARRs). To obtain ARRs for state space representation such as (1), it is necessary to manipulate the model to eliminate the unobserved variables (i.e., the state x).

As has been defined in [16], an ARR is a constraint deduced from the system model which contains only observed variables, and which can therefore be evaluated from any observation. It is noted as $r = 0$, where r is called the residual of the ARR.

Given the model defined in (1) with observed variables y_k and u_{k}, consistency tests can be derived from an ARR by generating a computational residual in the following way:

$$r_i = \Psi_i(y_k, u_k) = 0 \tag{2}$$

where Ψ_i is called the residual ARR expression. The set of ARR can be represented as

$$\mathcal{R} = \{r_i = \Psi_i(y_k, u_k) = 0,\ i = 1, \ldots, n_r\} \tag{3}$$

where n_r is the number of obtained ARRs.

In CIS, these ARRs can be derived following the structural approach, using the algorithms proposed in [9, 25, 54]. In particular, given a set of model equations $\mathcal{M}$, residuals can be obtained from the over-determined part of the model $\mathcal{M}^+$.

Each residual is obtained from a subset of redundant equations in $\mathcal{M}^+$. The minimal subset of redundant equations that are related to a residual r_i is called Minimal Structural Overdetermined (MSO) set [25] and is represented by $\omega_i \subseteq \mathcal{M}^+$. Given a set of model equations, the set of all possible MSO sets is: $\Omega = \{\omega_1, \omega_2, \ldots, \omega_r\}$.

Example 1 Given a model represented by:

$$\begin{aligned} e_1 &: x_1 + y_1 = 0 \\ e_2 &: x_1 + x_2 - y_2 = 0 \\ e_3 &: x_1 - x_2 + y_3 = 0 \\ e_4 &: x_2 - y_4 = 0 \end{aligned}$$

where y_i are known variables, the set of equations is $\mathcal{M} = \{e_1, e_2, e_3, e_4\}$. In this example, the over-determined part is $\mathcal{M}^+ = \mathcal{M}$. Applying the algorithm proposed by Krysander et al. [25] four possible MSO sets are found, $\Omega = \{\omega_1, \omega_2, \omega_3, \omega_4\}$, where their structure and computational form is given by:

$$
\begin{aligned}
r_1 &= y_4 - y_1 - y_2 \rightarrow \omega_1 = \{e_1, e_2, e_4\} \\
r_2 &= y_3 - y_4 - y_1 \rightarrow \omega_2 = \{e_1, e_3, e_4\} \\
r_3 &= y_3 - 2y_1 - y_2 \rightarrow \omega_3 = \{e_1, e_2, e_3\} \\
r_4 &= 2y_4 - y_3 - y_2 \rightarrow \omega_3 = \{e_2, e_3, e_4\}
\end{aligned}
$$

3.3 Fault Detection

Let $\mathcal{F}$ be the set of faults that must be monitored. For example, in a water distribution domain some faults could be pipe leakage or valve blocking.

Definition 3.1 *Detectable fault.* A $f \in \mathcal{F}$ is detectable if its occurrence can be observed, or at least one of the residuals in the residual set (3) satisfies $r_i \neq 0$.

A detectable fault can also be characterized in the structural analysis framework [26]. Without loss of generality, it is assumed that a single fault $f \in \mathcal{F}$ can only violate one equation, denoted by $e_f \in \mathcal{M}$. Then, a fault $f \in \mathcal{F}$ is structurally detectable if there exists at least one MSO $\omega \in \Omega$ such that

$$e_f \in \omega \tag{4}$$

Using the set of computable ARR residuals (3), the fault detection module must check at each time instant whether they are consistent with the observations. Under ideal conditions, residuals are zero in the absence of faults and non-zero when a fault is present.

However, as previously discussed in Sect. 2.1, modeling errors, disturbances and noise in complex engineering systems are inevitable, and hence a need of applying robust fault detection algorithms arises.

In the literature, there are different approaches to solve this problem. For example, statistical decision methods [3] can be used when unknown dynamics and measurement noise are stochastically modeled. In many practical situations, this assumption is not realistic, being more natural to assume that disturbances/ model errors and measurement noise are bounded and their effect is propagated to the residuals using, for example, interval methods [37]. Taking into account bounded uncertainties, the residual of the ARR is monitored by:

$$R_i = \{r_i | r_i = \Psi_i(y_k, u_k, \delta_k), \delta_k \in D\} \tag{5}$$

where D is the interval box $D = \left\{\delta \in \mathbb{R}^{n_\delta} \middle| \underline{\delta} \leq \delta \leq \overline{\delta}\right\}$, that includes all the bounded uncertainties.

Fault detection is formulated as ARR consistency checking using a set-membership approach [53].

Definition 3.2 *ARR consistency checking.* Given a system described by (5) and a sequence of measured inputs u_k and outputs y_k of the real system at time k, one ARR is consistent with those measurement and the known bounds of uncertain parameters and noise if there exist a set of sequences $\delta_k \in$ D which satisfies the ARR.

According to Definition (3.2) a residual of ARR r_i is consistent when zero is included in the interval bounding the residual, that is, $0 \in R_i$.

Definition 3.3 *Fault detection.* Given a sequence of observed inputs u_k and outputs y_k of the real system, a fault is said to be detected at time k if there does not exist a set of sequences $\delta_k \in D$ to which the set of ARRs is consistent.

According to Definition (3.2), a fault is detected when $0 \notin R_i$. The information provided by the consistency checking is stored as fault signal $\phi_i(k)$:

$$\phi_i(k) = \begin{cases} 0 \text{ if } 0 \in R_i \\ 1 \text{ if } 0 \notin R_i \end{cases} \tag{6}$$

3.4 Fault Isolation

While a single residual is sufficient to detect faults, a set (or a vector) of residuals is required for fault isolation [21]. Once the jth residual has been generated, it is evaluated in order to detect normal or abnormal behaviors. In general, a fault f affects a subset of ARRs.

Definition 3.4 *Isolable fault.* A fault f_i is isolable from a fault f_j if the occurrence of f_i can be observed independently of the occurrence of f_j.

Fault isolability can also be characterized in the structural analysis framework [26]. A fault f_i is structurally isolable from a fault f_j if there exists at least one MSO set $\omega \in \Omega$ such that

$$e_{f_i} \in \omega \wedge e_{f_j} \notin \omega \tag{7}$$

Let n_r be the number of residuals available and l the number of detectable faults, a fault signature matrix (FSM) that relates ARRs and faults is defined. This matrix has as many rows as residuals and as many columns as faults are considered. An element Σ_{ij} of this matrix is equal to 1 if the jth fault appears in the expression of the ith residual generator, otherwise it is equal to 0.

Example 2 Given the following FSM

	f_1	f_2	f_3
r_1	1	1	0
r_2	0	1	1

f_1 is isolable from f_3 but not from f_2, because f_2 can not be independently observed from the others.

In model based FDI, the fault effects on the residual can be expressed in terms of the residual fault sensitivity that leads to the residual internal form [21]. In case of considering for example additive faults f_1 and f_2, the internal form of residual r_1 can be expressed as follows:

$$r_1(k) = S_{f_1}(q^{-1})f_1(k) + S_{f_2}(q^{-1})f_2(k) \tag{8}$$

where, $S_{f_1}(q^{-1})$ and $S_{f_2}(q^{-1})$ are the residual fault sensitivity transfer functions that characterize the fault effect on the residual.

The fault isolation module proposed in this chapter is a generalization to a CIS of the one used in [36] (see Fig. 3). The first component is a memory that stores information about the fault signal occurrence history and the fault detection module updates it cyclically. The pattern comparison component compares the memory contents with the stored fault patterns. The classical Boolean fault signature matrix concept [21] is generalized since the binary interface is extended to take into account more fault signal properties. The last component represents the decision logic part of the method whose aim is to propose the most probable fault candidate.

3.4.1 Memory Component

The memory component consists of a table in which events in the residual history are stored. When $\phi_i = 1$, for each row, the first column stores the occurrence time k_o, the second one stores the maximum nominal residual r_i^o computed according to (5) considering the center of the uncertainty intervals δ_o

$$r_{i,\max} = \max_{k \in [k_o, k_o + T_w]} \left(\left| r_i^o(k) \right| \right) \tag{9}$$

for every residual signal, and the last one stores the *sign* of the residual. If the fault detection component detects a new fault signal, it updates the memory by filling out all those fields. The problem of different time instant appearances of the fault

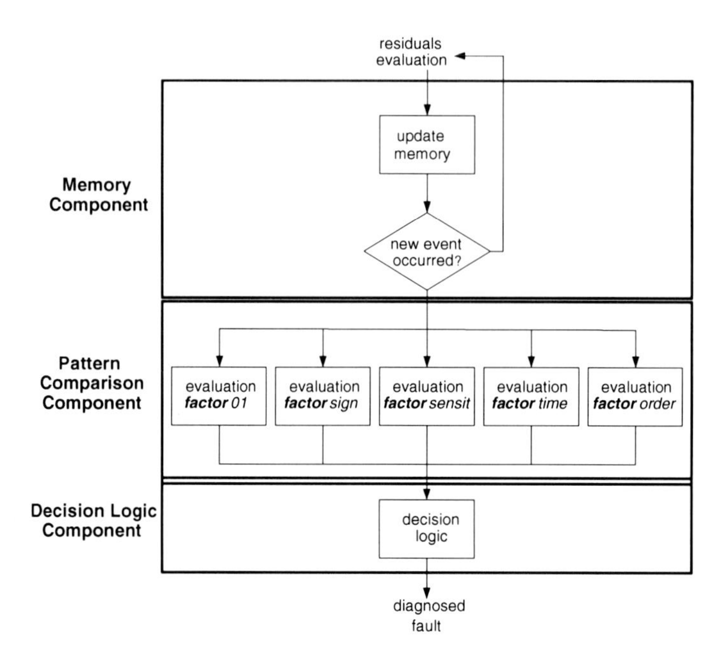

Fig. 3 Fault detection and isolation logic scheme

signal $\phi_i(k)$ is solved not indicating the isolation decision until a prefixed waiting time T_w has elapsed from the first fault signal appearance. This T_w is calculated from the larger transient time response from a non-faulty situation to any faulty situation. After this time has elapsed, a diagnosis is proposed and the memory component is reset in order to be ready to start the diagnosis of a new fault. Following [15], inside this diagnosis window, the maximum activation value of the memory-table $r_{i,\max}$ and for one residual i changes only if the current nominal residual is superior to the previous ones. Due to the max-operator activation values can only rise. Using this strategy the effect of noise and non-persistence fault indicators are filtered because just the peaks of activation are stored. The memory table makes the residual history accessible for later computation by explicitly storing that data. In this way, time aspects of fault isolation can be treated in a very easy and straightforward way.

3.4.2 Pattern Comparison Component

The pattern comparison component compares the memory contents with the stored fault patterns. Fault patterns are organized according to a theoretical ***FSM***. This interpretation assumes that the occurrence of f_j is observable in r_i, hypothesis known as *fault exoneration* or no *compensation*, and that f_j is the only fault affecting the monitored system. Five different fault signature matrices are considered in the evaluation task: Boolean fault signal activation (***FSM*01**), fault signal signs (***FSMsign***), fault residual sensitivity (***FSMsensit***), and, finally, fault signal occurrence order (***FSMorder***) and time (***FSMtime***). Theses matrices can be obtained from the analysis of residual fault sensitivity (8). Details on the general rules to obtain those matrices from (8) can be found in [29].

3.4.3 Decision Logic Component

The decision logic algorithm starts when the first residual is activated (that is, $\phi_i = 1$) and lasts T_w time instants or till all fault hypotheses except one are rejected because they do not fulfill the observed residual activation order/time or because an unexpected activation signal has been observed according to those fault hypotheses. Rejection is based on using the results of $\boldsymbol{factor01}_j$, $\boldsymbol{factorsign}_j$ and $\boldsymbol{factororder}_j$. If any of these factors is 'zero' for a given fault hypothesis, it will be rejected. Every factor represents some kind of a filter, suggesting a set of possible fault hypotheses. At the end of the time window T_w, for each non-rejected fault hypothesis, a fault isolation indicator is calculated using $\boldsymbol{factorsensit}_j$ and $\boldsymbol{factortime}_j$ factors. Thus, the biggest fault isolation indicator will determine the diagnosed fault. The fault isolation indicator associated to the fault hypothesis f_j is determined as it follows:

$$d_j = \max\left(\left|\boldsymbol{factorsensit}_j\right|, \boldsymbol{factortime}_j\right) \tag{10}$$

So, the final diagnosis result can be expressed as a set of fault candidates with their associated fault isolation indicator.

4 FTC in CIS

4.1 Motivation

Fault-tolerant control is an incipient research area in the automatic control field [9]. One way of achieving fault-tolerance is to employ a fault detection and isolation (FDI) scheme on-line. This system will generate a discrete event signal to a supervisor system when a fault is detected and isolated. The supervisor, in turn will activate some accommodation action in response, which can be pre-determined for each fault or obtained from real-time analysis and optimization. Due to the discrete-event nature of fault occurrence and reconfiguration/accommodation, an FTC system is a hybrid system by nature. Therefore, the analysis and design of FTC systems is not trivial. For design purposes of these systems, traditionally the hybrid nature has been neglected in order to facilitate a simple design, reliable implementation, and systematic testing. In particular, the whole FTC scheme can be

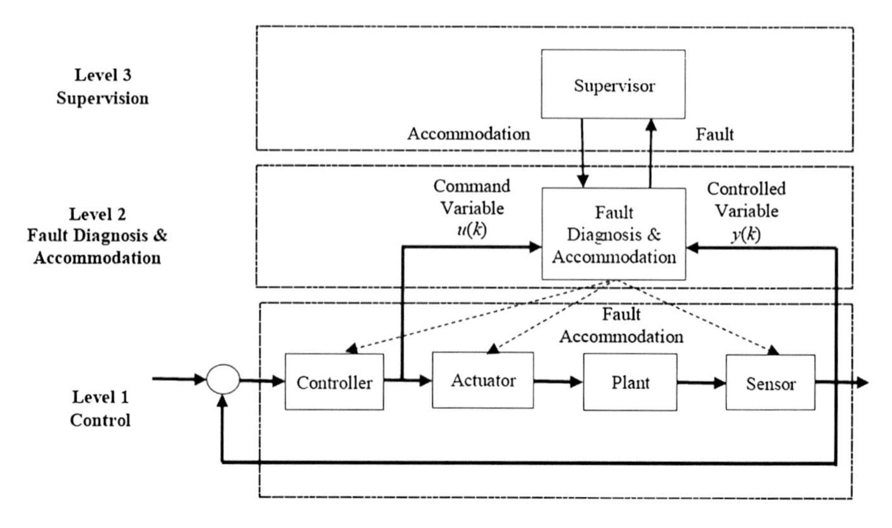

Fig. 4 Fault-tolerant control architecture

expressed using the three-level architecture proposed by Blanke et al. [9] (see Fig. 4).

- Level 1: "*Control Loop*". This level comprises a traditional control loop with sensor and actuator interfaces, signal conditioning and filtering and the controller.
- Level 2: "*Fault Diagnosis and Accommodation*". The second level comprises a given amount of detectors, usually one per each fault effect which will be detected, and effectors that implement the desired reconfiguration or other remedial actions given by the autonomous supervisor.
- Level 3: "*Supervision*". The supervisor is a discrete-event dynamical system (DEDS) which comprises state-event logic to describe the logical state of the controlled object. The supervisor functionality includes an interface to detectors for fault detection and demands remedial actions to accommodate the fault.

4.2 Fault-Tolerant MPC

Model Predictive Control (MPC) has become the accepted standard for CIS [32]. This control strategy computation is based on the implementation of a receding horizon control strategy that poses and solves an optimal control problem at each time k [11].

$$\min_{\tilde{u}_k} J(\tilde{x}_k, \tilde{u}_k, \tilde{d}_k) \tag{11}$$

subject to:

$$\begin{cases} x(k|j+1) = f(x(k|j), u(k|j), d(k|j), \theta(k)) \\ u(k|j) \in \mathbb{U} \quad j = 0, \ldots, H_p - 1 \\ x(k|j) \in \mathbb{X} \quad j = 1, \ldots, H_p \end{cases}$$

where:

$$\mathbb{U} = \left\{ u \in \mathbb{R}^m | u^{\min} \leq u \leq u^{\max} \right\}$$
$$\mathbb{X} = \left\{ x \in \mathbb{R}^n | x^{\min} \leq x \leq x^{\max} \right\}$$

and

$$\tilde{u}_k = (u(k|j))_{j=0}^{H_{p-1}} = \left(u(k|0), u(k|1), \ldots, u(k|H_{p-1}) \right)$$
$$\tilde{x}_k = (x(k|j))_{j=1}^{H_p} = \left(x(k|1), x(k|2), \ldots, x(k|H_p) \right)$$
$$\tilde{d}_k = (d(k|j))_{j=0}^{H_{p-1}} = \left(d(k|0), d(k|1), \ldots, d(k|H_{p-1}) \right)$$

According to this algorithm, at each time step, a control input sequence $\tilde{u}_k$ of present and future values is computed to optimize the performance function $J(\tilde{x}_k, \tilde{u}_k, \tilde{d}_k)$, according to a prediction of the system dynamics over the horizon H_p. However, only the first control input $u_{k|0}$ is actually applied to the system. The same procedure is restarted at time $k + 1$, using the new measurements obtained from sensors that allow estimating the actual value of system states to initialize the optimization problem (11). In this way, feedback from the telemetry system is used that allows the optimal control strategy to be re-computed at each time k.

At the next time step, the computation is repeated starting from the new state and over a shifted horizon, leading to a moving horizon policy. The solution relies on a linear dynamic model, that respects all input and output constraints, and optimizes a quadratic performance index. Thus, as much as a quadratic performance index together with various constraints can be used to express true performance objectives, the performance of MPC is excellent. Over the last decade a solid theoretical foundation for MPC has emerged so that in real-life large-scale MIMO applications controllers with non-conservative stability guarantees can be designed routinely and with ease [38, 41].

Fault-tolerance against faults can be embedded in MPC relatively easily [28]. This can be done in several ways:

(a) Changing the constraints in order to represent certain kinds of faults, being especially "easy" to adapt the algorithms for faults in actuators.
(b) Modifying the internal plant model used by the MPC in order to reflect the fault influence over the plant.
(c) Relaxing the initial control objectives in order to reflect the system limitations under fault conditions.

However, each way relies on an associated assumption:

(a) The nature of the fault can be located and its effects modeled.
(b) The internal model of the plant can be updated, essentially in an automatic manner.
(c) The set of control objectives defined in the MPC design process can be left unaltered once the fault has occurred.

5 Integration of FDI and FTC

One important question to emphasize is that the key to any FDI and FTC mechanisms is the existence of system redundancies (physical or analytical). Physical redundancies always come at additional cost for extra components and with added inconveniences, such as increased weight, size and cost of maintenance. Analytical redundancies necessitate building precise analytical models of the system and adapting these models to the operating condition changes of the CIS. Clearly, one has to seriously analyze the problem at hand to justify the use of FDI and/or FTC of CIS.

In the literature, the research and technological efforts on active fault tolerant control systems have been focused on fault detection/isolation/estimation mechanisms and on control system reconfiguration schemes. As a matter of fact, a significant amount of research and techniques have been accomplished separately in these areas. In comparison, relatively little work [24] has been done to integrate the developed methodologies from these areas together to produce an efficiently active fault tolerant control system.

Concerning the controller reconfiguration, a lot of work has been carried out under the assumption that a perfect model of the faulty system is already available and the task of reconfigurable control is simply to stabilize the faulty system and to recover the original system performance as much as possible based on the exact fault system model. Certainly, the availability in real-time of the fault system model is a big assumption. The uncertainty of the fault diagnosis module to isolate the real fault from the limited number of sensors in CIS, the difficulty to properly estimate the real magnitude of the fault in the system and the time delay to detect, isolate and estimate the fault system model are three important questions to take into account in the design of a fault tolerant control mechanism.

For actuator faults in a multivariable system such as a CIS, the FDI and FTC modules can be integrated switching the actuator suspicious of abnormal behavior by another safety actuator without waiting for the estimation of the fault magnitude in the suspicious actuator. In [39], a fault tolerant controller for a PEM fuel system has been designed considering that one actuator (the compressor) is switched off because it is suspicious to be faulty, and another actuator (fuel valve) is used to control the excess of oxygen in the PEM fuel cell system. This strategy is

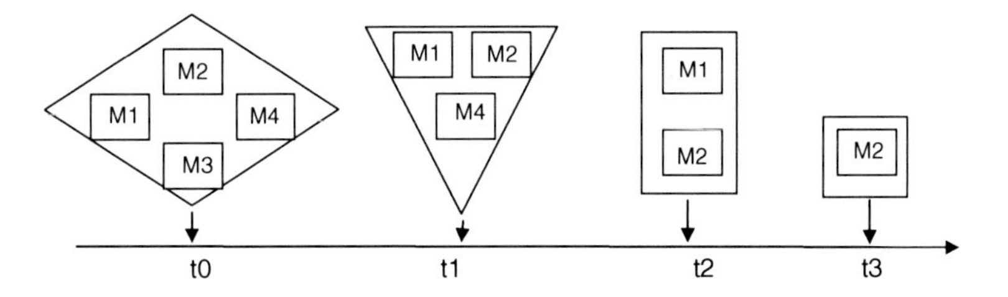

Fig. 5 The falsification model method in one simple example

quite conservative, because the mechanisms switch off a suspicious component that could be partly used should the actuator not be broken.

Another strategy is to take into account the dynamic knowledge of the fault situation in real time using model falsification method by [13, 22] which uses a different philosophy. Rather than identifying the most likely model of the faulty system, models that are not compatible with the input/output data are invalidated, thus avoiding the computation of decision thresholds. Figure 5 shows how the model falsification method works for a simple example. Assume that at a time four possible models M1, M2, M3 and M4 exist for a given real system. We are interested in deciding which model (if any) is able to explain the input/output data sequence that we are obtaining from the sensors and actuators measurements. Therefore, assume that, at a given time, t_0, all four models are plausible, as depicted in Fig. 5. Further, suppose that at time t_1, model M3 is invalidated, i.e., sensor measurements cannot be explained by model M3. Moreover, consider that at time t_2 model M4 is invalidated, and finally, model M1 is invalidated at time t_3. Then at time t_3, we conclude that the only model capable of explaining the sensor-actuator measurements by the system is model M2.

The fault tolerant control strategy based on this falsification model method requires the design of a set of passive fault tolerant controllers obtained from a large number of fault models (M1, M2, M3 and M4 in the example) to only one fault model (M2 in the example). Obviously, as the number of potential failure scenarios increase, the overall performance of the controller becomes less and less effective for each fault.

The philosophy of an integrated FDI and FTC scheme is essential to integrate all the tasks (detection/isolation/estimation of the fault and reconfiguration) of an active fault tolerant control in one design cycle. It is shown in the previous section that MPC is a very suitable method. The reason is because it integrates the estimation of the model parameters, and actuators/sensors new limits and adequate in real time the control actions to the new model of the system taking into account explicitly the new sensor/actuators restrictions. And for this reason, we select the MPC as the most interesting strategy to integrate the FDI and FTC for CIS.

6 Sensor and Actuator Placement

6.1 Background

In spite of their simplicity, structural models can provide much useful information for FDI and FTC design which in turn makes solving the sensor and actuator placement problem easier. In the literature, several researchers have addressed the optimal sensor and actuator placement problem. Bagajewicz et al. [2] formulate a mixed-integer linear program based on a digraph representation of the fault propagation behaviour. The goal is to design a sensor network for a chemical process that provides a good estimate of the state of the system and such that a set of faults can be detected and isolated.

In [26], the sensor placement problem is addressed by analyzing the analytical redundancy properties of a structural model based on the Dulmage-Mendelsohn (DM) decomposition, an efficient tool to analyze bipartite graphs. A comparative study on different sensor placement strategies is done in [46].

The problem of sensor placement in water networks has been addressed in several works. In [44] the cheapest sensor configuration is chosen such that some leak detection and isolability properties are satisfied, whereas in [45] sensors are chosen such that the leak isolability property is maximized while satisfying a budgetary constraint on the sensor configuration cost. In [6], water safety and security are addressed taking into account the network topology. Sensors are placed in order to assure the early detection of maliciously injected contaminants.

Transport networks have also been the focus of sensor placement strategies. In [7], the minimum number and location of counting points is determined in order to infer all traffic flows. Anderson et al. [1] develop a sensor location methodology aiming for homeland security. The goal is to optimally place monitoring devices for detecting malicious entities or materials transported on roadways around urban regions.

Sensor placement based on fault tolerant properties has been addressed in [52]. Sensors are chosen such as a redundancy degree is guaranteed against sensor failures when a reconfiguration strategy is used. The same concept of redundancy degree is applied in [42] to evaluate the fault tolerant properties of a system under actuator faults. Based on these ideas an actuator placement methodology could be easily developed.

6.2 General Problem Formulation

Let $\boldsymbol{I}$ be the set of candidate instruments. This set contains all sensors and actuators that are eligible for installation.

Let F be the set of faults that can affect process components. In general, these faults can concern components that are already installed in the system, as well as candidate instruments that are chosen by the sensor and actuator placement algorithm.

The fault diagnosis and fault tolerance properties will be stated in terms of F. Some examples of such properties are: the set of detectable faults, the set of isolable fault pairs, or the redundancy degree to actuator or sensor fault tolerance. Assume that a function $\mathcal{P}$ exists that evaluates whether a given set of fault diagnosis and fault tolerance properties T would be satisfied if an instrument configuration $I \subseteq \mathbf{I}$ was chosen for installation. Thus:

$$\mathcal{P}(\mathcal{M}, T, I) = \begin{cases} 1, & \text{if } T \text{ is fulfilled} \\ 0, & \text{otherwise} \end{cases} \tag{12}$$

Remark that this function requires a system model $\mathcal{M}$. In our approach this will be a structural model.

Function $\mathcal{P}$ induces a two-class partition in the set $2^{\boldsymbol{I}}$ as follows:

$$\begin{aligned} \left[2^I\right]^+ &= \{I \subseteq \mathbf{I} | \mathcal{P}(\mathcal{M}, T, I) = 1\} \\ \left[2^I\right]^- &= \{I \subseteq \mathbf{I} | \mathcal{P}(\mathcal{M}, T, I) = 0\} \end{aligned}$$

Thus, $[2^{\mathbf{I}}]^+$ contains all instrument configurations that satisfy T. Among these, an instrument configuration I^* exists that minimizes a certain cost function J. Therefore, the optimal sensor and actuator placement problem can be formally stated as follows:

$$\begin{aligned} &\min_{\forall I \in \mathbf{I}} J(\mathcal{M}, I) \\ &\text{subject to:} \mathcal{P}(\mathcal{M}, T, I) = 1 \end{aligned} \tag{13}$$

Remark that J also requires, in general, the process knowledge through model $\mathcal{M}$.

6.3 Solving the Sensor Placement Problem

Several approaches exist that solve the optimal problem defined in (13). In this section, the approach proposed in [44] is recalled, where an optimal sensor placement problem is solved based on a branch and bound strategy.

The cost function J in (13) is stated in terms of the cost of the sensor configuration S, $C(S)$. This cost can concern different issues, such as economical price, reliability, precision and ease of installation.

The set of system properties T in (13) is stated through two sets:

$$\begin{aligned} F_D &= \{f \in F | f \text{ is detectable}\} \\ F_I &= \left\{(f_i, f_j) \in F \times F \middle| f_i \text{ is isolable from } f_j\right\} \end{aligned} \tag{14}$$

Thus, fault detectability and isolability specifications are stated.

The recursive *Algorithm 1 searchOp* solves the sensor placement problem based on a depth-first search with backtracking. Function *isFeasible* corresponds to function $\mathcal{P}$ in (13), and it is implemented based on the DM decomposition of the structural model. The fault detectability property is implemented based on Eq. (4), whereas the fault isolability property is determined applying Eq. (7).

Algorithm 1 S^*:= searchOp(*node*, S^*, $C(\cdot)$, M, F_D, F_I)

```
for all s ∈ node.R ordered in decreasing cost do
    childNode.S := node.S\{s}
    node.R := node.R\{s}
    childNode.R := node.R
    if C(childNode.S\childNode.R) < C(S*) and
    isFeasible(childNode.S, M, F_D, F_I) then
        if C(childNode.S) < C(S*) then
            S* := childNode.S
        end if
        S* := searchOp(node, S*, C(·), M, F_D, F_I)
    end if
end for
return S*
```

A tree is built during the search, where each node in the tree consists of two sensor sets:

- *node.S*: a candidate sensor configuration to test
- *node.R*: a set of sensors that can be individually removed to create the corresponding child nodes in the tree.

Throughout the search, the best solution is updated in S^*, whenever a feasible solution with a lower cost than the current best one is found. A branch exploration is terminated at some node whenever any of the following two conditions is fulfilled:

- The lowest sensor configuration cost among the descendants of the current node is not lower than the cost of the current best solution.
- The fault detectability and isolability specifications for the current sensor configuration are not satisfied.

Algorithm 1 is initialized with the candidate sensor set as the root node of the search tree, the candidate sensor set as the current best sensor configuration, the cost corresponding to the candidate sensors, the fault-free system model and the desired fault detectability and isolability specifications.

6.4 Guidelines to Solve Other Sensor and Actuator Placement Problems

In the previous section, the solution to a particular sensor placement problem has been described. In this section, some guidelines will be given on how to address other alternative sensor and actuator problems derived from (13).

Algorithm 1 searches the lowest cost sensor configuration that fulfills the given fault diagnosis specifications (14). However, ensuring such diagnosis specifications might lead to an optimal solution with a large sensor configuration cost that could exceed the budget assigned by the company. A more appealing approach for a company is the one developed by Sarrate et al. [45]. A key contribution of this work is the concept of the isolability index as a measurement of the fault diagnosis performance achievable in a system. This index measures the number of fault isolability pairs in F_I (i.e., $|F_I|$). Then, this measurement allows to set up a sensor placement problem based on a fault diagnosis performance maximization under the constraint of a given maximum sensor configuration cost. Thus, this new formulation becomes appropriate in CIS with a bound in the budget assigned to instrumentation.

The fault tolerance property of a system critically depends on the set of sensors and actuators available. Thus, faults affecting either sensors or actuators can jeopardize its fault tolerance capability. In [9], the structural observability and controllability are characterized. Observability and controllability are key system properties to address fault tolerance. Observability states whether a given set of measurements (i.e., sensors) are sufficient to estimate the system state, whereas controllability states whether a given set of control variables (i.e., actuators) are sufficient to bring the system to a target state.

In [52], the concept of redundancy degree concerning sensors is introduced. For a given sensor configuration the sensor redundancy degree states how many sequential sensor faults can be tolerated such that the system fault tolerance capability is preserved. *Algorithm 1* could be adapted to address the fault tolerant specifications by reformulating the function *isFeasible* such that, for a given sensor configuration, it checks whether a nominal sensor redundancy degree is satisfied. The fault tolerance capability could be measured based on the structural observability, which could be evaluated by applying the DM decomposition.

The concept of redundancy degree can also be applied to actuators. For a given actuator configuration, the actuator redundancy degree states how many sequential actuator faults can be tolerated such that the system fault tolerance capability is preserved [42]. In such a framework, a similar algorithm to *Algorithm 1* could be designed to address the fault tolerant specifications. The algorithm would search for an actuator configuration such as a nominal actuator redundancy degree is satisfied. Function *isFeasible* should be reformulated so that it measures the actuator fault tolerance capability based on the structural controllability, which could be evaluated by applying the DM decomposition.

7 Application Case Study: Barcelona Water Network

7.1 Description of Network

7.1.1 Transport Network

The Barcelona water network supplies water to approximately 3 million consumers, distributed in 23 municipalities in a 424 km^2 area. Water can be taken from both surface and underground sources. The most important ones in terms of capacity and use are Ter, which is a surface source, and Llobregat, where water can be taken from one surface source and one underground source. Water is supplied from these sources to 218 demand sectors through around 4645 km of pipes. The complete transport network has been modelled using: 63 storage tanks, 3 surface sources and 7 underground sources, 79 pumps, 50 valves, 18 nodes and 88 demands. The network is controlled through a SCADA system (Fig. 6) with sampling periods of 1 h. For the predictive control scheme a prediction horizon of 24 h is chosen. This record is updated at each time interval.

In Fig. 7, the whole network representation using elements of the modelling and optimal control tool used is shown. It is a simplified model of the real system:

- Each demand is actually another more detailed network of connections to hundreds or thousands of users.
- Each actuator can integrate several pumps or valves working in parallel.

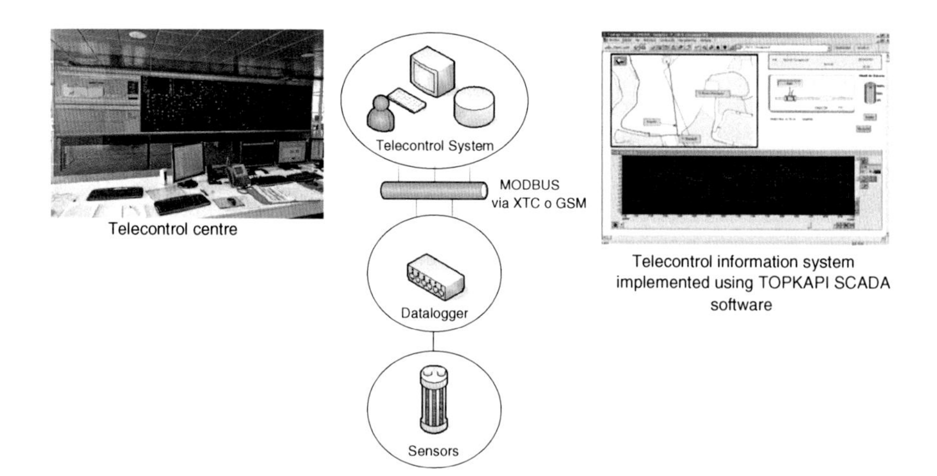

Fig. 6 Telecontrol of Barcelona water distribution system

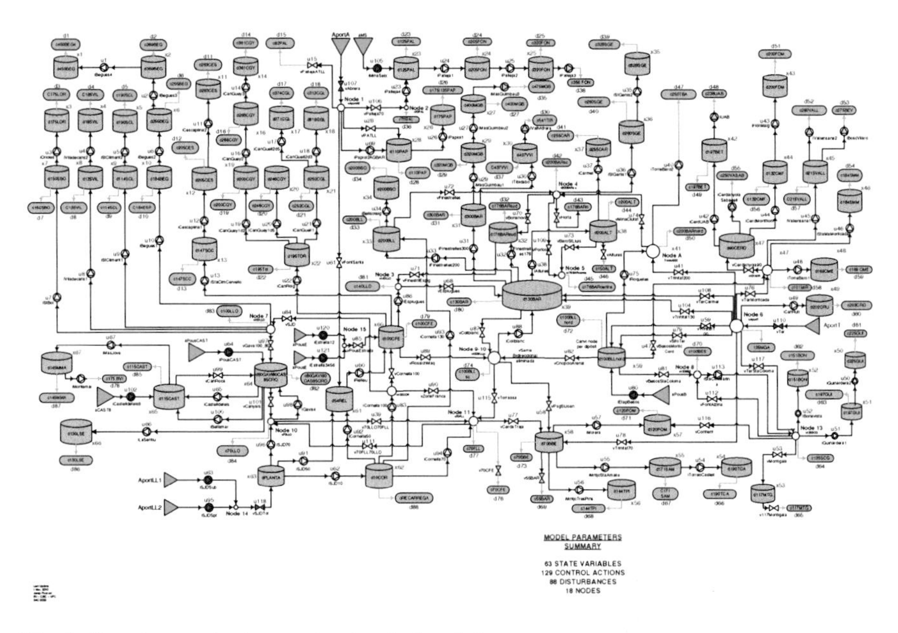

Fig. 7 Barcelona water transport network description

7.1.2 Distribution Network

The distribution network is organised as a District Metered Area (DMA). The DMA network used as the case study in this work is located in the Barcelona area (see Fig. 8) located at pressure floor 55 of the water transport network. It has 883 nodes and 927 pipes. The network consists of 311 nodes with demand (RM type), 60 terminal nodes with no demand (EC type), 48 nodes hydrants without demand (HI type), 14 dummy valve nodes without demand (VT type) and 448 dummy nodes without demand (XX type). The network has two inflow inputs modeled as two additional reservoir nodes.

7.2 Mathematical Model

A water network system will generally contain a number of flow or pressure-control elements, located at the supplies, at the water treatment plant inlets or within the network, and controlled through the telecontrol system. The dynamic model of a water network is represented as in (1). This general model could be specified by basic relationships. Function g represents the mass and energy balance in the water network and relates the stored volume in tanks, x, with the manipulated tank inflows and outflows; it can be written as the difference equation plus a saturation

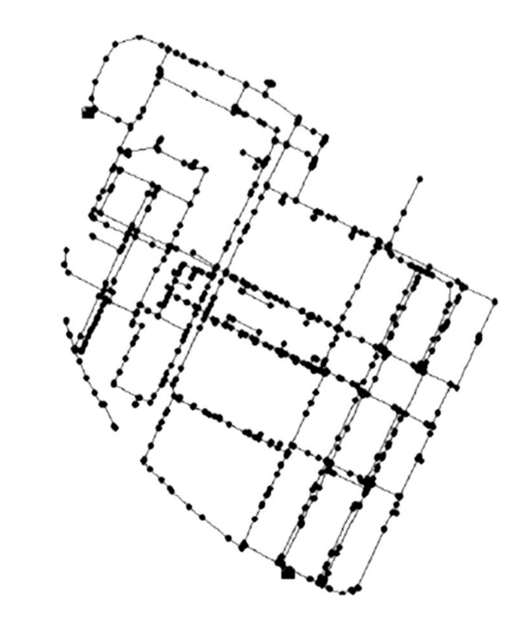

Fig. 8 Barcelona DMA network

$$
\begin{aligned}
& x_i(k+1) = x_i(k) + \left(\sum_i q_{in,i}(k) - \sum_j q_{out,j}(k)\right) \\
& x^{\min} \le x \le x^{\max}
\end{aligned} \tag{15}
$$

where $q_{in,i}(k)$ and $q_{out,j}(k)$ correspond to the i-th tank inflow and the j-th tank outflow, respectively, given in m^3/s, and x_{min} and x_{max} denote the minimum and the maximum volume capacity, respectively, given in m^3.

In a water network, nodes correspond to intersections of mains. Function f represents the static equation that expresses the mass conservation in these elements can be written as

$$
\sum_i q_{in,i}(k) - \sum_j q_{out,j}(k) = 0 \tag{16}
$$

where $q_{in,i}(k)$ and $q_{out,j}(k)$ correspond to the i-th node inflow and the j-th node outflow, respectively, given in m^3/s. Input and outputs flows in a the transport network are controlled using actuators (pumps and valves) while in the distribution network those flows depend of the difference of pressure between nodes

$$
q_l = c\,\mathrm{sgn}(p_i - p_j)\left(|p_i - p_j|\right)^{\gamma} \tag{17}
$$

7.3 FDI in the Barcelona Water Network

The case study used to illustrate the FDI methodology proposed in this chapter is based on part of this network. It includes two subsystems, known as Orioles and Cervello (Fig. 9). This part of the network includes the following elements:

- Tanks: d150SBO, d175LOR, d147SCC, d205CES, d263CES
- Actuators with sensor flows: iStBoi, iOrioles, iStaClmCervello, iCesalpina1, iCesalpina2
- Demands with sensor flows: c157SBO, c175LOR, c147SCC, c205CES, c263CES
- Sensor levels: d150SBO, xd175LOR, xd147SCC, xd205CES, xd263CES

In this study case the aim is to detect faults in actuators and sensors.

The case study model has 28 equations, being the set of unknown variables $\mathcal{X} = \{x_i, q_{in,i}, q_{out,i}, d_i\}$ and the set of known variables $\mathcal{O} = \{u_i, y_j\}$ for $i = 1,..,5$

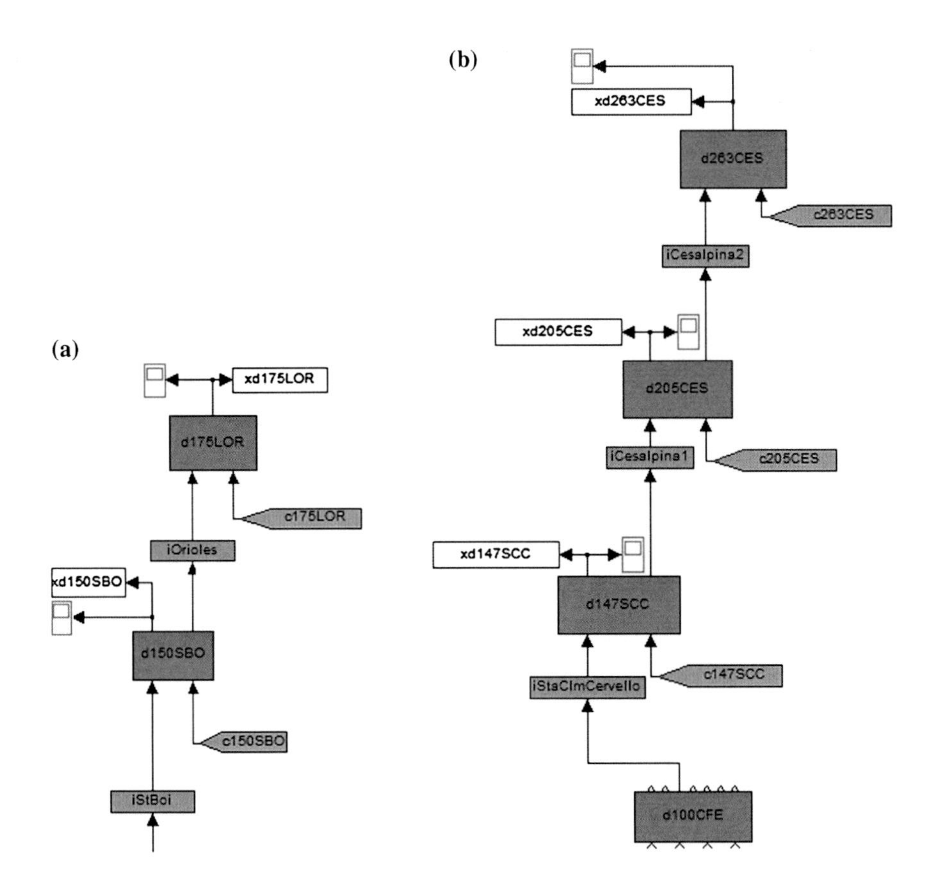

Fig. 9 Case study subsystems. **a** Orioles subsystem. **b** Cervello subsystem

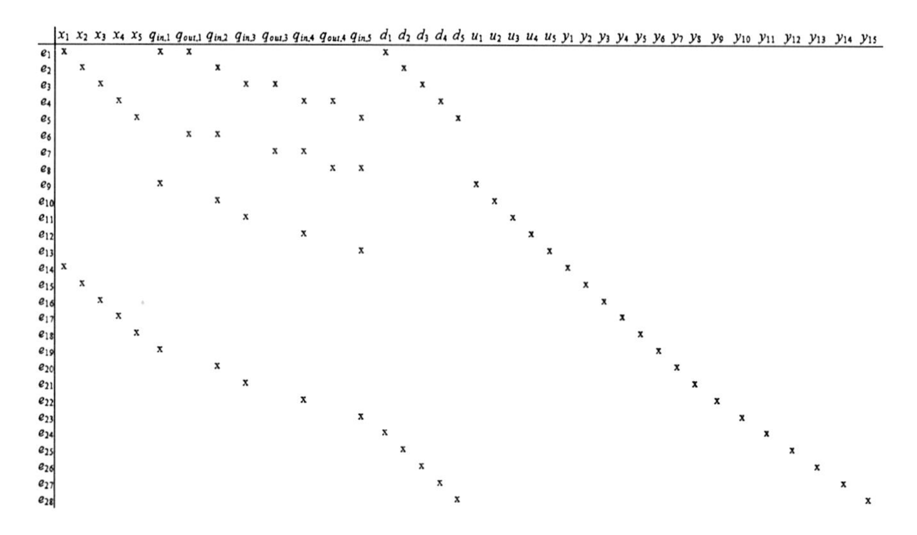

Fig. 10 Biadjacency matrix of the study case

and $j = 1,..,15$, where d_i is the demand and y_j means all the measured variables, which include the sensors described below. The structural model has been represented by a biadjacency matrix (Fig. 10).

Applying the algorithm proposed by Rosich et al. [43], 21 possible MSO sets are found. The computational form of each ARR residual could be implemented by a non-linear observer when the ARR includes tanks dynamics or a parity equation otherwise.

Considering fault in the actuators, f_{Pi}, flow transducers, f_{Fi}, level transducers, f_{Li}, and demand transducers f_{di}, for $i = 1,\ldots,5$, the fault signature matrix shown in Fig. 11 is derived. This fault signature matrix includes binary and sign information.

According to Definitions 3.1 and 3.4 all considered faults are detectable and only f_{Pi} and f_{Fi} are isolable. For instance, in Fig. 11 $\{f_{\mathrm{Li}}, f_{\mathrm{di}}\}$ can not be isolated because both are not observed independently. But if information about the residual sign is taken into account, both can be distinguished. Notice that the information provided for both transducers, $\{f_{\mathrm{Li}}, f_{\mathrm{di}}\}$, are essential for computing residuals; they can be considered as critical sensors. A failure in one of these sensors modifies the ARR sets, $\mathcal{R}$, resulting to an undetectable fault.

The fault detection and isolation procedure described in Sect. 3 has been applied in a simulation case. Figure 12 shows the first 8 residuals and fault signal evolution when a drift in sensor iOrioles flow, f_{P2}, is introduced at hour 362.

The time evolution of *factor*01 and *factor*sign are plotted at every time instant in Fig. 13. It can be seen that both factors indicate as a maximum fault hypothesis f_{P2}. There are also others activated factors but with a less indication.

	f_{P1}	f_{P2}	f_{P3}	f_{P4}	f_{P5}	f_{F1}	f_{F2}	f_{F3}	f_{F4}	f_{F5}	f_{L1}	f_{L2}	f_{L3}	f_{L4}	f_{L5}	f_{d1}	f_{d2}	f_{d3}	f_{d4}	f_{d5}
r_1	(-)1					(+)1														
r_2		(-)1					(+)1													
r_3		(+)1										(+)1					(-)1			
r_4							(+)1					(+)1					(-)1			
r_5	(+)1	(-)1									(+)1					(-)1				
r_6		(-)1				(+)1					(+)1					(-)1				
r_7	(+)1						(-)1				(+)1					(-)1				
r_8						(+)1	(-)1				(+)1					(-)1				
r_9			(-1)1					(+)1												
r_{10}				(-)1					(+)1											
r_{11}					(-)1					(+)1										
r_{12}			(+)1	(-)1									(+)1					(-)1		
r_{13}				(-)1				(+)1					(+)1					(-)1		
r_{14}			(+)1						(-)1				(+)1					(-)1		
r_{15}								(+)1	(-)1				(+)1					(-)1		
r_{16}				(+)1	(-)1									(+)1					(-)1	
r_{17}					(-)1				(+)1					(+)1					(-)1	
r_{18}				(+)1						(-)1				(+)1					(-)1	
r_{19}									(+)1	(-)1				(+)1					(-)1	
r_{20}					(+)1										(+)1					(-)1
r_{21}										(+)1					(+)1					(-)1

Fig. 11 Theoretical fault signature matrix FSM using *binary* and *sign* information

7.4 FTC in the Barcelona Water Network

Four tests/analyses have been made over the case study. Figure 14 shows the sequence of tests performed. The first test consists of finding the critical network actuators by means of its structural analysis. These critical actuators are those which, without them (outage), the connectivity of a path is lost. Results of this test, collected in Table 1, show an important number of critical actuators within the network. This fact is caused by the topology and the way the network elements are connected: most actuators (either valves or pumps) are just connected between tanks and demands. Therefore, if an actuator fault occurs, the corresponding demand will not be satisfied.

The second analysis performed on the Barcelona DWN focuses on the determination of those actuators whose physical constraints limit the capacity of water transport thought a certain path. Notice that this analysis does not consider any fault in these actuators. The analysis, performed by using an optimization algorithm, also results in the determination of several alternative paths through which the water transport is possible (or even mandatory) given the constraints of the paths for supplying demands. Results of this latter analysis yields the determination of another critical actuator: actuator 52 (namely *vBesos-MontCerd*). Notice that the increase in the amount of constraints is not significant.

The third analysis naturally focuses on the set of optimal paths including the objective function of the MPC controller and the system constraints (in states and

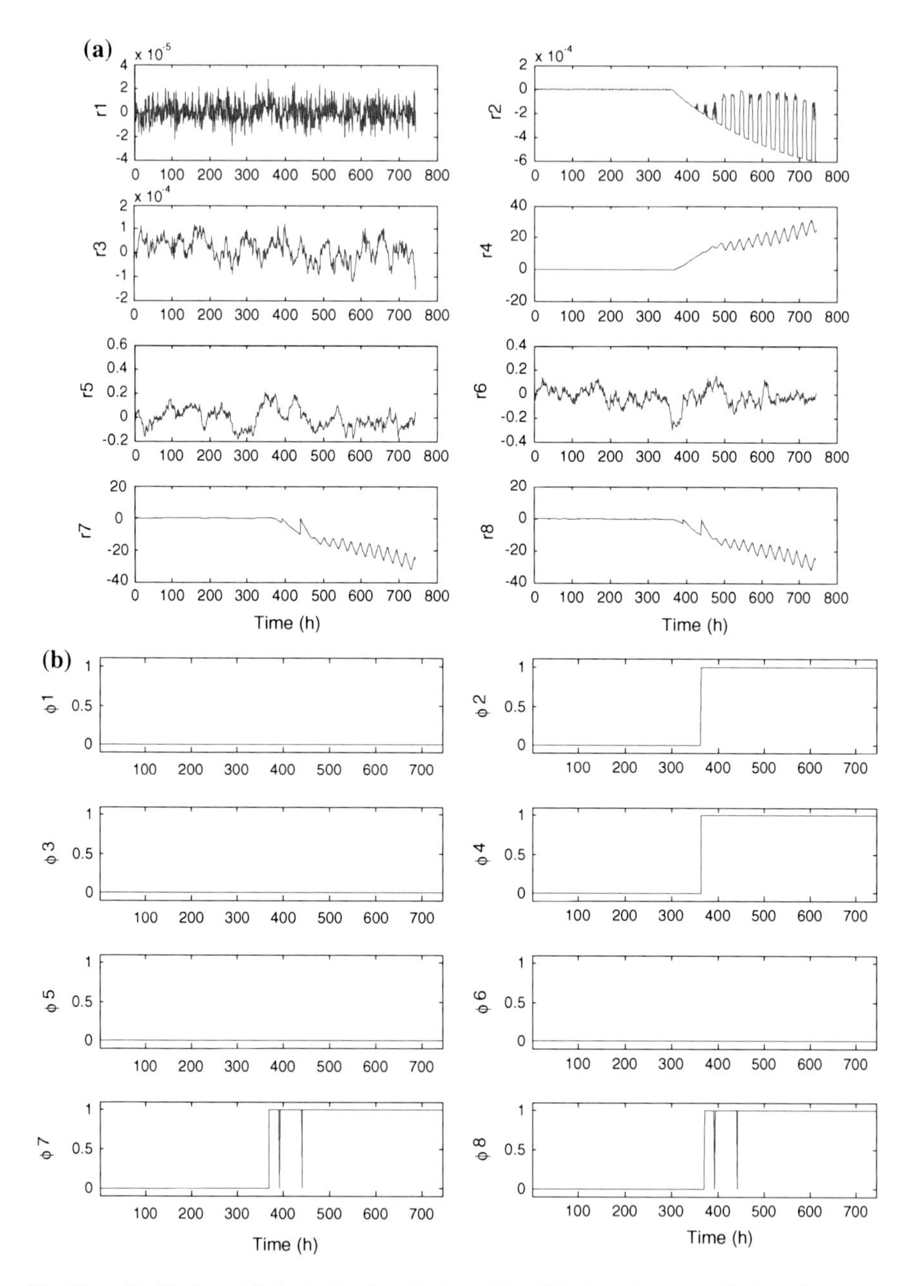

Fig. 12 **a** Residuals and **b** fault signal evolution with a drift fault in sensor iOrioles flow

actuators). Path details are not provided here, but the total economic cost of maintaining a functional network with satisfaction of all the demands is 502.25 e.u.

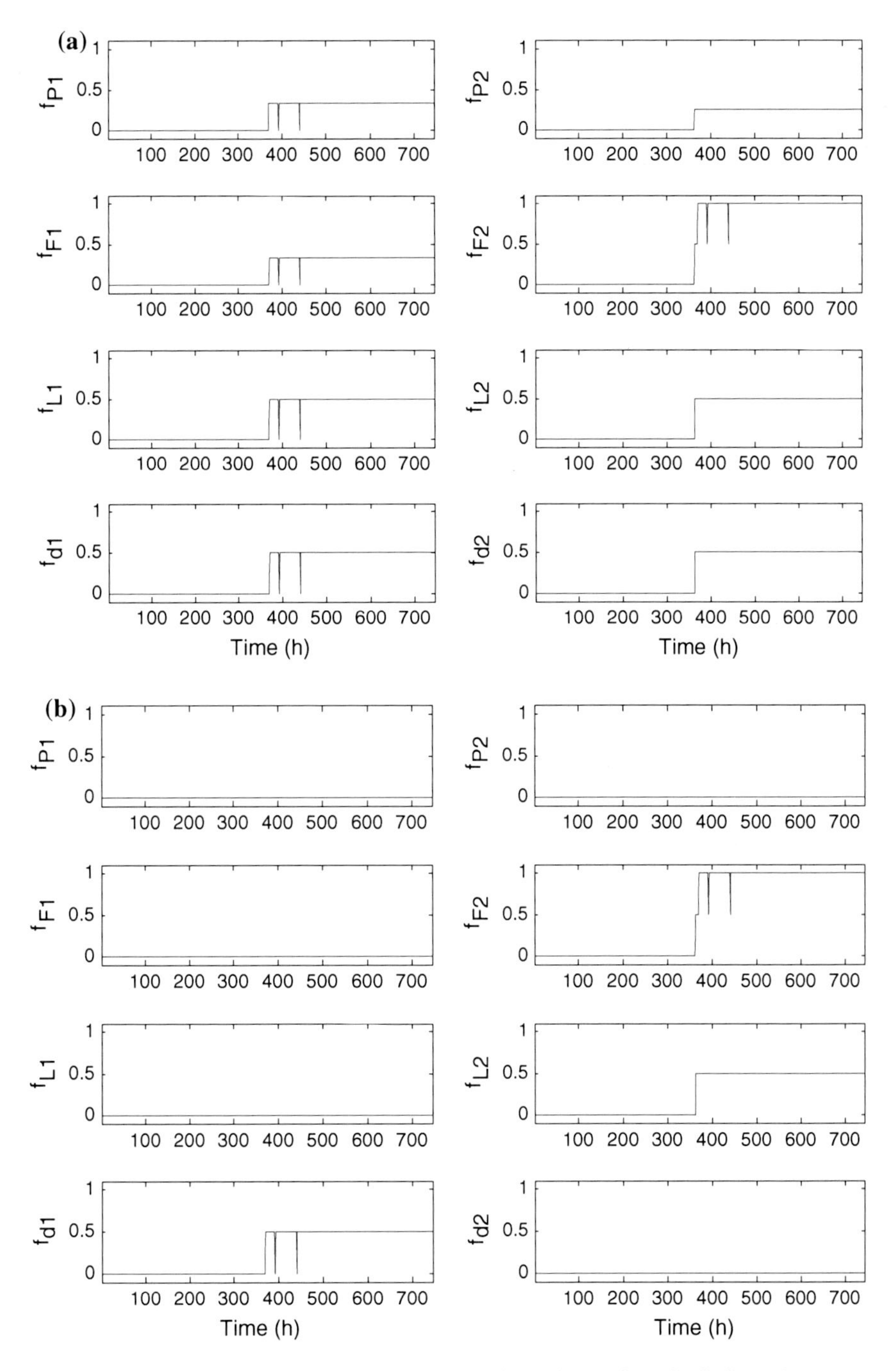

Fig. 13 Fault signal analysis based on *factor*01 and *factor*sign. **a** *factor*01, **b** *factor*sign

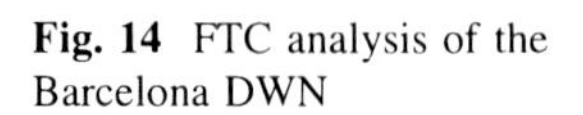

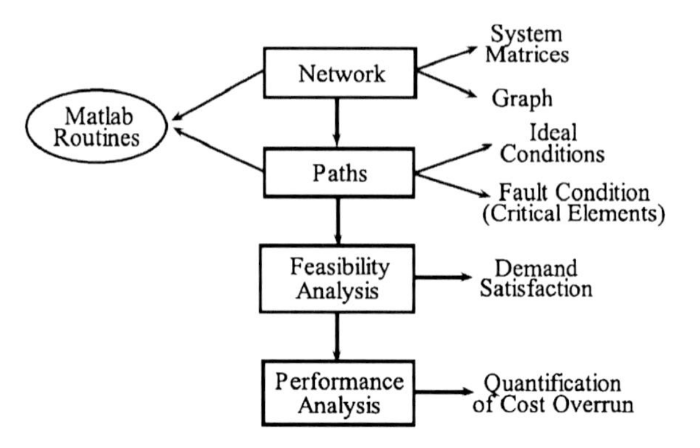

Fig. 14 FTC analysis of the Barcelona DWN

Table 1 Structural critical actuators

Number	Name	Number	Name
122	iAltures	68	iMntjeStaAmalia
10	iBegues 1	69	iMntjeTres Pins
6	iBegues2	3	iOrioles
2	iBegues 3	23	iPalleja1
1	iBegues 4	24	iPalleja2
31	iBellsoleig	26	iPalleja4
61	iBonavista	30	iPapiol1
20	iCanGüell	88	iSJD10
17	iCanGüell2d3	7	iStBoi
16	iCanGüell2d2	9	iStCliment1
18	iCanGüell1d2	5	iStCliment2
19	iCanGüell1d5	40	iStGenis 1
15	iCanGüell2	38	iStGenis2
14	iCanGüell3	13	iStaClmCervello
21	iCanRoig	45	iStaMaMontcada
57	iCanRuti	35	iTibidabo
37	iCarmel	56	iTorreBaro 1
43	iCerdMontflorit	65	iTorreoCastell
12	iCesalpina1	44	iVallensana1
11	iCesalpina2	8	iViladecans 1
82	iCornella100	4	iViladecans2
39	iFlorMaig	25	vAbrera
109	iFinestrelles300	54	vCerdanyola90
62	iGuinardera1	63	vMontigala
60	iGuinardera2	27	vPalleja70
101	iLaSentiu	90	vSJD
34	iMasGimbau1	104	vSJDTot
31	iMasGimbau2	58	vTer
100	iMasJove	59	vTerStaColoma

Table 2 Cost overrun analysis in some actuator faulty scenarios

Actuator no	Faulty price (e.u.)	Cost overrun (%)
41	514.44	2.43
47	515.94	2.73
74	528.05	5.14
78	557.62	11.03
86	515.08	2.55
89	556.22	10.74
97	510.49	1.64
102	539.87	7.49
103	552.21	9.95

Finally, performance analysis has been performed as a fourth test. The test has been performed using the objective function in the MPC controller, where the b array is formed with the real values for the actuator constraints. Several simulations using MATLAB have been performed, including a faulty component at a time; if a feasible solution is found, the cost of maintaining an operational network is computed and compared with costs obtained in previous analyses.

The difference in prices shows the impact that a single malfunctioning actuator has in the entire network. Results from this analysis are collected in Table 2. Notice that all comparisons are preformed taking into account an optimal functioning price (under faultless conditions) of 502.25 e.u. Moreover, the faulty price denotes the functioning price under faulty conditions.

According to the analysis made in this system, some actuators do not have a significant impact in the total performing cost (overrun over 1 %, e.g., actuators 28, 29, 33, 64, 71, 80, 81, 85, 87, 94, 107, 108, 113), but there are some others (such as 78 or 89) that induce an important increase in the price, taking into account the daily estimation. These latter actuators are shown in Table 2. Degradation in prices obtained with this analysis can be the foundation for the introduction of redundant actuators in the network or an alternative way of fault tolerant strategy.

7.5 Sensor Placement for Leakage Detection and Isolation

The DMA network is originally represented as a directed graph $G = (N,L)$ where pipe junctions are nodes, N, and pipes are edges, L. Each node represents, at the same time, a pressure variable and a flow balance equation. Similarly, each edge represents a flow variable and a pipe equation.

Now, the structural model of the water network can be defined as the bipartite graph involving the equation node set M and the unknown variable node set X. Let M_N be the set of flow balance equations and M_L be the set of pipe equations, so

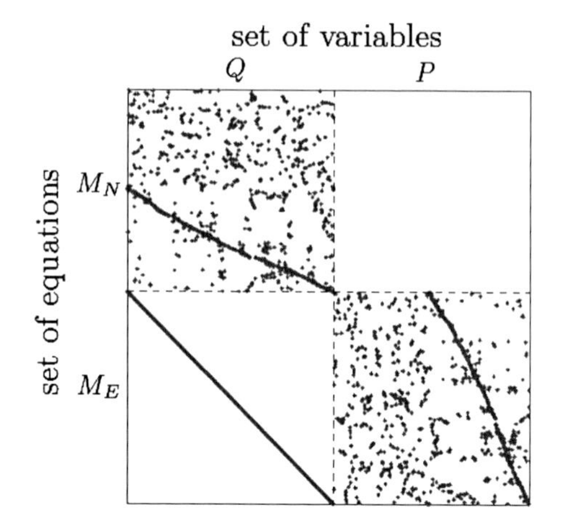

Fig. 15 Structural model of the DMA network

$M = M_N \cup M_L$. Note that there are as many equation in M_N as nodes in G and as many equations in M_L as edges in G. Thus, 1810 equations for the Barcelona DMA network are used here. On the other hand, let Q be the set of flow variables and P be the set of pressure variables, then it holds that $X = Q \cup P$. Therefore, the number of unknown variables is 1810. The edges of the structural model are defined from graph G (see Sect. 2). In Fig. 15, the resulting structural model is depicted in biadjacency matrix form where the equation set corresponds to rows and the variable set corresponds to columns. A dot in the *(i,j)* element indicates that there exists an edge incident to equation $e_i \in M$ and variable $x_j \in X$, i.e., $(e_i, x_j) \in A$. Note that the structural model of the DMA network is a just-determined model where all unknown variables can be computed, i.e., the model could be used for simulation.

It should be noted that when a leak is present in a dummy node (XX type), the corresponding Eq. (16) does no longer hold. Indeed, a term q_f should be added to the equation so that the model becomes consistent with the faulty water network. However, since detecting inconsistencies in the equation is the objective of model-based diagnosis, the term q_f is omitted and the set of faults, or leaks, is now represented as the subset of structural model equations in M_N related to dummy nodes. Therefore, the following set of fault equations is defined as

$$M_F = \{e \in M_N | e \, \text{comes from an XX type node}\}$$

The set of sensors is characterized by the subset of pressure variables in P such that its corresponding node is an RM type node. When a sensor measuring pressure p_i is placed, an equation of the form of $p_i = \hat{p}_i$ is added to the structural model, where $\hat{p}_i$. is the known measured value.

Before the sensor placement problem is solved, the maximum leakage detection and isolation specifications must be determined. This can be straightforwardly done by placing all candidate sensors in the model and then performing diagnosability analysis according to the detectability and isolability definitions in Sect. 3. The results obtained from this analysis are that all leaks can be detected, and 417 out of 448 leaks can be completely isolated from any other leak. Moreover, there are 16 leaks that are pair-wise non-isolable (they can be isolated from any other leak except the paired one), two sets of three non-isolable leaks each, a set of 4 non-isolable leaks and a set of 5 non-isolable leaks. In conclusion, there are 31 out of 448 leaks that cannot be completely isolated and in the worst case, when one of the 5 non-isolable leaks is present, we will not be able to isolate the correct one, among the 5 leaks.

The optimal sensor placement algorithm presented in Sect. 6 is now applied to the Barcelona DMA water network. Since all candidate sensors are of the same type (all of them measure pressure), the cost of installing a sensor is assumed to be equal for all candidate sensors. Therefore, solving the sensor placement problem involves finding the minimum cardinality sensor set that satisfies the maximum leakage detection and isolation specifications.

The optimal sensor configuration is found in just 55.5 s after applying Algorithm 1. The 12 nodes corresponding to the optimal sensors are labeled in Fig. 16. The optimal solution seems reasonable since it involves sensors related to nodes located in peripheral loop-free branches in the graph. This is mainly due to the fact that redundancy concerning nodes located in loop-free branches is more difficult to attain than redundancy concerning nodes located in loops.

Fig. 16 DMA optimal sensor location

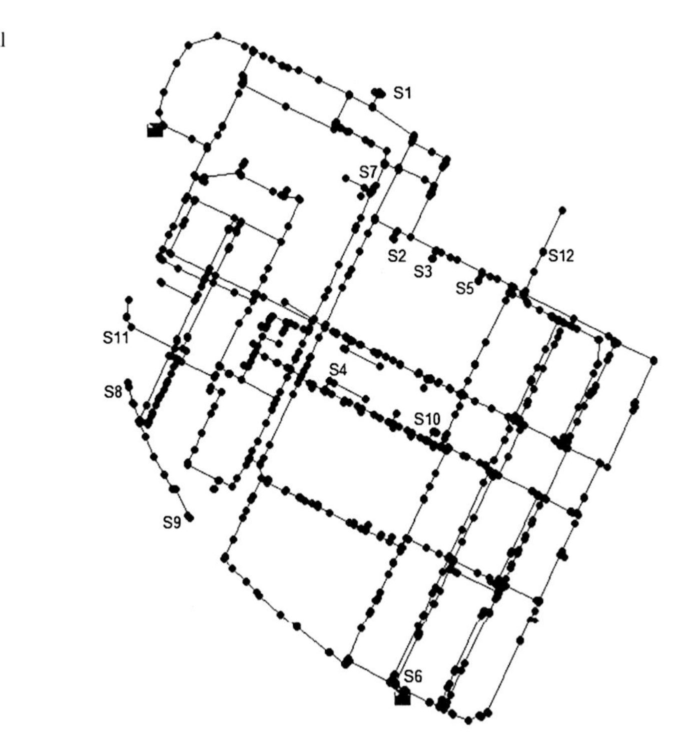

Algorithm 1 has been proven to be highly efficient. Given a set of 311 candidate sensors, the sensor placement algorithm searches a solution among 2^{311} potential sensor configurations. Obviously, checking all of them would be unaffordable. However, Algorithm 1 just needs to traverse 714,392 sensor configurations. Of these, the fault diagnosis specifications are only verified against 4,740 sensor configurations.

Note that the same diagnosis capabilities obtained by installing all 311 candidate sensors, are now achieved by just installing 12 sensors, which a significant improvement is regarding the investment that the water company should confront.

8 Conclusions

Critical infrastructure systems (CIS) are complex large-scale systems which in turn require highly sophisticated supervisory-control systems to ensure that high performance can be achieved and maintained under adverse conditions. The global RTC need of operating in adverse conditions involve, with a high probability, sensor and actuator malfunctions (faults). This problem calls for the use of an on-line FDI system able to detect such faults and correct them (if possible) by activating fault tolerant mechanisms. FTC techniques such as the use of soft sensors or using the embedded tolerance of the controller, prevent the global RTC system from stopping every time a fault appears. To exemplify the FDI and FTC methodologies in CIS, the Barcelona drinking water network is used as the case study.

Acknowledgments This work has supported by CICYT SHERECS DPI- 2011-26243 and CICYT WATMAN DPI- 2009-13744 of the Spanish Ministry of Education, by iSense grant FP7-ICT-2009-6-270428 and by EFFINET grant FP7- ICT-2012-318556 of the European Commission. The authors acknowledge the help of CETAQUA and AGBAR (Barcelona Water Company).

References

1. Anderson, L.B., Atwell, R.J., Barnett, D.S., Bovey, R.L.: Application of the maximum flow problem sensor placement on urban road networks for homeland security. Homel. Secur. Aff. **3**(3), 1–15 (2007)
2. Bagajewicz, M., Fuxman, A., Uribe, A.: Instrumentation network design and upgrade for process monitoring and fault detection. AIChE J. **50**(8), 1870–1880 (2004)
3. Baseville, M., Nikiforov. I.: Detection of abrupt changes–Theory and Applications. Prentice Hall Information and Systems.Sciences Series (1993)
4. Bayoudh, M., Travé-Massuyès L., Olive, X.: On-line analytic redundancy relations instantiation guided by component discrete-dynamics for a class of non-linear hybrid systems. In: Proceedings of CDC, pp. 6970–6975. Shanghai, PR China (2009)
5. Bemporad, A., Morari, M.: Control of systems integrating logic, dynamics, and constraints. Automatica **35**(3), 407–427 (1999)

6. Berry, J.W., Fleischer, L., Hart, W.E., Phillips, C.A., Watson, J.P.: Sensor placement in municipal water networks. J. Water Res. Plann. Manag. **131**(3), 237–243 (2005)
7. Bianco, L., Confessore, G., Reververi, P.: A network based model for traffic sensor location with implications on O/D matrix estimates. Trans. Sci. **35**(1), 52–60 (2001)
8. Biswas, G., Simon, G., Mahadevan, N., Narasimhan, S., Ramirez, J., Karsai, G.: A robust method for hybrid diagnosis of complex systems. In: Proceedings of 5th Symposium on Fault Detection, Supervision and Safety for Technical Processes, pp 1125–1131. Washington, DC (2003)
9. Blanke, M., Kinnaert, M., Lunze, J., Staroswiecki, M.: Diagnosis and Fault-Tolerant Control, 2nd edn. Springer, Heidelberg (2006)
10. Calafiore, G., Campi, M. C., El Ghaoui, L.: Identification of reliable predictor models for unknown systems: A data-consistency approach based on learning theory. In: Proceedings of the 15th IFAC World Congress (2002)
11. Camacho, E.F., Bordons, C.: Model Predictive Control, Second Edition. Springer-Verlag (2004)
12. Carrozza, G., Cotroneo, D., Russo, S.: Software faults diagnosis in complex OTS based safety critical systems. In: Proceedings of Seventh European Dependable Computing Conference, pp. 25–34 (2008)
13. Casau, P., Rosa, P., Tabatabaeipour, S.M., Silvestre, C., Stoustrup, J.:A Set-Valued Approach to FDI and FTC of Wind Turbines, Control Systems Technology, IEEE Transactions on , (in press)doi: 10.1109/TCST.2014.2322777
14. Cembrano, G., Quevedo, J., Salamero, M., Puig, V., Figueras, J., Martı́, J.: Optimal control of urban drainage systems: a case study. Control Eng. Pract. **12**(1), 1–9 (2004)
15. Combastel, C., Gentil, S., Rognon, J.P.: Toward a better integration of residual generation and diagnostic decision. In: Proceedings of IFAC Safeprocess'03, Washington, USA (2003)
16. Cordier, M.-O., Dague, P., Levy, F., Montmain, J., Staroswiecki, M., Trave-Massuyes, L.: Conflicts versus analytical redundancy relations: a comparative analysis of the model based diagnosis approach from the artificial intelligence and automatic control perspectives. IEEE Trans. Syst. Man Cybern B Cybern **34**(5), 2163–2177 (2004)
17. Ding, S.X.: Model-Based Fault Diagnosis Techniques. Springer, Berlin (2008)
18. Dulmage, A.L., Mendelsohn, N.S.: Coverings of bipartite graphs. Can. J. Math **10**(4), 517–534 (1958)
19. El Ghaoui, L., Calafiore, G.: Robust filtering for discrete-time systems with bounded noise and parametric uncertainty. IEEE Trans. Autom. Control **46**(7), 1084–1089 (2001)
20. Gelormino, M., Ricker, N.: Model-predictive control of a combined sewer system. Int. J. Control **59**, 793–816 (1994)
21. Gertler, J.: Fault Detection and Diagnosis in Engineering Systems. Marcel Dekker, New York (1998)
22. Ingimundarson, A., Sanchez, R.: Using the unfalsified control concept to achieve fault tolerance. In: A: 17th World Congress of the International Federation of Automatic Control (IFAC). 17th IFAC World Congress, pp. 1236–1242. IFAC (2008)
23. Isermann, R.: Fault Diagnosis Systems: An Introduction from Fault Detectionto Fault Tolerance. Springer, New York (2006)
24. Jiang, J.: Fault-tolerant control systems —an introductory overview. Automatica SINCA. Vol. 13(1), pp. 161-174, (2005)
25. Krysander, M., Åslund, J., Nyberg, M.: An efficient algorithm for finding minimal over-constrained sub-systems for model-based diagnosis. IEEE Trans. Syst. Man Cybern. A **38**(1), 197–206 (2008)
26. Krysander, M., Frisk, E.: Sensor placement for fault diagnosis. IEEE Trans. Syst. Man Cybern. A **38**(6), 1398–1410 (2008)
27. Marinaki, M., Papageorgiou, M.: Optimal Real-time Control of Sewer Networks. Springer London (2005)
28. Maciejowski, J.M.: Predictive Control with Constraints. Prentice Hall, Essex (2002)

29. Meseguer, J., Puig, V., Escobet, T.: Fault isolation module implementation using a timed discrete-event approach based on interval observers. In: 7th IFAC Symposium on Fault Detection, Supervision and Safety of Technical Processes, pp. 1563–1568 (2009)
30. Nejjari, F., Perez, R., Escobet, T., Travé-Massuyès, L.: Fault diagnosability utilizing quasi-static and structural modelling. Math. Comput. Model. **45**(5–6), 606–616 (2006)
31. Ocampo-Martínez, C., Ingimundarson, A., Puig, V., Quevedo, J.: Objective prioritization using lexicographic minimizers for MPC of sewer networks. IEEE Trans. Control Syst. Technol. **16**(1), 113–121 (2008)
32. Ocampo-Martinez, C., Puig, V., Cembrano, G., Quevedo, J.: Application of predictive control strategies to the management of complex networks in the urban water cycle. IEEE Control Syst. Mag. **33**(1), 15–41 (2013)
33. Petti, T.F., Klein, J., Dhurjati, P.S.: Diagnostic model processor: using deep knowledge for process fault diagnosis. AIChE J. **36**, 565 (1990)
34. Pleau, M., Colas, H., Lavallée, P., Pelletier, G., Bonin, R.: Global optimal real-time control of the Quebec urban drainage system. Environ. Model. Softw. **20**, 401–413 (2005)
35. Ploix, S., Adrot, O., Ragot, J.: Bounding apprcomplex networks in the urbanoach to the diagnosis of a class of uncertain static systems. In: SAFEPROCESS2000. Budapest, Hungary (2000)
36. Puig, V., Quevedo, J., Escobet, T., Pulido, B.: A new fault diagnosis algorithm that improves the integration of fault detection and isolation. In: ECC-CDC'05, Sevilla, Spain (2005)
37. Puig, V., Quevedo, J., Escobet, T. Nejjari, F., de las Heras, S.: Passive Robust Fault Detection of Dynamic Processes Using Interval Models, IEEE Transactions on Control Systems Technology, Vol. 16(5), pp. 1083–1089, Sept. (2008)
38. Qin, S.J., Badgwell, T.A.: A survey of industrial model predictive controltechnology. Control Engineering Practice, vol. 11(7), pp. 733 –764 (2003)
39. Quevedo, J.: Polynomial approach to design a virtual actuator for a fault tolerant control: application to a Pem fuel cell. In: 5th Workshop on Advanced Control and Diagnosis, Grenoble France (2007)
40. Ragot, J., Maquin, D.: Fault measurement detection in an urban water supply network. J. Process Control **16**(9), 887–902 (2006)
41. Rawlings, J.B., Mayne, D.Q.: Model Predictive Control: Theory and Design, Nob Hill Publishing, (2009)
42. Robles, D., Puig, V., Ocampo Martínez, C., Garza, L.E.: Methodology for actuator fault tolerance evaluation of linear constrained MPC: application to the Barcelona water network. In: Proceedings of the 20th Mediterranean Conference on Control and Automation, pp 518–523 (2012)
43. Rosich, A., Frisk, E., Åslund, J., Sarrate, R., Nejjari, F.: Fault diagnosis based on causal computations. IEEE Trans. Syst. Man Cybern A **42**(2), 371–381 (2012)
44. Rosich, A., Sarrate, R., Nejjari, F.: Optimal sensor placement for leakage detection and isolation in water distribution networks. In: Proceedings of the 8th IFAC International Symposium on Fault Detection, Supervision and Safety of Technical Processes, pp. 776–781. México city, México (2012)
45. Sarrate, R., Nejjari, F., Rosich, A.: Sensor placement for fault diagnosis performance maximization under budgetary constraints. in Proceedings of the 2nd International Conference on Systems and Control, pp. 178–183. Marrakech, Morocco (2012a)
46. Sarrate, R., Nejjari, F., Rosich, A.: Model-based optimal sensor placement approaches to fuel cell stack system fault diagnosis. In: Proceedings of the 8th IFAC International Symposium on Fault Detection, Supervision and Safety of Technical Processes, pp. 96–101. México city, México (2012b)
47. Schilling, W., Anderson, B., Nyberg, U., Aspegren, H., Rauch, W., Harremoës, P.: Real-time control of wastewater systems. J. Hydraul. Res. **34**(6), 785–797 (1996)
48. Schütze, M., Butler, D., Beck, B.: Modelling, Simulation and Control of Urban Wastewater Systems. Springer, London (2002)

49. Schütze, M., Campisanob, A., Colas, H., Schilling, W., Vanrolleghem, P.: Real time control of urban wastewater systems: where do we stand today? J. Hydrol. **299**, 335–348 (2004)
50. Šiljak, D.D.: Decentralized control of complex systems. Academic Press, New York (1991)
51. Staroswiecki, M., Cassar, J. P., Declerk , P.: A structural framework for the design of FDI system in large scale industrial plants. In Issues of Fault Diagnosis for Dynamic Systems. Patton, R.J., Frank, P.M., Clark, R.N. (eds). Springer-Verlag. (2000)
52. Staroswiecki, M., Hoblos, G., Aitouche, A.: Sensor network design for fault tolerant estimation. Int. J. Adapt. Control Signal Process. **18**, 55–72 (2004)
53. Tornil-Sin, S., Ocampo-Martinez, C., Puig, V., Escobet,T.: Robust fault detection of non-linear systems using set-membership state estimation based on constraint satisfaction. Eng. Appl. Artif. Intell. 25 (1), pp. 1–10, (2012)
54. Travé-Massuyès, L., Escobet, T., Olive, X.: Diagnosability analysis based on component supported analytical redundancy relations. IEE Trans. Syst. Man Cybern. A **36**(6), 1146–1160 (2006)
55. Venkatasubramanian, V., Rengaswamy, R., Kavuri, S.: A review of process fault detection and diagnosis—part i: Quantitative model-based methods. Comput. Chem. Eng. **27**, 293–311 (2003)
56. Venkatasubramanian, V., Rengaswamy, R., Kavuri, S.: A review of process fault detection and diagnosis—part ii: qualitative models and search strategies. Comput. Chem. Eng. **27**, 313–326 (2003)
57. Venkatasubramanian, V., Rengaswamy, R., Kavuri, S.: A review of process fault detection and diagnosis—part iii: process history based methods Comput. Chem. Eng. **27**, 327–346 (2003)

Wireless Sensor Network Based Technologies for Critical Infrastructure Systems

Attila Vidács and Rolland Vida

Abstract This chapter presents an overview on how wireless sensor networks (WSN) can be used in critical infrastructure systems (CIS). First, the architecture of a typical sensor node is presented, and it is shown how these nodes can be grouped to form a network. Then, the requirements of a traditional WSN (such as low complexity, energy efficiency, scalability, or self-organization) are given, and those special requirements are highlighted that are characteristic to CIS: reliability, availability, real-time operations, security, etc. Currently available sensing and communication technologies that best fit these special requirements are also given. In the second part of the chapter some specific use cases are described: WSN for water system monitoring, electric power system monitoring, and nuclear reactor monitoring.

Keywords Sensor motes · Wireless sensor networks · WSN technology · WSN requirements · Water system monitoring · Electric power system monitoring · Nuclear reactor monitoring

1 Intelligent Sensors and Sensor Networks

Sensors are everywhere around us. Most of them are discretely hidden and unnoticed, but they do their job, monitor their physical environment (e.g., temperature, light conditions, radioactivity). Such sensors are used in our everyday life since they are deployed in our cars, homes, and workplaces. Devices such as mercury thermometers, watt-hour meters, barometers, carbon monoxide detectors, or seismographs are well known and intensely used.

A. Vidács (✉) · R. Vida
Department of Telecommunications and Media Informatics, Budapest University of Technology and Economics, Budapest, Hungary
e-mail: vidacs@tmit.bme.hu

R. Vida
e-mail: vida@tmit.bme.hu

E. Kyriakides and M. Polycarpou (eds.), *Intelligent Monitoring, Control, and Security of Critical Infrastructure Systems*, Studies in Computational Intelligence 565,
DOI 10.1007/978-3-662-44160-2_11

These sensor devices are passive; they are merely the input devices to collect information for the system. Usually there are more than one sensor connected to the controlling entity. Taking one step forward towards intelligent sensors, we can have sensor devices equipped with some kind of intelligence (i.e., processor and memory) so that they can be capable of data collection, (pre-)processing and even self-configuration regarding how the sensing task is performed. For example, the intelligent thermometer can monitor the inside and outside temperature, taking notes on the daily minimum and maximum values, and it can adaptively set its sampling interval to save battery when the operating temperature of the device is too low.

Taking one more step forward we can implement communication capabilities to the sensors in order to be able to share information. They can do it either by communicating to a master device, being simple slave nodes, or—more interestingly—they can communicate with each other as sensor peers. Assuming bidirectional peer-to-peer communication among sensor nodes, the vision of a network of intelligent sensor devices is attained, also referred simply as a sensor network.

Although most of the actual sensors are wired constructions, when we talk about sensor networking the wireless communication is almost immediately assumed. The reason behind this is that most application scenarios assume a (very) large number of nodes distributed over a (relatively) large geographical area. This excludes the possibility of manual node configuration, making it impossible to wire each of the nodes. This is why we talk about wireless sensor networks (WSNs). Note, however, that wireless here means not just wireless communication, but operation without wired energy supply.

Many different kinds of sensor networks exist, and many more could be imagined and developed. The enabling sensor networking technology does not seem to be the bottleneck any more; a wide range of general building blocks and technological solutions are already available. The technological progress made it possible to manufacture sensors at a microscopic scale using the MEMS (Micro-electro-mechanical Systems) technology. The goal is to build sensors that provide significantly higher speeds and sensitivity, but on the other hand are cheap and tiny. Their size and low price permits then to deploy hundreds or thousands of them inside the area that has to be monitored. Nevertheless, these cheap and tiny devices have limited resources (memory, processing power, energy); therefore, special operation and communication schemes have to be designed and implemented in order to cope with these limitations. The real challenge is to design and develop solutions using sensor network technology to a particular application area of interest, such as critical infrastructure monitoring.

1.1 Sensor Node Architecture

A typical sensor node, also called as *mote*, consists of five main parts. One or more *sensors* sense and collect data from the environment. The sensors can be built-in, or they can be installed on a separate sensor board. The *central unit* in the form of a micro-controller manages the operation tasks. A *transceiver* (radio) module is used for communicating with other devices, and some *memory* is used to store measured data and also the application software. Energy supply is solved using *batteries*. Besides the main components, some LEDs and buttons, and a serial communication interface can be integrated as well.

1.2 Sensor Network Architectures

A wireless sensor network (WSN) can be regarded as a sensing system: sensing the physical world and providing the collected information to the users (humans or control system intelligence). The monitored data gathered by the (typically many) sensor nodes is sent to one or more sinks which are connected to other networks (e.g., Internet) through a gateway. Thus, a typical network installation has many simple nodes and a few special ones called sink nodes (or base stations) which are connected to the Internet. The measurement data can be transmitted either directly or through intermediate relay nodes (called multi-hop), to one or several sinks. The nodes can be static or mobile, and different sensing motes (i.e., heterogeneous) can be considered as well. A traditional homogeneous single-sink WSN topology can be seen in Fig. 1 as an example.

1.3 Information Gathering and Communication

Data gathering

Information gathering in sensor networks can follow different patterns, depending mostly on the specific needs of the applications. In a time-driven scenario all sensors send data periodically to the sink. As opposed to this, in the event-driven case sensors start communicating with the sink only if sensing an event, i.e., a situation that is worth reporting. Finally, in a query-driven scenario a sensor transmits its data only if the sink asks for it.

Communication

If the covered area is small, sensors can send their data directly to the sink. In larger setups—e.g., covering a large geographical area—the range of the sensors' radio is in general quite short when compared to the network size. In this case, multi-hop communication is used: the sensors forward each others' data toward the sink.

Routing

In general, it can be said that the majority of routing protocols assume homogeneous and static sensor nodes. There are literally hundreds of proposals for routing protocols in WSNs (see for example [1] for an excellent survey).

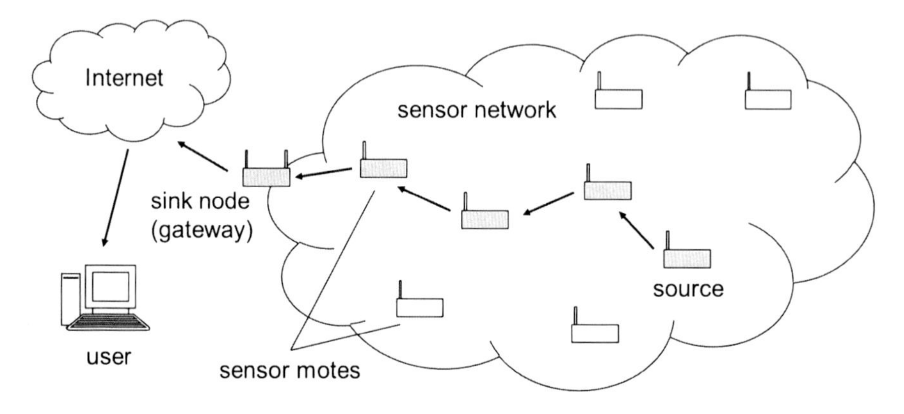

Fig. 1 Typical wireless sensor network architecture

2 WSN for CIS

Critical Infrastructure System monitoring and protection requires continuous monitoring to detect either accidental failures or intentional attacks. The technology that can achieve this can be the wireless sensor networks. On one hand, equipment control and diagnosis of CIS as the targeted application sets requirements that drive the selection of WSN technologies and protocols. On the other hand, the inherent features of WSN and radio communication pose constraints to the application's performance. As a result of this, there is a mismatch between the "traditional" requirements and the definition of WSN in the literature when compared to wireless sensors used for CIS. WSNs for CIS have some additional requirements (e.g., security, reliability) and some of the generally assumed properties (e.g., self-organized, ad hoc) are no longer (or not fully) adequate. Similarities and differences are highlighted next.

2.1 Traditional WSN Requirements

Low cost, low complexity

Wireless networks have the cost advantage over wired networks since installation and maintenance costs tend to be lower. Upgrading and replacing wireless devices is also easier and more flexible. On the other hand, low complexity results in limited capabilities.

Energy efficiency

To assure long network lifetime while operating unattended, energy efficiency in all parts of the system is crucial. This is true at all levels: hardware components, physical layer, MAC, routing, and application protocols. For example, at physical and MAC layers the nodes should operate with low duty cycles by spending most of their time in sleep mode. However, sleep modes make it harder to synchronize nodes and special attention is needed when designing the communication protocols and end-to-end networking.

The design of energy efficient communication and operation solutions is a must. When a node is in an active state transmitting or receiving information over the radio it consumes much more energy than when being active but just processing information, not to mention the passive state when the node sleeps. The ratio of energy consumption when transmitting compared to processing is much larger than one, it is rather more than one hundred or even one thousand in most platforms [3]. For this reason, the communication protocols need to be designed to be highly energy efficient, or in other words, minimize packet transmissions over the air interface and, equally important, avoid idle listening and overhearing (i.e., receiving packets not intended to that particular node). But energy efficiency is not just about minimizing radio transmissions. It can easily happen that the processing task of a sensor node becomes extensive and takes much longer time than it is needed to transmit the resulting information bits over the radio. If that is the case (e.g., when data encryption algorithms are running), care should be taken to design computationally efficient algorithms and protocols for the specific application.

Scalability

The simplest single sink topology lacks scalability, as well as the direct communication limits the network size. Although allowing multiple sinks provides scalability in theory, this is not a trivial extension of a single sink solution from the engineering point of view. The network topology and communication protocols are necessarily more complicated in the multi-sink case, giving rise to more distributed networking solutions. Besides the additional complexity, the distributed networking provides many advantages from the applications point of view as well. The requirements on scalability are in-line with the fact that the large-scale nature of CIS calls for a scalable and low cost technology.

Self-Organization and Self-Healing

Wireless networks in general have the ability to organize and reconfigure themselves into (cost) effective communication networks.

2.2 WSN Requirements for CIS

The application requirements of CIS monitoring and protection strongly affects the choice of WSN technology. Effective protection of critical infrastructures requires an increased innovation and development of advanced, intelligent detection and sensor systems for both physical and cyber aspects [18].

Reliability and availability

A powerful centralized control center together with a robust information infrastructure with high-speed and reliable communication links are essential for CIS operation and management. However, these centralized elements are especially vulnerable during serious attacks or damages to the system. Besides protecting the centralized control elements, a distributed intelligent control system would be desirable in critical situations to ensure at least partial operation of the whole system to avoid total system failures. One solution to provide such a distributed monitoring and control system is to use WSN technologies for CIS protection. The communication must be robust to interference and failures as well.

Real-time operation

Timely monitoring and alerting is a must if a problem is detected or is likely going to happen soon. Both proactive and reactive operations can be desirable for the WSN. The sensor network can be deployed on the spot when needed in critical situations. It can be seen as a self-powered, redundant and robust monitoring and communication system, collecting and providing real-time information of the system and its surrounding physical environment.

Integrity and security

Robustness against attacks is among the primary requirements. This should be even more emphasized since a WSN can be highly vulnerable against attacks if no special attention is given to it. In CIP applications solutions are needed to protect the WSN itself. In [4] the authors give an overview of the security challenges and the approaches to ensure dependability of WSN for CIP at three layers: the node architecture, the networking protocols, and the service layer. One proposed solution in [2] is a special key management tool for security.

Two-way communication

The two-way communication is required if not only sensors, but actuators are used to control the system as well.

Wired-in

In CIS monitoring and protection the wireless nodes are to be connected to a wired network to allow remote monitoring, management and control.

3 WSN Technologies for CIS

The application requirements (e.g., Critical Infrastructure Protection (CIP)) strongly affect the choice of the most promising wireless sensor technology to be used. This section briefly describes WSN technologies available today to choose from.

We also direct the reader's attention towards the project results of the Wireless Sensor and Actuator Networks for Critical Infrastructure Protection (WSAN4CIP) Project. The goal of WSAN4CIP (see http://www.wsan4cip.eu/ for more details) is to advance the technology of Wireless Sensor and Actuator Networks (WSANs) beyond the current state of the art, in order to improve the protection of Critical Infrastructures (CIS).

3.1 Sensor Mote Technology

Sensors monitor physical properties of the surrounding environment and convert them to measurable signals. Thus, the sensors can be regarded as the interface between the physical world and the virtual world. Actuators function the other way around: their task is to convert electric signals into a physical phenomenon.

Various motes have been developed and made available by different vendors. Here we list some of the most widely used ones based on survey results reported in [11], but the complete list would be much longer. For a more detailed survey of WSN hardware please refer to [11] and references therein.

MICA2, MICAz

The processor board of MICA motes is based on Atmel's ATmega128L low-power micro-controller. A variety of sensors and data acquisition boards (e.g., light, temperature, RH, barometric pressure, acceleration/seismic, acoustic, magnetic, etc.) are available for the motes which can be connected to the standard expansion connector, and custom boards are also available. The sensor mote can also function as a base station when interfaced with a serial or USB interface for both programming and data communication [9, 10]. The main difference between the MICA2 and MICAz platforms is the radio board. The MICA2 is equipped with a 315, 433 or 868/916 MHz multi-channel transceiver with a communication range of 150 m [9]. The later-developed MICAz has a IEEE 802.15.4 compliant RF transceiver on board, a direct sequence spread spectrum radio in the 2.4–2.48 GHz globally compatible ISM band, that is more resistant to RF interference and provides high speed (250 kbps) communication and inherent (hardware) data security (AES-128) [10].

Telos B

The Telos B motes have an IEEE 802.15.4/ZigBee compliant radio that operates in the 2.4–2.48 GHz ISM band and enables 250 kbps data speed. It includes Texas Instrument MSP430 micro-controller with 10 kB RAM, 1 MB external flash, and an integrated on-board antenna. Optional light, humidity and temperature sensors can be integrated as well. The Telos B motes are said to be more suitable for test-bed deployments in lab experimentation.

IRIS

The IRIS 2.4 GHz mote has three-times improved radio range, over 300 m outdoor and more than 50 m indoor range, when compared to the MICA and Telos B motes. It uses XM2110CA processor board that is based on the Atmel ATmega1281. Several sensor boards can be attached to the mote via the standard 51 pin connector [7]. The IRIS 2.4 GHz mote was designed specifically for deeply embedded sensor networks. The mote can also function as a base station. The IRIS OEM Edition was specially designed for mesh networking; it utilizes XMesh(TM) software technology for low-power reliable mesh networking.

IMote

The IMote 2.0 platform uses the CC2420 IEEE 802.15.4 (ZigBee) radio from Texas Instruments that supports 250 kbps data rate in the 2.4 GHz ISM band. Its surface-mounted antenna supports a nominal range of 30 m. The mote utilizes Intel's PXA271 processor at 13–416 MHz with a DSP coprocessor, and has 32 MB SRAM and 32 MB Flash RAM. It also has a camera interface [6].

SunSPOT

The Sun Small Programmable Object Technology (SPOT) solution is based on Java platform. It uses ARM 920T micro-controller, and an IEEE 802.15.4 ZigBee compliant radio. It possesses 512 kB RAM for programs and data and 4 MB external Flash.

Stargate Gateway

Gateways serve as the connection point for a WSN towards the Internet. Many hardware solutions exist, here we name only one. The Stargate gateway was developed by a joint effort of many research groups, and finally licenced to Crossbow Technologies for commercial production [8]. It works on an embedded Linux platform running on a 400 MHz Intel PXA255 processor, and has various connecting options such as RS232, 10/100 Ethernet, USB host and JTAG. It is compatible with Crossbow's MICA motes family.

3.2 Radio (Hardware and Software) Technology

IEEE 802.15.4

The IEEE 802.15 TG4 was chartered to investigate a low data rate solution with multi-month to multi-year battery life and very low complexity. It is operating in an unlicensed, international frequency band [17]. The key features of IEEE 802.15.4 are: 16 channels in the 2.4 GHz ISM band, 10 channels in the 915 MHz and one channel in the 868 MHz band; data rates of 20–250 kbps; support for critical latency devices; automatic network establishment by the coordinator; fully handshaked protocol for transfer reliability; power management to ensure low power consumption. The main field of application of this technology is the implementation of WSNs. The IEEE 802.15.4 Working Group focuses on the standardization of the bottom two layers (physical and data link layers) of the ISO/OSI protocol stack. There are two options for the upper layers: the ZigBee protocols defined by the ZigBee Alliance (http://www.zigbee.org/) and the 6LoWPAN defined by IETF (http://6lowpan.net). IPv6 uses 128-bit addresses. With these, IPv6 can be flexible in address allocation and in routing. The most problematic part is that the sensor motes have limited resources. A common mote has 2–256 kilobytes of ROM and the size of the RAM varies between 128 bytes and 16 kilobytes; for comparison, a normal IPv6 packet's size is 1280 bytes.

Bluetooth 4.0

IEEE Bluetooth v4.0 is the most recent version of Bluetooth wireless technology. It introduced low energy technology to the Bluetooth Core Specification, enabling new Bluetooth Smart devices that can operate for months or even years on tiny, coin-cell batteries. Key features of Bluetooth low energy wireless technology include: ultra-low peak, average and idle mode power consumption; ability to run for years on standard, coin-cell batteries; low cost; multi-vendor interoperability; enhanced range (http://www.bluetooth.com).

UWB

Ultra-wideband (UWB) radio is a fast emerging technology with uniquely attractive features that has attracted a great deal of interest lately. The most widely used definition of UWB is a signal with spectral occupancy in excess of 500 MHz or a fractional bandwidth of more than 20 %. One of the most promising UWB techniques for WSN applications is named Impulse Radio-UWB (IR-UWB). The IR-UWB technique relies on ultra-short (nanosecond scale) waveforms that can be free of sine-wave carriers and do not require IF processing because they can operate at baseband. The IR-UWB technique has been selected as the PHY layer of the IEEE 802.15.4a Task Group for WPAN Low Rate Alternative PHY layer [3].

4 Use Cases

This section presents some real examples for the use of wireless sensor networks to monitor critical infrastructure systems such as water distribution systems, electric power grids, telecommunication systems, nuclear reactors, facility protection systems, or systems deployed to monitor the structural health of bridges, roads or buildings. Besides presenting the specific hardware components usually employed for these monitoring tasks, the different challenges and considerations that have to be taken into account when deploying and operating a wireless sensor network tailored for the needs of a given application scenario are presented.

4.1 WSN for Water System Monitoring

Water system monitoring is a term that covers several different applications and services. On one hand, it includes water quality monitoring and management, which relates to monitoring the quality of the water in households or the water used in industrial processes, be that in a natural or an artificial environment. But, on the other hand, it also includes the monitoring of the water distribution system in terms of potential leakages, quality or efficiency.

4.1.1 Water Quality Monitoring

For water quality monitoring, there are several sensing devices that are able to monitor a wide range of physical, chemical and biological parameters, such as water flow velocity, temperature, evaporation rate, turbidity, salinity (measured through electrical conductivity), chlorophyll concentration, and sensors for detecting the presence of different bacteria and algae [13]. As an example, Fig. 2

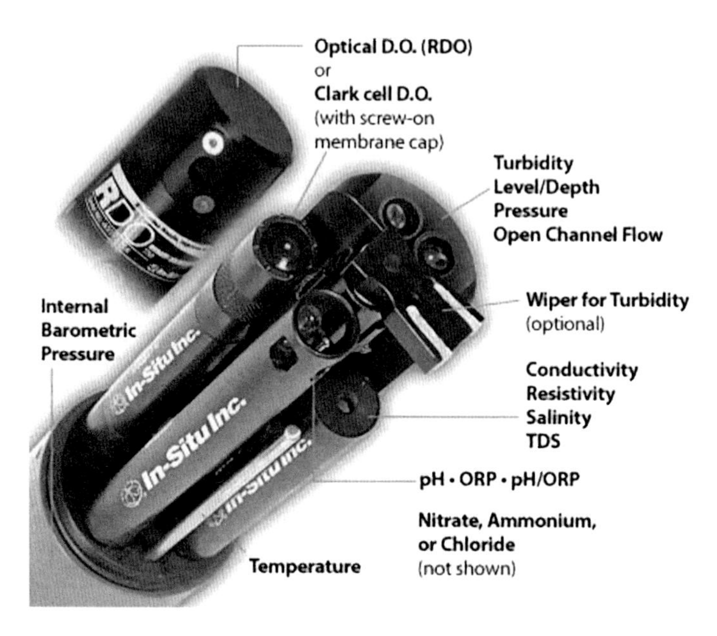

Fig. 2 The MP-TROLL 9500 multi-parameter water quality measurement sensor developed by In Situ Inc (All rigths reserved)

shows a multi-parameter water quality measurement sensor called MP-TROLL 9500 developed by In Situ Inc.

These sensors, if enhanced with communication capabilities and grouped in networks, can be used for many possible applications. In case of beach water monitoring, sensors floating in the water along popular beaches can continuously monitor water quality, and send out warning messages when they detect, or predict a deterioration. Similar water quality monitoring can be performed for lakes, rivers, groundwater wells, wetlands or tidal areas, having different monitoring objectives. Probably one of the best-known WSN-based water monitoring systems was developed in the GBROOS (Great Barrier Reef Ocean Observing System), where the goal was to understand the water quality parameters that influence coral bleaching and other coral diseases. However, these water monitoring applications are more related to environment monitoring; thus, they will not be addressed here in detail. The focus will be more on the monitoring of (urban) water distribution systems that are part of the critical infrastructure.

4.1.2 Monitoring Water Distribution Systems

Water distribution systems are continuously facing a deterioration of their infrastructure due to aging pipes and continuous repairs. Water losses associated with the inefficiency of the underground water distribution system result in important economic losses, but also in the inefficient use of the water reserves, which is especially important in areas where water shortage starts to become an important problem. In order to monitor, localize and quantify possible leakages in the system, parameters such as water flow speed, pressure, soil and air moisture, soil and air temperature can all be monitored, together with the water consumption in the different daily periods. In this context, deploying a wireless sensor network to perform real-time monitoring and data gathering seems to be the straightforward solution.

A pilot water distribution monitoring system was developed by the University of Cyprus, in the research project Urban Water Distribution Modelling, Simulation and Optimization of Leakage Detection via Sensing Technologies [5]. Moisture, sound and pressure sensors were deployed along an experimental pipe-line system, to monitor soil and air moisture and temperature, water pressure and flow, rainfalls and sound. Decagon Echo dielectric sensors were used to measure soil moisture at different locations between 0.5 and 1 m below the pipelines. Water flow was measured using Elster's KENT V100 water meters, while Gutterman's Zonescan-800 leak intelligence units were used to detect leak noise. Sensors were connected to Mica2 motes using MDA100 data acquisition boards, housed in waterproof casing, and communicated on the 433 MHz frequency with a Stargate gateway that was responsible for gathering the data and sending it to a remote base station through a GPRS connection.

The Stargate gateway was also able to update and fine-tune the different operational parameters of the sensors, such as the sampling rate or the sampling duration. Note that most of the deployed sensors typically generate low data rates, as compared to the acoustic sensor that operates at much higher rates. As the wireless communication module consumes a significant portion of the mote's limited energy resources, it was wise to operate initially the sensor network so as to gather data only from the low-rate nodes. Then, if the probability of a leakage was detected, the acoustic sensor was activated to localize the leak based on some standard correlation algorithms.

Another example of such water distribution system monitoring is the Wireless Water Sentinel project in Singapore (WaterWiSe@SG), deployed to monitor a 60 sq-km area in downtown Singapore, containing more than 20.000 pipes and 19.000 junctions. Sensor nodes were located at an average distance of 1 km, and included a pressure transducer and a hydrophone (sampling at a frequency of 2 kHz, 2000 samples per second), a flowmeter (sampling at a frequency of 1 Hz) and some water quality sensors (sampling at 0.33 Hz). Data is temporarily stored on a 2 GB memory card, but is also continuously transmitted through a USB 3G modem towards a central database. Each node is also equipped with a GPS unit that is not used for localisation, but mainly for high accuracy clock-synchronisation, to enable burst location schemes based on the relative arrival times of the transient pressure wave front at different points in the network [16]. All these capabilities (high frequency sampling, large data storage, 3G communication and GPS operation) come at a price of high energy consumption, which was however handled by installing solar panels that were linked to the sensor nodes.

Several other WSN-based water distribution monitoring systems exist, but the goal of this section is not to give an exhaustive survey of all these solution, but rather to just show the viability of the approach and present the directions that are currently envisaged in this area.

4.2 WSN for Electric Power System Monitoring

Although the power grid infrastructure deployed in the developed world is extremely large and complex, it is at the same time quite old, rendering it vulnerable to faults. Most of the infrastructure is more than 50 years old, and its difficulty to cope with stress situations results in high energy costs and serious economic losses. However, despite these problems, governments are often reluctant to invest in the maintenance of the existing power infrastructure, while building new power lines becomes increasingly difficult. Therefore, using wireless sensor networks to monitor the electric power systems becomes a very interesting and useful application scenario.

Overhead conductor sag, tension and temperature measurement help in retrieving real-time data about the power line thermal capacity and its power carrying capacity. For overhead conductor sag measurements there are several techniques, such as inclinometers, sagometers, tension measurement, etc. Dynamic thermal rating systems include power donuts and power line sensors that are directly deployed on the power line to measure temperature, wind, or inclination, but also conductor replicas that are placed near the line, not on the line, to evaluate weather conditions without generating physical modifications to the line [19]. In Fig. 3 Sentient Energy's Advanced Monitoring Platform is shown as an example of a line-mounted power sensor.

As opposed to overhead cables, underground power lines experience fewer interruptions, but there are also a number of failures that might affect them. Their drawback is that fault localization and repairing might be more expensive and time consuming. Finally, besides monitoring the conductor lines, the structural health and mechanical strength of the towers and poles has also to be monitored.

A simulation-based analysis of WSN-based electric distribution system monitoring is provided in [14]. The described network includes wireless sensor nodes that measure power quality (harmonics, voltage sags and voltage swells) regularly, once every second. These sensors are placed on pole transformers, forming a sparse network with nodes situated at hundreds of meters from each other. The nodes communicate with each other using WiFi (802.11b) radios at a 2.4 GHz frequency. As many of the pole transformers are placed at quite larger distances from the substation (which plays the role of a sink node), they can not communicate directly, so a multi-hop routing solution has to be deployed to relay the information. Since we deal with a static topology, it is relatively easy to build then a least-cost spanning tree along which data will be forwarded thereafter. Note that in case of energy distribution systems we do not have to worry about the limited

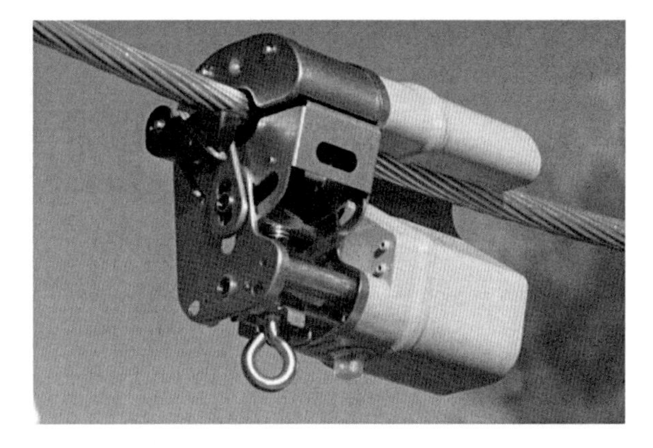

Fig. 3 Sentient Energy's Advanced Monitoring Platform (All rights reserved)

energy resources of the sensors, as they can be directly powered by the monitored network. Therefore, even the high energy demand of using WiFi technology to send data to large distances can be easily supported.

Obviously, if the aim is to monitor a real, large scale electric distribution system that covers several substations, data has to be relayed from the substations to a monitoring center. As this is usually located far from the substations, communication at this level should be ensured through wired, highly-reliable high speed Internet connections. However, in [14] authors address only the wireless data connection subsystem, and evaluate the efficiency of the proposed approach through simulations.

4.3 WSN for Nuclear Reactor Monitoring

Inside nuclear plants, a continuous monitoring of equipment, processes and nuclear material is needed. Wireless sensor networks can very well be deployed for such purposes, especially in areas where human intervention is not advised, due to the harmful radioactivity. Gathering data in a wireless manner, without installing cables in protected areas, is also important as it is much easier to deploy, and the risk of accidents is reduced. Moreover, costs are significantly reduced as well, due to sparing the cabling and deployment, but also because expensive cable shielding against corrosion and leakage is not needed anymore.

Nuclear reactors are however placed not only in nuclear plants, with the aim of producing electricity, but also in research labs, to serve different research goals related to cancer detection and treatment, or electronic equipment production, to name just a few. These research reactors are significantly different from nuclear reactors in terms of size, heat generation, or the type of activity carried out inside the premises that host the reactor. Therefore, the wireless sensor networks deployed for monitoring them differ as well, both in terms of the used hardware, the deployment strategy, or the expected robustness and fault tolerance.

4.3.1 Radiation Sensors

One of the most important parameters to monitor in the proximity of a nuclear reactor is of course the level of radioactivity. The best-known solution for that is the Geiger-Mueller (GM) detector, that contains a gas-filled chamber, with a central wire that acts as an anode if a voltage is applied, while the chamber wall acts a cathode. If the detector is exposed to radioactivity, an electronic pulse is created, which can then be measured. These are portable, hand-held detectors, providing very reliable measurements. However, their main drawback is that human intervention is needed in the potentially contaminated area. Also, they provide just a snapshot of the situation at a given point in time and at a given location, without any long-term history or aggregation of the accumulated data.

Another solution that does not need human intervention is the use of Thermal Luminescent Dosimeters, which are passive badges not requiring any battery power. If heated after being exposed to radiation, they will emit light proportional to the amount of radiation received. These badges are very cheap and reliable, but their main drawback is that they cannot provide immediate feedback about a radiation hazard; they have to be sent to outside laboratories for analysis which can last several days or even weeks. Moreover, they do not provide time-stamped information, the results of the analysis reflect only the level of radiation accumulated while the badge was exposed.

As opposed to these traditional methods, radiation sensor boards could be attached to wireless motes to provide long-term, fine-grained, real-time monitoring. The Radiation Sensor Board produced by Libelium and shown in Fig. 4 is an example of such a device. It contains an autonomous and wireless Geiger counter and it works as follows [12]. The node sleeps most of the time, to save its battery, that is always a critical resource in WSNs. At periodic intervals however it wakes up for a very short, one-minute interval and it counts the number of pulses that are generated in the Geiger tube during this interval. If the result is below a well defined threshold, the result is time-stamped and simply sent to a local gateway node through the Zigbee interface of the attached mote. If, however, the result is above the threshold, besides logging locally the data, an immediate alarm is sent through the GPRS connection to the concerned security personnel. If the board is not statically placed at a given location, its GPS coordinates might also be sent along with the alarm, to allow fast detection of the radiation source.

4.3.2 Specific WSN Challenges in Nuclear Reactors

Nuclear reactors are not placed in traditional buildings, but dedicated ones, specifically built to limit the possibilities of accidents and radiation leakage. Concrete walls and floors are for example exceptionally thick (about 1.5 m or above, usually), which helps protecting the external environment, but is especially harmful for the propagation of electromagnetic waves, which will be affected by

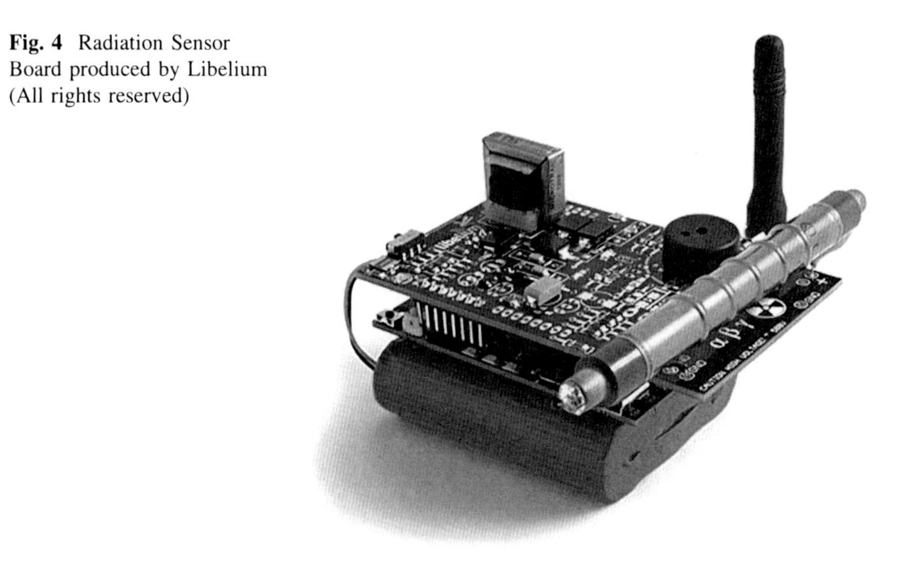

Fig. 4 Radiation Sensor Board produced by Libelium (All rights reserved)

reflections, diffraction or scattering. Thus, to prevent rapid signal attenuation, the placement of the sensors and the sink node has to be carefully chosen.

In case the radiation level monitoring system is based on a wireless technology, it has to be made certain that the system is very robust and reliable. If the radio communication system is disturbed or fails, this could potentially affect the lives of the people operating the reactor. Different techniques could be implemented to increase reliability, but they should also think about sparing the wireless resources, which leads to a prolonged network lifetime. Data aggregation techniques can be very well used for example, both at the node and the network levels. If everything is normal, there is no need to flood the network with useless redundant information. However, if the radiation level approaches the critical threshold, very detailed information is needed from as many sensors as possible. These readings have then to be prioritized when processed and transmitted in the network, prohibiting data aggregation. Fur further reading, a very detailed description of the critical issues related to the deployment and operation of a wireless sensor network in the McMaster Nuclear Reactor is given in [15].

References

1. Al-Karaki, J., Kamal, A.: Routing techniques in wireless sensor networks: a survey. Wireless Communications, IEEE **11**(6), 6–28 (2004). doi:10.1109/MWC.2004.1368893
2. Alcaraz, C., Roman, R.: Applying key infrastructures for sensor networks in cip/ciip scenarios. In: Proceedings of the First International Conference on Critical Information Infrastructures Security, CRITIS'06, pp. 166–178. Springer-Verlag, Berlin, Heidelberg (2006)

3. Buratti, C., Conti, A., Dardari, D., Verdone, R.: An overview on wireless sensor networks technology and evolution. Sensors **9**(9), 6869–6896 (2009). doi:10.3390/s90906869, http://www.mdpi.com/1424-8220/9/9/6869
4. Buttyan, L., Gessner, D., Hessler, A., Langendoerfer, P.: Application of wireless sensor networks in critical infrastructure protection: challenges and design options [security and privacy in emerging wireless networks]. Wireless Communications, IEEE **17**(5), 44–49 (2010). doi:10.1109/MWC.2010.5601957
5. Christodoulou, S., Agathokleous, A., Kounoudes, A., Milis, M.: Wireless sensor networks for water loss detection. European Water, Springer (2010)
6. Crossbow IMote2 data sheet. Tech. rep
7. Crossbow IRIS data sheet. Tech. rep
8. Crossbow MIB600 data sheet. Tech. rep
9. Crossbow MICA2 data sheet. Tech. rep
10. Crossbow: MICAz data sheet. Tech. rep
11. Desai, U.B., Jain, B.N., Merchant, S.N.: Wireless sensor networks: technology roadmap. Tech. rep
12. Gascon, D., Yarza, M.: Wireless sensor networks to control radiation levels. Libelium. http://www.libelium.com/wireless_sensor_networks_to_control_radiation_levels_geiger_counters/
13. Goldman, J., Ramanathan, N., Ambrose, R., Caron, D.A., Estrin, D., Fisher, J.C., Gilbert, R., Hansen, M.H., Harmon, T.C., Jay, J., Kaiser, W.J., Sukhatme, G.S., Tai, Y.C.: Distributed sensing systems for water quality assessment and management (2007)
14. Lim, Y., Kim, H.M., Kang, S.: A design of wireless sensor networks for a power quality monitoring system. Sensors **10**(11), 9712–9725 (2010). doi:10.3390/s101109712, http://www.mdpi.com/1424-8220/10/11/9712
15. Merizzi, N.: Sensor network deployment in the mcmaster nuclear reactor. McMaster University (2005). http://books.google.hu/books?id=b36zMAAACAAJ
16. Srirangarajan, S., Iqbal, M., Whittle, A.J., Allen, M., Preis, A., Lim, H.B.: Water main burst event detection and localization, Chap. 118, pp. 1324–1335. doi:10.1061/41203(425)119, http://ascelibrary.org/doi/abs/10.1061/41203%28425%29119
17. (TG4), I.W.T.G.: http://www.ieee802.org/15/pub/tg4.html
18. US: 2004 National critical infrastructure protection research and development plan. Tech. rep., US (2004)
19. Yang, Y., Lambert, F., Divan, D.: A survey on technologies for implementing sensor networks for power delivery systems. In: IEEE, Power Engineering Society General Meeting, 2007, pp. 1–8 (2007). doi:10.1109/PES.2007.386289

System-of-Systems Approach

Massoud Amin

Abstract Energy, telecommunications, transportation, and financial infrastructures are becoming increasingly interconnected, thus, posing new challenges for their secure, reliable and efficient operation. All of these infrastructures are, themselves, complex networks, geographically dispersed, non-linear, and interacting both among themselves and with their human owners, operators, and users. No single entity has complete control of these multi-scale, distributed, highly interactive networks, nor does any such entity have the ability to evaluate, monitor, and manage them in real time. In fact, the conventional mathematical methodologies that underpin today's modeling, simulation, and control paradigms are unable to handle the complexity and interconnectedness of these critical infrastructures.

Keywords System-of-systems · Critical infrastructure interdependencies · Complex systems · Smart self-healing infrastructure · Security · Risk · Uncertain dynamic systems

1 Introduction

Virtually every crucial economic and social function depends on the secure, reliable operation of energy, telecommunications, transportation, financial, and other infrastructures. Indeed, they have provided much of the good life that the more developed countries enjoy. However, with increased benefit has come increased risk. As these infrastructures have grown more complex to handle a variety of demands, they have become more interdependent. The Internet,

M. Amin (✉)
Technological Leadership Institute, University of Minnesota, 290 McNamara Alumni Center, 200 Oak Street Southeast, Minneapolis, MN 55455, USA
e-mail: amin@umn.edu

E. Kyriakides and M. Polycarpou (eds.), *Intelligent Monitoring, Control, and Security of Critical Infrastructure Systems*, Studies in Computational Intelligence 565,
DOI 10.1007/978-3-662-44160-2_12

computer networks, and our digital economy have increased the demand for reliable and disturbance-free electricity; banking and finance depends on the robustness of electric power, cable, and wireless telecommunications. Transportation systems, including military and commercial aircraft and land and sea vessels, depend on communication and energy networks [1, 2]. Links between the power grid and telecommunications and between electrical power and oil, water, and gas pipelines continue to be a lynchpin of energy supply networks. This strong interdependence means that an action in one part of one infrastructure network can rapidly create global effects by cascading throughout the same network and even into other networks.

Modeling interdependent infrastructures (e.g., the electric power, together with telecommunications, oil/gas pipelines and energy markets) in a control theory context is especially pertinent since the current movement toward deregulation and competition will ultimately be limited only by the physics of electricity and the topology of the grid. In addition, mathematical models of complex networks are typically vague (or may not even exist); existing and classical methods of solution are either unavailable, or are not sufficiently powerful. For the most part, no present methodologies are suitable for understanding their behavior.

There is reasonable concern that national and international energy and information infrastructures have reached a level of complexity and interconnection which makes them particularly vulnerable to cascading outages, initiated by material failure, natural calamities, intentional attack, or human error. The potential ramifications of network failures have never been greater, as the transportation, telecommunications, oil and gas, banking and finance, and other infrastructures depend on the continental power grid to energize and control their operations. Although there are some similarities, the electric power grid is quite different from gas, oil or water networks—phase shifters rather than valves are used, and there is no way to store significant amounts of electricity. To provide the desired flow on one line often results in "loop flows" on several other lines.

In the aftermath of the tragic events of September 11th and recent natural disasters and major power outages, there are increased national and international concerns about the security, resilience and robustness of critical infrastructures in response to evolving spectra of threats. Secure and reliable operation of these networks is fundamental to national and international economy, security and quality of life.

Our work in this area draws from methods in statistical physics, complex adaptive systems, discrete-event dynamical systems, and hybrid, layered networks. Modeling complex systems is one of three main areas in our ongoing work. The others are measurement—to know what is or will be happening and develop measurement techniques for visualizing and analyzing large-scale emergent behavior—and management—to develop anticipatory distributed management and control systems to keep power and energy infrastructures robust and operational. From a broader viewpoint, *agility* and *robustness/survivability* of smart grids as large-scale dynamic networks that face *new* and *unanticipated* operating conditions is presented.

2 Definition of Critical Infrastructure

Executive Order 13010, signed by President Clinton in 1996, defined critical infrastructures as "so vital that their incapacity or destruction would have debilitating impact on the defense or economic security of the United States" and included "telecommunications, electrical power systems, gas and oil storage and transportation, banking and finance, transportation, water supply systems, emergency services and continuity of government".

The U.S. Department of Homeland Security (DHS) in the National Infrastructure Protection Plan has expanded the concept to include "key resources" and added food and agriculture, health and healthcare, defense industrial base, information technology, chemical manufacturing, postal and shipping, dams (including locks and levees), government facilities, commercial facilities, critical manufacturing and national monuments and icons [3].

The Board on Infrastructure and the Constructed Environment (BICE) has argued that five "lifeline" infrastructures are the most critical because all the others depend on them for survival. These are power, telecommunications, transportation, water and wastewater systems [4]. This chapter focuses on these five, agreeing with the BICE, but expecting that the results will ultimately apply to many of the 18 sectors identified by DHS. The focus, as an example, will be the critical infrastructures of the United States (US), as well as the current practices and investments (or lack of) taking place.

3 Consequences of Aging Infrastructures

The infrastructures provide the lifelines on which our communities and our economy depend. Many of these, however, are aging in place, imposing risks of fatalities, serious injuries and massive economic disruptions. For example, in the United States:

3.1 Highways

The vast bulk of our major highways were built in the 1950s and 1960s and have been maintained largely at the level of hot-patching even as the number of vehicles has increased enormously. (New highway construction, however, has continued.) Between 1970 and 2002, passenger travel doubled, with a growth of another 67 % by 2022. The Federal Highway Administration (FHWA) currently considers one-third of the nation's roads to be "poor," "mediocre" or "fair"—all three categories requiring investment. The American Automobile Association estimates that as many as 12,000 lives (of a total of 44,000) could be saved if highways were improved by adding lighting and guardrails and straightening dangerous curves. The Road Information Program estimated costs as much as $220 billion a year, which includes car repairs ($65B), congestion ($78B) and accidents ($77B)—the equivalent of nearly 2 % points of the national economy.

3.2 Bridges

The age of famous bridges is illuminating: New York's Brooklyn Bridge is over 120 years old and its George Washington Bridge is 75+ years; St. Louis's Eads Bridge is 130+; San Francisco's Bay Bridge and Louisiana's Huey P. Long Bridge are both 70+. The FHWA rates 13.1 % of America's highway bridges "structurally deficient" and an additional 13.6 % as "functionally obsolete". Most remain open to traffic. One of these was the I-35 bridge over the Mississippi River that collapsed, killing 13 people in 2007. Others were the I-95 Mianus River Bridge in Greenwich, Connecticut, where three people died in 1982, and the New York Throughway Bridge near Amsterdam, N.Y., where ten people perished in 1987. More than 1,500 bridges failed between 1966 and 2005, 60 % due to soil erosion around the bridge supports, a potential weakness seldom checked in inspections.

3.3 Dams and Levees

Hoover Dam is 74 years old; the Wilson Dam, 84; the Grand Coulee Dam, 66; and all of the dams over the Tennessee River are more than 60 years old. The number of "high hazard" dams, whose failure would endanger human life, has increased from 9,281 in 1998 to 10,213 in 2007. In the past 2 years, more than 67 dam incidents, including 29 dam failures, were reported to the National Performance of Dams program. States report more than 3,500 "unsafe" dams with conditions that could cause them to fail. Seepage has been noted under the 55-year-old Wolf Creek Dam, forcing its water level to be lowered to avoid devastating failure of the dam holding the largest man-made reservoir east of the Mississippi and the flooding of Nashville and neighboring communities. Precautions came too late for the levees protecting New Orleans after Hurricane Katrina in 2005, killing almost 1,500 and making thousands refugees, many still far from returning to their homes.

3.4 Electric Power

Modern life depends on electricity. The transmission of electric power is largely based on technologies installed more than 50 years ago. From 1988 to 1998, US electricity demand rose by nearly 30 % while the transmission network's capacity grew by only 15 %. The Electric Power Research Institute (EPRI) anticipated that the disparity would further increase during the period 1999–2009: The Institute projects demand to grow by 20 % and system capacity to increase by just 3.5 % [5]. The Northeast Blackout of 2003 [6], caused by human error and transmission lines contacting improperly trimmed trees, resulted in the shutdown of more than 100 power plants, denied power to 40 million Americans (one-seventh of the national population) and ten million Canadians (one-third of its national population), with a cost of more than six billion dollars. Water and wastewater plants were idled, transportation by all modes slowed to a stop, communications and industry were largely stopped and at least eleven fatalities were reported. In a CNN interview about the Blackout, Governor Bill Richardson of New Mexico, former U.S. Secretary of Energy, described the U.S. as "a major superpower with a third-world electrical grid" [7]. In the 1960s and before, blackouts lasting more than more than a few hours were rare but, by the mid-1990s, their frequency rose to yearly or greater. An EPRI survey of industry which tracked the cost of blackouts over this time indicated a rise from insignificant levels to $100 billion/year. The White House's National Energy Policy of 2001 suggests that the nation will need 1,300–1,900 new power plants in the next two decades, but this ignores the investments needed in transmission and distribution.

3.5 Water and Wastewater

Many cities, especially in the east and northwest, have water systems containing components more than 100 years old, including asbestos-cement pipes, lead pipes, even wooden pipes and storage tanks. New York City's fresh water is transmitted through two tunnels, built in 1917 and 1936 (a third will be completed by 2020 if the schedule holds). A survey by EPA found that in systems that serve more than 100,000 people, about 30 % of the pipes are between 40 and 80 years old and about 10 % of the pipes are more than 80 years old. Some systems treat as much as 100 % more water than is consumed due to high rates of leakage from old transmission and distribution pipes. The U.S. Conference of Mayors reports that more than half of a 330-city survey suffer annual or more frequent water main breaks—some as many as 50 or more per year. During the 2003 Northeast Blackout, emergency generators in New York City failed and some 30 million gallons of raw sewage were dumped into the East River. In New York and many other cities, raw sewage mixes into waterways with every significant rainstorm. Sanitary sewer overflows caused by blocked or broken pipes result in the release of as much as 10 billion gallons of raw sewage annually, according to the EPA. During blackouts, pressure drops in water mains, impeding the cooling of high-rise buildings, severely limiting the ability to control fires, and risking introduction of contaminants into drinking water.

The consequences of aging infrastructures are dire and add to the congestion costs of inadequate infrastructure capacity in parts of the country undergoing economic or population growth. A large and growing sentiment among the American public cries out for government at all levels and the private utilities that provide these services to invest in new, renewal and replacement infrastructure necessary to avoid or mitigate such consequences.

4 Recent and Near-Term Infrastructure Investment Rates

Hurricane Katrina and the collapse of Minneapolis's I-35 bridge have stimulated public awareness of the necessity for accelerated programs of new investment, replacement, rehabilitation and renewal. Most Americans are unaware of the magnitude or trends in American investment in infrastructure. In the sources cited below, the definition of investments included in "infrastructure" varies widely. Where possible, the text or figure labels will make clear what is included.

Most public infrastructure investment is in transportation and water, while most private sector infrastructure is in energy and telecommunications. Public investment in infrastructure has grown steadily since World War II (Fig. 1a), but, as a portion of Gross Domestic Product, has declined more or less steadily since the late 1950s (Fig. 1b).

Relative to its international competitors, the U.S. ranks 13th in infrastructure investment among OECD nations (Fig. 2a). Both China and India, not OECD members, also rank higher than the U.S. This is not a recent trend. U.S. infrastructure investment has been well below the average of the leading 17 countries in every decade since at least the 1970s (Fig. 2b, c).

In a more detailed look, capital investment by public and private sectors in selected infrastructures in 2004, the last year of complete data, totaled more than $300 billion (Table 1), split about evenly between the public (mostly in transportation and water/wastewater) and the private sector (mostly in energy and telecommunications).

5 The Costs of Renewal or Replacement

In brief, vast sums of money are being and will be invested. The American Society of Civil Engineers (ASCE), however, has judged that it will not be nearly enough, estimating that $2.2 *trillion* dollars will be required to restore the United States infrastructure to a sound level [8]. Some examples illustrate this pressing need:

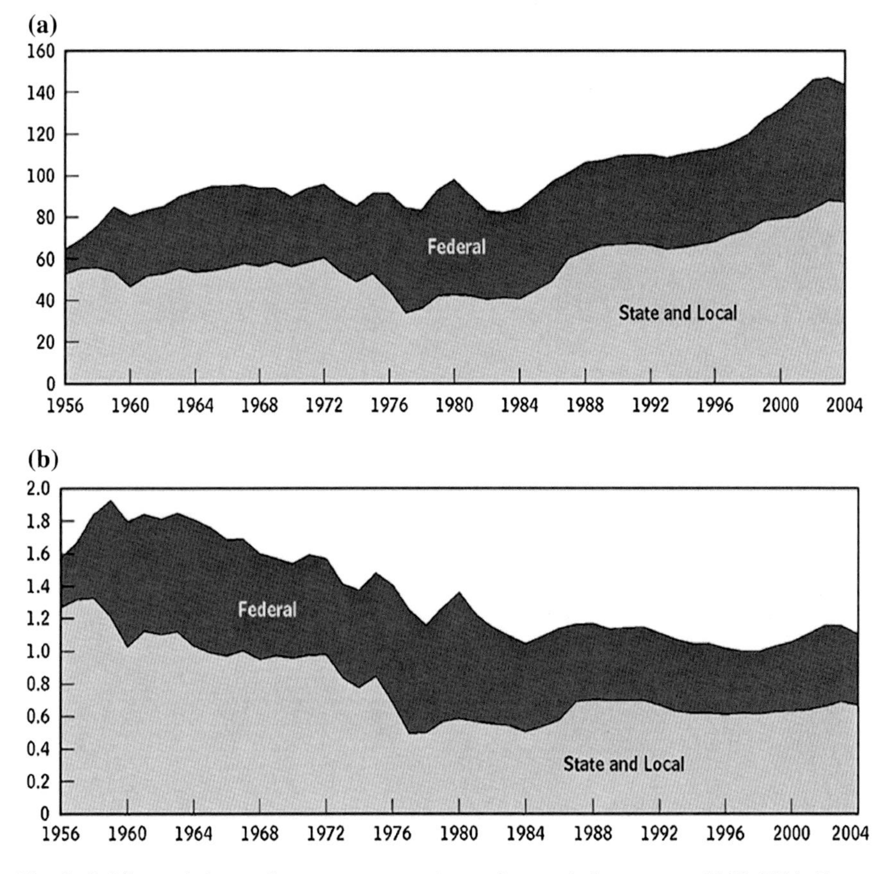

Fig. 1 Public capital spending on transportation and water infrastructure, 1956–2004 (*Source* CBO, 2008). **a** Billions of 2006 dollars. **b** Percentage of gross domestic product

5.1 Highways

The American Association of State Highway and Transportation Officials, Federal Highway Administration and ASCE all point to the need for massive increases in capital outlays by all levels of government to reach the cost-to-maintain level, and roughly double that to reach the cost-to-improve level. ASCE is calling for $186 billion annually, compared to current actual outlays, well below even maintenance level.

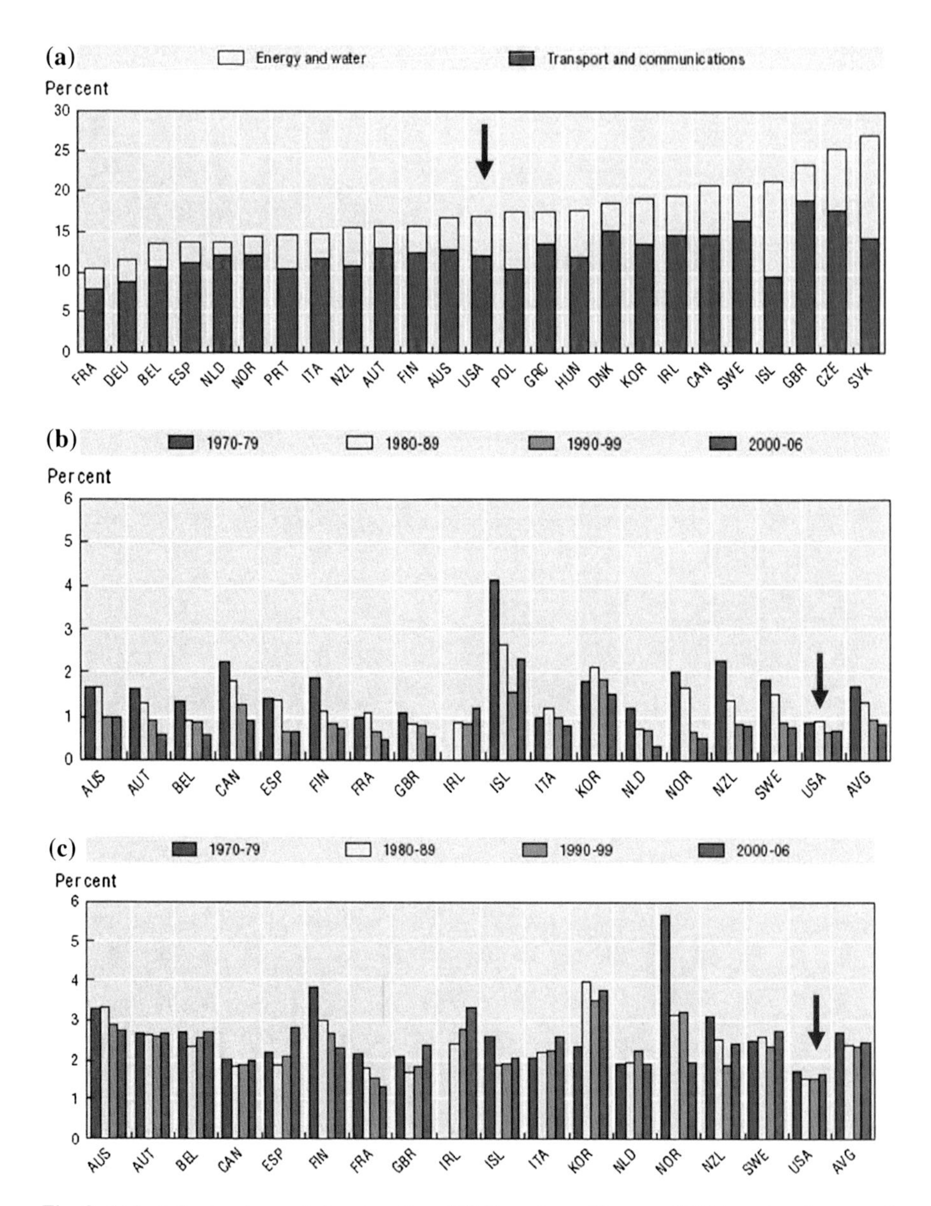

Fig. 2 U.S. infrastructure relative to other OECD nations (*Source* OECD, going for growth 2009). **a** Infrastructure investment as a percentage of total fixed investment, averages over latest 5 years. **b** Electricity, gas and water investment as a percentage of GDP. **c** Transport and communications investment as a percentage of GDP

Table 1 Capital spending on infrastructure in 2004, by category (Billions of 2004 dollars; columns may not add due to rounding)

	Public[a]		Total			Amer. recovery and reinvestment act[b]
Transportation	Federal	State	Public	Private	Total	
Highways	30.2	36.5	66.7	n.a.	66.7	27.5
Mass transit	7.6	8.0	15.5	0	15.5	8.4
Railroads	0.7	0	0.7	6.4	7.1	9.3
Aviation	5.6	6.8	12.4	2.0	14.4	2.1
Water transportation	*0.7*	*1.7*	*2.4*	*0.1*	*2.5*	*0.2*
Total transportation	44.7	53.0	97.7	8.5	106.2	47.5
Drinking water and wastewater	2.6	25.4	28.0	n.a.	28.0	7.4
Energy	1.7	7.7	9.4	69.0	78.4	36.8
Telecommunications	3.9	n.a.	3.9	68.6	72.5	7.2
Pollution control and waste disposal	0.8	1.8	2.6	3.6	6.2	0.8
Water and other natural resources	*7.1*	*4.3*	*11.3*	*n.a.*	*11.3*	6.1
Total, non-transportation	16.1	39.2	55.2	141.2	196.4	58.3
Total	60.8	92.2	152.7	149.7	302.6	105.8

[a] Peter R. Orszag, Director, Congressional Budget Office, "Investing in Infrastructure," Testimony before the Committee on finance, U.S. Senate, July 10, 2008, obtained from http://www.cbo.gov/ftpdocs/95xx/MainText.1.2.shtml, accessed on June 22, 2009. The original table included investments for schools ($99.7 billion), prisons ($2.9 billion) and postal facilities ($0.9 billion), for a total of $406.1 billion. These were deleted as not being parts of the infrastructures of interest in this chapter

[b] Interpreted from U.S. Congress, House Committee on Appropriations, "Summary: American Recovery and Reinvestment—Conference Agreement," February 13, 2009

5.2 Bridges

ASCE estimates that it will cost $17 billion per year over 20 years to eliminate bridge deficiencies compared to the $10.5 billion currently being spent.

5.3 Dams and Levees

ASCE did not estimate the cost of restoring American dams to safety, but states that $100 billion is needed to repair and rehabilitate U.S. levees.

5.4 *Electrical Power*

ASCE estimates that $1.5 trillion in new investments will be needed by 2030.

5.5 *Water and Wastewater*

For drinking water, EPA estimates potential funding gaps as high as $263 billion by 2019. For wastewater, EPA and ASCE estimate $390 billion is needed to replace existing wastewater infrastructure systems and to build new ones.

6 Interdependent Infrastructures

Valuation of infrastructure investments is complicated by the dependencies and interdependencies *among* the constituent system elements of some infrastructures (e.g., the electricity system of power plants, substations, and transmission and distribution lines; the transportation system of ports, rail, barges, roads, highways, airports, bridges and tunnels; water storage dams and water supply). Interdependencies also exist *between* infrastructures, such as:

- Cooling of power plants uses about 40 % of the fresh water withdrawals, as much as all the nation's irrigation requirements;
- Water-related energy requirements are reported in California to account for 19 % of all power generated in the state;
- The Northeast Blackout of 2003 and Hurricane Isabel of 2005 illustrated how loss of electricity can cause loss or diminution of communications, water and sanitation services, automotive fuel pumping, rail transportation, traffic control, food distribution, building cooling, fire protection, hospital services and numerous other life-essential services—even power generation itself where cooling water relies on electrically powered pumps;
- The failure of one infrastructure in New Orleans, the levee system, caused the near total collapse of all the infrastructures—indeed, of basic human viability—following Hurricane Katrina.

Such interdependencies can and do result in "cascading failures". These interdependencies make it difficult for decision-makers in the responsible institutions and organizations to consider *all* the consequences as they decide where to allocate their severely limited funds for infrastructure replacement and restoration. Often, whole metropolitan areas or even multi-state regions can be impacted by a single failure, such as a poorly trimmed tree in the wrong place. The interactions among infrastructures necessitate a "system-of-systems" analytic approach in which the assets and networks of each infrastructure are assessed in the context of their consequences on all the impacted infrastructures and communities.

7 Infrastructure Investment Decision-Making

Infrastructure investment is a major societal challenge but the United States is currently ill-equipped to make the needed priority and resource allocation decisions. Rationally allocating such funds is complicated by long-standing, ingrained practices—"earmarks" and "pork-barrel" funding for special projects (some necessary, others perhaps less so); "trust-fund" single-purpose financing of some infrastructures (e.g., roads) and the absence of such funds for others (e.g., drinking water); formulaic allocation of block grants based on criteria at best loosely related to infrastructure requirements on the ground. Congressional committees and subcommittees authorize and appropriate funds that flow directly to specialized Federal agencies and, from there, often to state and local specialized agencies, creating "stove-pipes" from funds source to sink, from concept to spade tip, without any challenge of comparative assessment of the benefits and costs of the investments. The absence of a central clearing point or set of standard metrics makes essential comparisons for rational optimization impossible [9].

Powerful vested interests have a stake in maintaining the current jumble of allocation schemes because they are in the position to exercise power, take credit, and/or receive funding. Starting with Congressional earmarks and horse-trading, through trust funds and federal agency formula grants, clear through to state and local elected and appointed officials' final decisions, there is little or no comprehensive, comparative analysis of value, sustainability, risk or resilience of potential infrastructure projects. Furthermore, certain individuals at each level profit from those arrangements. Availability of a competent, standardized technology for valuing and prioritizing infrastructure investments would permit the adaptation of existing processes to provide a more rational analytic underpinning and more nearly optimal allocation of investment funds.

Further, there are few incentives for the needed design and advocacy to be undertaken. Infrastructures are, almost by definition, networked and highly interdependent, so an improved valuation and selection method requires contributions from diverse disciplines and industries that seldom collaborate, making it unlikely that the needed research will be undertaken without Federal support.

To date, the combination of political and functional barriers has resulted in there being no comprehensive, comparative method for rationally undertaking these critically important decisions. Politically robust forces benefit materially from the current jumble of allocation processes and the requirements for constructing new, more competent methods cover too broad and diverse a set of disciplines to be feasible in the absence of a concentrated Federal effort. The opportunity cost of continuing with the present system will be enormous as the

United States rushes to make investments that will entail jobs, while potentially wasting tens or hundreds of billions of dollars.

According to a distinguished bipartisan commission that included two sitting Senators, two former Senators, three sitting governors and a number of former cabinet members and ambassadors, convened by the Center for Strategic and International Studies (CSIS), "America's economic well-being and physical security depend on safe and reliable... infrastructure... But we are both under-investing in infrastructure and investing in the wrong projects: *new investments are critically needed, but we lack the policy structures to make the correct choices and investments*... A centralized infrastructure project approval process would force all infrastructure modes to be evaluated using common methods and parameters" (emphasis in original) [10]. The commission was not specific as to a particular set of "common methods and parameters". Establishment of objective, transparent methods that use "common methods and parameters" that yield directly comparable estimates of benefits and costs of alternative investments is the *sine qua non* of rational allocation of limited resources.

Rehabilitating or renewing aging infrastructures often yields little near-term revenue or profit to companies and seldom reflects great credit on public officials faced with competing, more visible demands. Strong incentives encourage deferral of these investments. Accordingly, investment in older infrastructures lags behind the rate of deterioration. Conversely, most new infrastructure development conspicuously adds local employment, so tends to highlight politicians who can take credit, accounting for much of the "earmarking" for favored projects. Ribbon cutting for new, highly visible facilities attracts far more media attention than rehabilitating or upgrading existing structures.

The diversity among the infrastructures and the complexity of the interdependencies among them encourage decision-makers to evaluate their investment options from their own very limited perspective, even though the full consequences may have very far-reaching effects. Each infrastructure sector and each operator may use any perspective, methodology, metrics or level of rigor it chooses, within only broad guidelines, often prescribed by the oversight or funding authority. Numerous criteria and metrics for valuing benefits and costs are used, and usually, no attempt is made at optimizing the allocation of the resource base for aging infrastructures. Across all infrastructures, local to national in scale, this results in an absolute inability to compare one project's benefits or costs to competing project opportunities. At best, resource allocation decisions can be "sub-optimized" as each oversight or funding authority, functional agency, utility operator or company makes its choices by its own criteria within its own "stove-pipe". The overall national infrastructure portfolio cannot be optimized in any sense, so scarce resources are expended for less benefit than a more comprehensive approach could accomplish.

8 A Case Study: Interdependent End-to-End Power Infrastructure and its Couplings

Power, telecommunications, banking and finance, transportation and distribution, and other infrastructures are becoming more and more congested partially due to dramatic population growth, particularly in urban centers. These infrastructures are increasingly vulnerable to failures cascading through and between them. A key concern is the avoidance of widespread network failure due to cascading and interactive effects. Moreover, interdependence is only one of several characteristics that challenge the control and reliable operation of these networks. Other factors that place increased stress on the power grid include dependencies on adjacent power grids (increasing because of deregulation) by telecommunications, markets and computer networks. Furthermore, reliable electric service is critically dependent on the whole grid's ability to respond to changed conditions instantaneously. Dependencies of other infrastructures on electric power grid telecommunications, markets, and so much else...

Secure and reliable operation of complex networks poses significant theoretical and practical challenges in analysis, modeling, simulation, prediction, control, and optimization. The pioneering initiative in the area of complex interactive networks and infrastructure interdependency modeling, simulation, control and management was launched and successfully carried out its goals during 1998–2002, through the Complex Interactive Networks/Systems Initiative (CIN/SI) [5, 11], studied closely challenges to the interdependent electric power grid, energy, sensing and controls, communications, transportation, and financial infrastructures. It comprised of six university research groups consisting of 108 university faculty members and over 220 researchers involved in the joint Electric Power Research Institute (EPRI) and U.S. Department of Defense program. During 1998–2002, CIN/SI developed modeling, simulation, analysis, and synthesis tools for damage-resilient control of the electric power grid and interdependent infrastructures connected to it.

Earlier work by the author during the 1990s on damaged F-15 aircrafts in part provided background for the creation, successful launch, and management of research programs for the electric power industry, including the EPRI/DOD CIN/SI mentioned above, which involved six university research consortia, along with two energy companies, to address challenges posed by our critical infrastructures. This work was done during the period of 1998 to early 2002. CIN/SI laid the foundation for several on-going initiatives on the self-healing infrastructure and subsets focusing on smart reconfigurable electrical networks. These have now been under development for some time at several organizations, including programs sponsored by the U.S. NSF, DOD, DOE, and EPRI, including EPRI's "Intelligrid" program and the US Department of Energy's "Gridwise," "Modern Grid," and "Smart Grid" initiatives.

To provide a context for this, the EPRI/DOD CIN/SI aimed to develop modeling, simulation, analysis, and synthesis tools for robust, adaptive, and reconfigurable control of the electric power grid and infrastructures connected to it. In part, this work showed that the grid can be operated close to the limit of stability given adequate situational awareness combined with better sensing of system conditions, communication, and controls. A grid operator is similar to a pilot flying the aircraft, monitoring how the system is being affected, how the "environment" is affecting it and having a solid sense of how to steer it in a stable fashion by keeping the lines within their operating limits while helping an instantaneous balance between loads (demand) and available generation—grid operators often make these quick decisions under considerable stress. Given that in recent decades we have reduced the generation and transmission capacity, we are indeed flying closer to the edge of the stability envelope.

As an example, one aspect of the Intelligrid program aimed at enabling grid operators greater look-ahead capability and foresight into the road-ahead overcoming limitations of the current schemes which at best have over a 30-s delay in assessing system behavior—analogous to driving the car by looking into the rear-view mirror instead of the road ahead. This tool using advanced sensing, communication and software module was proposed during 2000–2001 and the program was initiated in 2002 [12]. This advanced simulation and modeling program promotes greater grid self-awareness and resilience in times of crisis in three ways: by providing faster-than-real-time, look-ahead simulations (analogous to master chess players rapidly expanding and evaluating their various options under time constraints) and thus avoiding previously unforeseen disturbances; by performing what-if analysis for large-region power systems from both operations and planning points of view; and by integrating market, policy and risk analysis into system models, and quantifying their integrated effects on system security and reliability.

Focusing on the electric power sector, the power outages and power quality disturbances cost the U.S. economy over $80 billion annually, and sometimes up to $188 billion in a single year. Transmission and distribution losses in the U.S. were about 5 % in 1970, and grew to 9.5 % in 2001, due to heavier utilization and more frequent congestion. Regarding the former, starting in 1995, the amortization/ depreciation rate exceeded utility construction expenditures. Since that crossover point in 1995, utility construction expenditures have lagged behind asset depreciation. This has resulted in a mode of operation of the system that is analogous to harvesting more rapidly than planting replacement seeds. As a result of these diminished "shock absorbers," the electric grid is becoming increasingly stressed, and whether the carrying capacity or safety margin will exist to support anticipated demand is in question.

To assess impacts one can use actual electric power outage data for the U.S., which are generally available from several sources, including from the U.S. DOE's Energy Information Administration (EIA) and from the North American Electric Reliability Corporation (NERC). In general, the EIA database contains more events, and the NERC database gives more information about the events. Both databases are extremely valuable sources of information and insight. In both databases, a report of a single event may be missing certain data elements, such as the amount of load dropped or the number of customers affected. In the NERC database, the amount of load dropped is given for the majority of the reported events, whereas the number of customers affected is given for less than half the reported events. Analyses of these data collected revealed that in the period from 1991 to 2000, there were 76 outages of 100 MW or more in the second half of the decade, compared to 66 such occurrences in the first half (Fig. 3) [13]. Furthermore, there were 41 % more outages affecting 50,000 or more consumers in the second half of the 1990s than in the first half (58 outages in 1996–2000 versus 41 outages in 1991–1995). In addition, between 1996 and 2000, outages affected 15 % more consumers than they did between 1991 and 1995 (the average size per event was 409,854 customers affected in the second half of the decade versus 355,204 in the first half of the decade). Similar results were determined for a multitude of additional statistics such as the kilowatt magnitude of the outage, average load lost, etc. These trends have persisted in this decade. NERC data show that during 2001–2005 we had 140 occurrences of over 100 MW dropped, and 92 occurrences of over 50,000 or more consumers affected.

The U.S. electrical grid has been plagued by ever more and ever worse blackouts over the past 15 years. In an average year, outages total 92 min per year in the Midwest and 214 min in the Northeast. Japan, by contrast, averages only 4 min of interrupted service each year [14]. The outage data excludes interruptions caused by extraordinary events such as fires or extreme weather.

Two sets of data, one from the U.S. Department of Energy's Energy Information Administration (EIA) and the other from the North American Electric Reliability Corp. (NERC) are analyzed below. Generally, the EIA database contains more events, and the NERC database gives more information about the events, including the date and time of an outage, the utility involved, the region affected, the quantity of load dropped, the number of customers affected, the duration of the outage, and some information about the nature of the event. The narrative data in the NERC (and also the EIA) databases are sufficient to identify factors, such as equipment failure or severe weather (or a combination of both!) that may have contributed to an outage. Establishment of the precise cause is beyond the scope of most of the narratives. Both databases are extremely valuable sources of information and insight.

In both databases, a report of a single event may be missing certain data elements such as the amount of load dropped or the number of customers affected. In the NERC database, the amount of load dropped is given for the majority of the reported events, whereas the number of customers affected is given for less than half the reported events. In the EIA database, the number of customers affected is reported more frequently than the amount of load dropped.

In both sets, each 5-year period was worse than the preceding one: According to data assembled by the U.S. Energy Information Administration (EIA) for most of the past decade, there were 156 outages of 100 MW or more during 2000–2004; such outages increased to 264 during 2005–2009. The number of U.S. power outages affecting 50,000 or more consumers increased from 149 during 2000–2004 to 349 during 2005–2009, according to EIA (Fig. 4).

In 2003 EIA changed their reporting form from EIA-417R to OE-417. Both forms were attached with descriptions of reporting requirements (page 3 and page

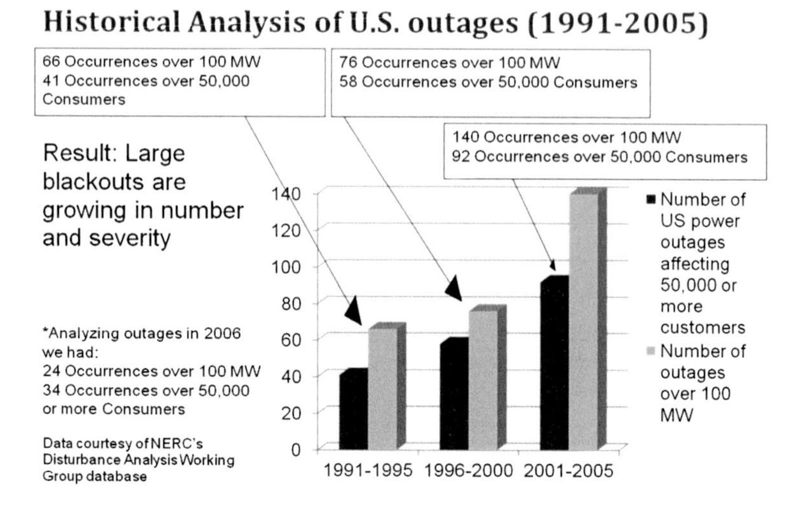

Fig. 3 U.S. electric power outages over 100 MW and affecting over 50,000 consumers (1991–2005)

6 respectively). In all, the reporting requirements are very similar, with OE-417 being a little more stringent. The main change in the requirement affecting the above figures is that all outages greater than 50,000 customers for 1 h or more be reported in OE-417, where it was only required for 3 h or more in EIA-417R prior to 2003. Adjusting for the change in reporting in 2003 (using all the data from 2000–2009 and only counting the outages that met the less stringent requirements of the EIA-417R form used from 2000–2002), there were 152 outages of 100 MW or more during 2000–2004; such outages increased to 248 during 2005–2009. The number of U.S. power outages affecting 50,000 or more consumers increased from 130 during 2000–2004 to 272 during 2005–2009 (Fig. 5).

In summary the number of outages adjusted for 0.9 % annual increase in load and adjusted for change in reporting in 2003 is:

	Occurrences of 100 MW or more	Occurrences of 50,000 or more consumers
2000–2004	152	130
2005–2009	248	272

As an energy professional and electrical engineer, I cannot imagine how anyone could believe that in the United States we should learn to "cope" with these increasing blackouts—and that we do not have the technical know-how, the political will, or the money to bring our power grid up to 21st century standards. Coping as a primary strategy is ultimately defeatist. We absolutely can meet the needs of a pervasively digital society that relies on microprocessor-based devices in vehicles, homes, offices, and industrial facilities. And it is not just a matter of

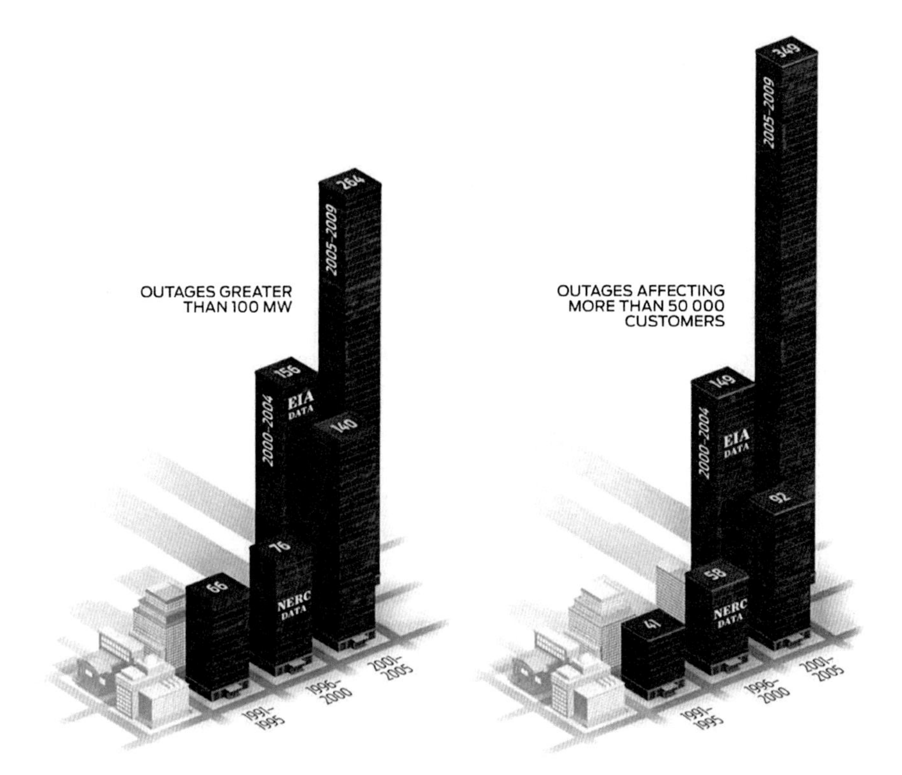

Fig. 4 Power outages have steadily increased [14]. (Research was supported by a grant from the NSF and a contract with the Sandia National Labs)

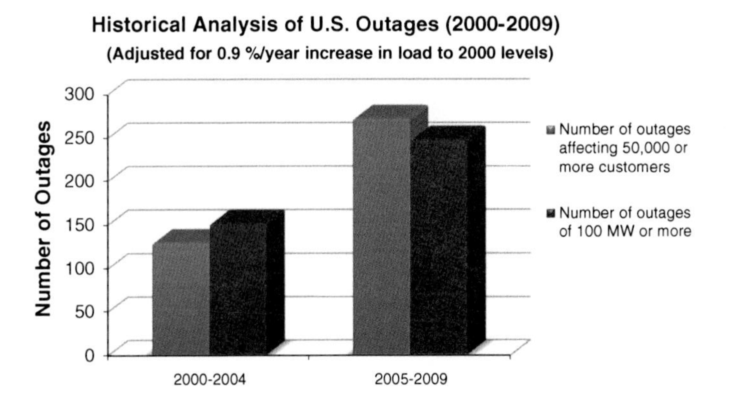

Fig. 5 U.S. electric power outages over 100 MW and affecting over 50,000 consumers during 2000–2009, adjusted for 0.9 % annual increase in load and adjusted for change in reporting in 2003 (using all the data from 2000–2009 and only counting the outages that met the less stringent requirements of the EIA-417R form used during 2000–2002)

"can". We must—if we want to continue on the road of technological advancement. However, it will not be easy or cheap [13, 15].

9 Background: Where Are We and How Did We Get Here?

The existing electricity infrastructure evolved to its technology composition today from the convolution of several major forces, only one of which is technologically based. Today opportunities and challenges persist in world-wide electric power networks. These include: reducing transmission congestion, increasing system/cyber security, increasing overall system and end-use efficiency while maintaining reliability—many other challenges engage those who plan for the future of the power grid: producing power in a sustainable manner (embracing renewable fuels while accounting for their scalability limitations, e.g., increased use of land and natural resources to produce higher renewable electricity will not be sustainable, thus not being able to lower emissions from existing generators), delivering electricity to those who do not have it (not just on the basis of fairness but also because electricity is the most efficient form of energy, especially for things like lighting), and using electricity more wisely as a tool of economic development, and pondering the possible revival of advanced nuclear reactor construction. To prepare for a more efficient, resilient, secure and sustainable electrical system it is helpful to remember the historical context, associated pinch-points and forcing functions.

The trends of worldwide electrical grid deployment, costing trillions of dollars and reaching billions of people, began very humbly. Some obvious electrical and magnetic properties were known in antiquity. In the 17th and 18th centuries, partially through scientific experiments and partially through parlor games, more was learned about how electric charge is conducted and stored. But only in the 19th century, with the creation of powerful batteries, and through insights about the relations between electric and magnetic force could electricity in wires service large scale industries—first the telegraph and then telephones.

And only in the 1880s did the first grids come into being for bringing electrical energy to a variety of customers for a variety of uses, at first mostly for illumination but later for turning power machines and moving trolley cars. The most important of these early grids, the first established big city grid in North America, was the network built by Thomas Edison in lower Manhattan. From its power station on Pearl Street, practically in the shadow of the Brooklyn Bridge, Edison's company supplied hundreds and then thousands of customers. Shortly thereafter, Edison's patented devices, and those of his competitors—devices such as bulbs, generators, switching devices, generators, and motors—were in use in new grids in towns all over the industrialized world.

From a historical perspective the electric power system in the U.S. evolved in the first half of the 20th century without a clear awareness and analysis of the system-wide implications of its evolution. In 1940, 10 % of the energy consumption in America was used to produce electricity. By 1970, this had risen to 25 %, and by 2002 it had risen to 40 %. (Worldwide, current electricity production is near 15,000 billion Kilowatt-hours per year, with The United States, Canada, and Mexico responsible for about 30 % of this consumption). This grid now underlies every aspect of our economy and society, and it has been hailed by the National Academy of Engineering as the 20th century's engineering innovation most beneficial to our civilization. The role of electric power has grown steadily in both scope and importance during this time and electricity is increasingly recognized as a key to societal progress throughout the world, driving economic prosperity, security and improving the quality of life. Still it is noteworthy that at the time of this writing there are about 1.4 billion people in the world with no access to electricity, and another 1.2 billion people have inadequate access to electricity (meaning that they experience outages of 4 h or longer per day).

Once "loosely" interconnected networks of largely local systems, electric power grids increasingly host large-scale, long-distance wheeling (movement of wholesale power) from one region or company to another. Likewise, the connection of distributed resources, primarily small generators at the moment, is growing rapidly. The extent of interconnectedness, like the number of sources, controls, and loads, has grown with time. In terms of the sheer number of nodes, as well as the variety of sources, controls, and loads, electric power grids are among the most complex networks made.

In the coming decades, electricity's share of total energy is expected to continue to grow, as more efficient and intelligent processes are introduced into this network [12]. Electric power is expected to be the fastest-growing source of end-use energy supply throughout the world. To meet global power projections, it is estimated by the U.S. DOE/EIA that over $1 trillion will have to be spent during the next 10 years. The electric power industry has undergone a substantial degree of privatization in a number of countries over the past few years. Power generation growth is expected to be particularly strong in the rapidly growing economies of Asia, with China leading the way.

The electric power grid's emerging issues include creating distributed management through using distributed intelligence and sensing; integration of renewable resources; use of active-control high-voltage devices; developing new business strategies for a deregulated energy market; and ensuring system stability, reliability, robustness, and efficiency in a competitive marketplace and carbon-constrained world. In addition, the electricity grid faces (at least) three looming challenges: its organization, its technical ability to meet 25-year and 50-year electricity needs, and its ability to increase its efficiency without diminishing its reliability and security.

As an example of historical bifurcation points, the 1965 Northeast blackout not only brought the lights down, it also marked a turn in grid history. The previous economy of scale, according to which larger generators were always more efficient than small machines, no longer seemed to be the only risk-managed option. In addition, in the 1970s two political crises—the Mideast war of 1973 and the Iranian Revolution in 1979—led to a crisis in fuel prices and a related jump in electric rates. For the first time in decades, demand for electricity stopped growing. Moreover, the prospects of power from nuclear reactors, once so promising, were now under public resistance and the resultant policy threats. Accidents at Brown's Ferry, Alabama in 1974 and Three Mile Island, Pennsylvania in 1979, and rapidly escalating construction costs caused a drastic turnaround in orders for new facilities. Some nuclear plants already under construction were abandoned.

In the search for a new course of action, conservation (using less energy) and efficiency measures (to use available energy more wisely) were put into place. Electrical appliances were re-engineered to use less power. For example, while on the average today's refrigerators are about 20 % larger than those made 30 years ago, they use less than half the electricity of older models. Furthermore, the Public Utility Regulatory Policy Act (PURPA) of 1978 stipulated that the main utilities were required to buy the power produced by certain independent companies which co-generated electricity and heat with great efficiency, providing the cost of the electricity was less than the cost it would take the utilities to make it for their own use.

What had been intended as an effort to promote energy efficiency, turned out, in the course of the 1980s and 1990s, to be a major instigator of change in the power industry as a whole. First, the independent power producers increased in size and in number. Then they won the right to sell power not only to the neighboring utility but also to other utilities further away, often over transmission lines owned by still other companies. With the encouragement of the Federal Energy Regulatory Commission (FERC), utilities began to sell off their own generators. Gradually the grid business, which for so long had operated under considerable government guidelines since so many utilities were effective monopolies, became a confusing mixture of regulated and unregulated companies.

Opening up the power industry to independent operators, a business reformation underway for some years in places like Chile, Australia, and Britain (where the power denationalization process was referred to as "liberalization"), proved to be a bumpy road in the US. For example, in 2001 in the state of California the effort to remove government regulations from the sale of electricity, even at the retail level, had to be rescinded in the face of huge fluctuations in electricity rates, rolling blackouts, and amid allegations of price-fixing among power suppliers. Later that year, Enron, a company that had grown immense through its pioneering ventures in energy trading and providing energy services in the new freed-up wholesale power market, declared bankruptcy.

Restructuring of the US power grid continues. Several states have put deregulation into effect in a variety of ways. New technology has helped to bring down costs and to address the need for reducing emission of greenhouse gases during the process of generating electricity. Examples include high-efficiency gas turbines, integrated "microgrids" of small generators (sometimes in the form of solar cells or fuel cells), and a greater use of wind turbines.

Much of the interest in restructuring has centered around the generation part of the power business and less on expanding the transmission grid itself. About 25 years ago, the generation capacity margin, the ability to meet peak demand, was between 25 and 30 %—it has now reduced to less than half and is currently at about 10–15 %. These "shock absorbers" have been shrinking; e.g., during the 1990s actual demand in the U.S. increased some 35 %, while transmission capacity has increased only 18 %. In the current decade, the demand is expected to grow about 20 %, with new transmission capacity lagging behind at under 4 % growth.

In the past, extra generation capacity served to reduce the risk of generation shortages in case equipment failed and had to be taken out of production, or in case there was an unusually high demand for power, such as on very hot or cold days. As a result, capacity margins, both for generation and transmission, are shrinking. Other changes add to the pressure on the national power infrastructure as well. Increasing inter-regional bulk power transactions strain grid capacity. New environmental considerations, energy conservation efforts, and cost competition require greater efficiency throughout the grid.

As a result of these "diminished shock absorbers," the network is becoming increasingly stressed, and whether the carrying capacity or safety margin will exist to support anticipated demand is in question. The most visible parts of a larger and growing US energy crisis are the result of years of inadequate investments in the infrastructure. The reason for this neglect is caused partly by uncertainties over what government regulators will do next and what investors will do next.

Growth, environmental issues, and other factors contribute to the difficult challenge of ensuring infrastructure adequacy and security. Not only are infrastructures becoming more complexly interwoven and more difficult to comprehend and control, there is less investment available to support their development. Investment is down in many industries. For the power industry, direct infrastructure investment has declined in an environment of regulatory uncertainty due to deregulation, and infrastructure R&D funding has declined in an environment of increased competition because of restructuring. Electricity investment was not large to begin with. Presently the power industry spends a smaller proportion of annual sales on R&D than do the dog foods, leather, insurance, or many other industries—less than 0.3 %, or about $600 million per year.

Most industry observers recognize this shortage of transmission capability, and indeed many of the large blackouts in recent years can be traced to transmission problems, either because of faults in the lines themselves or in the coordination of power flow over increasingly congested lines. However, in the need to stay "competitive," many energy companies, and the regional grid operators that work with them, are "flying" the grid with less and less margin for error. This means keeping costs down, not investing sufficiently in new equipment, and not building new transmission highways to free up bottlenecks.

10 A Stressed Infrastructure

From a broader view, the North American electricity infrastructure is vulnerable to increasing stresses from several sources. One stress is caused by an imbalance between growth in demand for power and enhancement of the power delivery system to support this growth. From 1988 to 1998, the United States electricity demand rose by nearly 30 %, but the capacity of its transmission network grew by only 15 %. This disparity increased from 1999 to 2009: demand grew by about 20 %, while planned transmission systems grow by only under 3.8 % [1]. Along with that imbalance, today's power system has several sources of stress:

- *Demand is outpacing infrastructure expansion and maintenance investments.* Generation and transmission capacity margins are shrinking and unable to meet peak conditions, particularly when multiple failures occur while electricity demand continues to grow.
- *The transition to deregulation is creating new demands that are not being met.* The electricity infrastructure is not being expanded or enhanced to meet the demands of wholesale competition in the industry; so connectivity between consumers and markets is at a gridlock.
- *The present power delivery infrastructure cannot adequately handle those new demands of high-end digital customers and 21st century economy.* It cannot support levels of security, quality, reliability, and availability needed for economic prosperity.
- *The infrastructure has not kept up with new technology.* Many distribution systems have not been updated with current technology including IT.
- *Proliferation of distributed energy resources (DER).* DER includes a variety of energy sources—micro turbines, fuel cells, photovoltaics, and energy storage devices—with capacities from approximately 1 kW to 10 MW. DER can play an important role in strengthening energy infrastructure. Currently, DER accounts for about 7 % of total capacity in the United States, mostly in the form of backup generation, yet very little is connected to the power delivery system. By 2020, DER could account for as much as 25 % of total U.S. capacity, with most DER devices connected to the power delivery system.
- *Return on investment (ROI) uncertainties are discouraging investments in the infrastructure upgrades.* Investing new technology in the infrastructure can meet these aforementioned demands. More specifically, according to a June 2003 report by the National Science Foundation, R&D spending in the U.S. as a percent of net sales was about 10 % in the computer and electronic products industry and 12 % for the communication equipment industry in 1999. Conversely, R&D investment by electric utilities was less than 0.5 % during the same period. R&D investment in most other industries is also significantly greater than that in the electric power industry (NSF 2003).
- *Concern about the national infrastructure's security* [13]; EPRI (2001). A successful terrorist attempt to disrupt electricity supplies could have devastating effects on national security, the economy, and human life. Yet power

systems have widely dispersed assets that can never be absolutely defended against a determined attack.

Competition and deregulation have created multiple energy producers that share the same energy distribution network, one that now lacks the carrying capacity or safety margin to support anticipated demand. Investments in maintenance and research and development continue to decline in the North American electrical grid. Yet, investment in core systems and related IT components are required to insure the level of reliability and security that users of the system have come to expect.

In addition, the power industry is just beginning to adapt to a wider spectrum of risks. Both the number and frequency of weather-caused major outages have increased from 2 to 5 during 1950s–1980s and even early 1990s to a range of 70–130 per year during 2008–2012—now accounting for 66 % root cause of disruptions affecting 178 million customers (meters), and accounting for about 1333 outages, comprising 78 % of all outages, during 1992–2011. We are at the early stages of this adaption process to implement the strategies, systems, technologies and practices that will harden the grid and improve restoration performance after a physical disturbance.

10.1 Smart Self-Healing Grid

An interdependent, secure, integrated, reconfigurable, electronically controlled system that operates in parallel with an electric power grid. An electric power grid can be defined as the entire apparatus of wires and machines that connects power plants with customers. Adding and utilizing a "smart" element—sensors, communications, monitors, optimal controls and computers—to the electric grid can substantially improve its efficiency and reliability. In particular, secure digital technologies added to the grid and the architecture used to integrate these technologies into the infrastructure make it possible for the system to be electronically controlled and dynamically configured. This gives the grid unprecedented flexibility and functionality as well as a self-healing capability. It can react to and minimize the impact of unforeseen events, such as power outages, so that services are more robust and always available.

In addition, a stronger and smarter grid, combined with massive storage devices, can substantially increase the integration of wind and solar energy resources into the generation mix. It can support a wide-scale system for charging electric vehicles. Utilities can use its technologies to charge variable rates based on real-time fluctuations in supply and demand, and consumers can directly configure their services to minimize electricity costs.

10.2 Smart Grids and the Consumer: Empowered Consumers

Throughout the history of electric power systems, grid operators have largely worked in the paradigm that supply (i.e., generators) exists to follow all variability in consumers' demand. This has profound implications for how we design and operate the grid: the size of system peak relative to average power influences how much capacity needs to be built and the levelized costs of existing capacity, short time scale (sub-daily) variability determines how much flexible generation is required to follow ramps in demand and forecast errors.

Also throughout the history of power systems, though electricity prices that follow *fixed* time-of-use schedules have been common, the price of electricity most consumers pay has been independent of the evolution of system conditions on a day-to-day or hour-to-hour basis. For example, if consumers demand more electricity and requires more expensive generation to do so, the price consumers pay would nonetheless remain the same. Moreover system operators have had very few tools to reduce demand from customers *with their permission*; emergency load shedding programs and rolling blackouts are among the system operators' bluntest tools, and they only use them in the most extreme conditions.

However, a number of changes in technology and society are inspiring a move away from the world of "inflexible loads" and into a world where loads become enabled for real-time responsiveness to system conditions. As we will mention below, some of this responsiveness (motivated by dynamic prices or reliability signals) has existed in parts of the commercial and industrial sector for some time and there are many practical lessons to be learned from that history. However, though we do not yet fully understand what the benefits would be, it is possible that the scale of consumer engagement in power systems could be many orders of magnitude greater than it is today.

The Smart Grid extends all the way from the source of fuel for the electric power production to the many devices that use electricity, such as a household refrigerator, a piece of manufacturing equipment or a city park's lighting fixtures.

The Smart Grid is essential to support the related goals of price transparency, clean energy, grid reliability and electrified transportation. For example, Smart Grids allow for charging variable rates for energy based on supply and demand. This will reduce peak usage and increase system reliability by providing an incentive for consumers to shift their heavy uses of electricity (such as for heavy-duty appliances or processes that are less time-sensitive) to times of the day when demand is low (called peak shaving or load leveling).

These technologies provide consumers with the ability to obtain information, control and options of their own energy consumption. This involves advanced metering and demand response technologies. Two way communications and consumer-ready interfaces are essential for the success of this technology group. Conditional information of the grid and utility systems are additional benefits.

The data interfaces and communications are essential pieces to link to the other areas. These combinations of technologies enable customer empowerment.

10.3 Smart Grids and the Supplier

The U.S. federal government recognized this potential by implementing the Energy Independence and Security Act (EISA) of 2007. Title XIII of the Act mandates a Smart Grid that is focused on modernizing and improving the information and control infrastructure of the electric power system. Among the areas being addressed in the Smart Grid are: transmission, distribution, home-to-grid, industry-to-grid, building-to-grid, vehicle-to-grid, integration of renewable and distributed energy resources (such as wind and solar, which are intermittent), and demand response.

In particular, the secure digital technologies added to the grid and the architecture used to integrate these technologies into the infrastructure make it possible for the system to be electronically controlled and dynamically configured. This gives the end-to-end grid unprecedented flexibility and functionality and self-healing capability. It can react to and minimize the impact of unforeseen events, such as power outages, so that services are more robust and always available.

Operationally focused technologies are utilized not only on the utility-side of the smart grid, but also at the consumer and asset management side. Together these technologies allow for the realization of a number of different key deliverables for the execution of the smart grid. They allow for a Distribution Management System that can enable two-way power flow on the system. They form an Advanced Outage Management System that can integrate Consumer technologies and Advanced Distribution Operational technologies to detect and diagnose stressed areas in the system. There is also an increasing number of microgrid integration projects that allow the local grids to determine (in time of need) if they should remain interconnected or become islanded.

Microgrids are small power systems of several MW or less in scale with three primary characteristics: distributed generators with optional storage capacity, autonomous load centers, and the capability to operate interconnected with or islanded from a larger grid. Storage can be provided by batteries, supercapacitors, flywheels, or other sources. Microgrid assemblies are groups of interconnected microgrids that are in some sense "near" one another so that interconnection distances are small. As a result, line characteristics for such assemblies are similar to those of distribution systems.

Certainly, the power grid backbone also needs to become increasingly efficient and smart grids are designed to do this by efficiently integrating renewable resources that reduce society's need for fossil-based resources, among other approaches. The upgraded backbone, combined with microgrids, will help us meet our goals for an efficient and eco-friendly electric power system.

The smart grid also has very important features that help the planet deal with energy and environmental challenges and reduce carbon emissions. Smart grids have the potential to substantially reduce energy consumption and CO_2 emissions. In fact CO_2 emissions alone could be reduced by 58 % in 2030, compared to 2005 emissions.

In summary a stronger and smarter grid, combined with massive storage devices, can substantially increase the integration of wind and solar energy resources into the generation mix. It can support a wide-scale system for charging electric vehicles. Utilities can use its technologies to charge variable rates based on real-time fluctuations in supply and demand, and consumers can directly configure their services to minimize electricity costs.

11 A New Energy Value Chain

In the past, power grids consisted of loosely connected networks of largely local systems. Today, however, they increasingly host the movement of wholesale power from one company to another (sometimes over the transmission lines of a third company) and from one region to another. Likewise, more and more distributed resources, primarily small generators, are connecting to the grid. The extent of interconnectedness, like the number of sources, controls, and loads, has grown with time.

As a result of these new technologies, players, regulatory environments that encourage competitive markets, a new energy value chain is emerging. In the past, much of the focus was on the supply side to enable competitive wholesale transactions. Changes in technology and the resulting economics have disrupted this traditional value chain and stimulated the adoption of distributed energy resources (DER). These distributed resources can assume many forms, but some key examples are distributed generation and storage, and plug-in hybrid electric vehicles (PHEVs).

In addition, because of competition and deregulation, an entire new area of energy services and transactions has been created around the demand side of the value chain. One of these new energy services is demand response (DR). DR enables load and other DER resources to provide capacity into the bulk power system in response to grid contingencies and market pricing signals. DR is an example of an energy service that requires the interaction and integration of multiple-party business systems and physical assets, resulting in both physical and financial transactions.

12 The Self-Healing Grid

A self-healing smart grid can be built to fulfill three primary objectives. The most fundamental is real-time monitoring and reaction [6]. An array of sensors would monitor electrical parameters such as voltage and current, as well as the condition of critical components. These measurements would enable the system to constantly tune itself to an optimal state.

The second goal is anticipation. The automated system constantly looks for potential problems that could trigger larger disturbances, such as a transformer that is overheating. Computers would assess trouble signs and possible consequences. They would then identify corrective actions, simulate the effectiveness of each action, and present the most useful responses to human operators, who could then quickly implement corrective action by dispatching the grid's many automated control features. This is a fast look-ahead capability to anticipate problems, to adapt to new conditions after an outage, or an attack, the way a fighter plane reconfigures its systems to stay aloft even after being damaged. This advanced feature enables resilience in times of crisis in three ways: by providing faster-than-real-time, look-ahead simulations (analogous to master chess players rapidly expanding and evaluating their various options under time constraints) and thus avoiding previously unforeseen disturbances; by performing what-if analysis for large-region power systems from both operations and planning points of view; and by integrating market, policy and risk analysis into system models, and quantifying their integrated effects on system security and reliability.

The third objective is rapid isolation. If failures were to occur, the whole network would break into isolated "islands," each of which must fend for itself. Each island would reorganize its power plants and transmission flows as best it could. Although this might cause voltage fluctuations or even small outages, it would prevent the cascades that cause major blackouts. As line crews repaired the failures, human controllers would prepare each island to smoothly rejoin the larger grid. The controllers and their computers would function as a distributed network, communicating via microwaves, optical fibers or the power lines themselves. As soon as power flows were restored, the system would again start to self-optimize.

13 Smart Grids and Security

The existing end-to-end energy and power-delivery system is vulnerable to natural disasters and intentional cyber-attacks. Virtually every crucial economic and social function depends on the secure, reliable operation of power and energy infrastructures. Energy, electric power, telecommunications, transportation, and financial infrastructures are becoming interconnected, thus posing new challenges for their secure, reliable, and efficient operation. All of these interdependent infrastructures are complex networks, geographically dispersed, non-linear, and interacting both among themselves and with their human owners, operators, and users.

Challenges to the security of the electric infrastructure include:

Physical security—The size and complexity of the North American electric power grid makes it impossible both financially and logistically to physically protect the entire end-to-end and interdependent infrastructure. There currently exist over 450,000 miles of 100 kV or higher transmission lines, and many more thousands of miles of lower-voltage lines. As an increasing amount of electricity is generated from distributed renewable sources, the problem will only be exacerbated.

Cyber security—Threats from cyberspace to our electrical grid are rapidly increasing and evolving. While there have been no publicly known major power disruptions due to cyber-attacks, public disclosures of vulnerabilities are making these systems more attractive as targets.

Security,[7] which includes privacy and cybersecurity, is fundamentally necessary for reliable grid operations and for customer acceptance of smart grids, and many in IEEE and the smart grid community are developing technologies and standards addressing this issue. What is most important, however, is that security is incorporated into the architectures and designs at the outset, not as an afterthought. For the microgrids it is necessary to employ security technologies for each equipment component that is used and for each customer application that is developed. If any part of the system is compromised, the system reconfigures to protect itself, localize and fend off the attacks.

Due to the increasingly sophisticated nature and speed of some malicious code, intrusions, and denial-of-service attacks, human response may be inadequate. Furthermore, currently more than 90 % of successful intrusions and cyber-attacks take advantage of known vulnerabilities and misconfigured operating systems, servers, and network devices. Technological advances targeting system awareness, cryptography, trust management and access controls are underway and continued attention is needed on these key technological solutions.

[7] Regarding interoperability and emerging security standards, the IEEE-SA has published an architectural framework for the smart grid, called IEEE 2030, which defines the interconnection and interoperability standards for the power, IT and communications technologies that will be used in smart grids. IEEE-SA is working actively on standardization with the NIST Smart Grid Interoperability Panel, which includes IEEE-SA standards in its catalog of smart grid standards. IEEE-SA also collaborates with many standards organizations that represent specific industries, countries or regions to help make sure that products that operate on smart grids are complementary and compatible with one another.

14 Smart Grid: Costs and Benefits

What are the costs/benefits and range of new consumer-centered services enabled by smart grids? What is the smart grid's potential to drive economic growth? To begin addressing these, the costs of full implementation for a nationwide Smart Grid range over a 20-year period (2010–2030):

- According to energy consulting firm Brattle Group, the necessary investment to achieve an overhaul of the entire electricity infrastructure and a smart grid is $1.5 trillion spread over 20 years (~$75 billion/year), including new generators and power delivery systems.
- A detailed study by the Electric Power Research Institute (EPRI) published in April 2011, finds that that the estimated *net* investment needed to realize the envisioned power delivery system of the future is between $338 and $476 billion. The new estimates translate into annual investment levels of between $17 and $24 billion over the next 20 years.

The costs cover a wide variety of enhancements to bring the power delivery system to the performance levels required for a smart grid. They include the infrastructure to integrate distributed energy resources and achieve full customer connectivity but exclude the cost of generation, the cost of transmission expansion to add renewables and to meet load growth and a category of customer costs for smart-grid-ready appliances and devices. Despite the costs of implementation, investing in the grid would pay for itself, to a great extent. Integration of the Smart Grid will result in:

(1) Costs of outages reduced by about $49B per year,
(2) Increased efficiency and reduced emissions by 12–18 % per year (PNNL report, January 2010),
(3) A greater than 4 % reduction in energy use by 2030; translating into $20.4 billion in savings,
(4) More efficient to move electrical power through the transmission system than to ship fuels the same distance. From an overall system's perspective, with goals of increased efficiency, sustainability, reliability, security and resilience, we need both:

    - Local microgrids (that can be as self-sufficient as possible and island rapidly during emergencies), and
    - Interconnected, smarter and stronger power grid backbone that can efficiently integrate intermittent sources, and to provide power for end-to-end electrification of transportation.

(5) Reduction in the cost of infrastructure expansion and overhaul in response to annual peaks. The demand response and smart grid applications could reduce these costs significantly.

(6) The benefit-to-cost ratios are found to range from 2.8 to 6.0. Thus, the smart gird definition used as the basis for the study could have been even wider, and yet benefits of building a smart grid still would exceed costs by a healthy margin. By enhancing efficiency, for example, the smart grid could reduce 2030 overall CO_2 emissions from the electric sector by 58 %, relative to 2005 emissions.
(7) Increased cyber/IT security, and overall energy security, if security is built in the design as part of layered defense system architecture.

On options and pathways forward, I am often asked "*should we have a high-voltage power grid or go for a totally distributed generation, for example with microgrids*?" We need both, as the "choice" in the question poses a false dichotomy. It is not a matter of "this OR that" but it is an "AND". To elaborate briefly, from an overall energy system's perspective (with goals of efficiency, eco-friendly, reliability, security and resilience) we need both (1) microgrids (that can be as efficient and self-sufficient as possible, and to island rapidly during emergencies), AND we need (2) a stronger and smarter power grid as a backbone to efficiently integrate intermittent renewable sources into the overall system.

The global investments so far in advanced metering infrastructure and the coming wave of investment in distribution automation are but the beginning of a multi-decade, multi-billion-dollar effort to achieve an end-to-end, intelligent, secure, resilient, and self-healing system. It is noteworthy that the cost-effective investments to harden the grid and support resilience will vary by region, by utility, by the legacy equipment involved and even by the function and location of equipment within a utility's service territory.

15 Options and Possible Futures: What Will It Take to Succeed?

Revolutionary developments in both information technology and material science and engineering promise significant improvement in the security, reliability, efficiency, and cost-effectiveness of all critical infrastructures. Steps taken now can ensure that critical infrastructures continue to support population growth and economic growth without environmental harm.

As a result of demand growth, regulatory uncertainty, and the increasing connectedness of critical infrastructures, it is quite possible that in the near future the ability, for example, of the electricity grid to deliver the power that customers require in real-time, on demand, within acceptable voltage and frequency limits, and in a reliable and economic manner may become severely tried. Other infrastructures similarly may be tested.

At the same time, deregulation and restructuring have added concern about the future of the electric power infrastructure (and other industries as well). This shift marked a fundamental change from an industry that was historically operated in a very conservative and largely centralized way as a regulated monopoly, to an industry operated in a decentralized way by economic incentives and market forces. The shift impacts every aspect of electrical power including its price, availability, and quality. For example, as a result of deregulation, the number of interacting entities on the electric grid (and hence its complexity) has been dramatically increasing while, at the same time, a trend towards reduced capacity margins has appeared. Yet, when deregulation was initiated, little was known about its large-scale, long-term impacts on the electricity infrastructure, and no mathematical tools were available to explore possible changes and their ramifications.

It was in this environment of concern that the smart self-healing grid was conceived. One event in particular precipitated the creation of its foundations: a power outage that cascaded across the western United States and Canada on August 10, 1996. This outage began with two relatively minor transmission-line faults in Oregon. But ripple effects from these faults tripped generators at McNary dam, producing a 500 MW-wave of oscillations on the transmission grid that caused separation of the primary West Coast transmission circuit, the Pacific Intertie, at the California-Oregon border. The result: blackouts in 13 states and provinces costing some $1.5 billion in damages and lost productivity. Subsequent analysis suggests that shedding (dropping) some 0.4 % of the total load on the grid for just 30 min would have prevented the cascading effects and prevented large-scale regional outages (note that load shedding is not typically a first option for power grid operators faced with problems).

From a broader perspective, any critical national infrastructure typically has many layers and decision-making units and is vulnerable to various types of disturbances. Effective, intelligent, distributed control is required that would enable parts of the constituent networks to remain operational and even automatically reconfigure in the event of local failures or threats of failure. In any situation subject to rapid changes, completely centralized control requires multiple, high-data-rate, two-way communication links, a powerful central computing facility, and an elaborate operations control center. But all of these are liable to disruption at the very time when they are most needed (i.e., when the system is stressed by natural disasters, purposeful attack, or unusually high demand).

Had the results of the CIN/SI been in place at the time of the August 1996 blackout, the events might have unfolded very differently. For example, fault anticipators located at one end of the high voltage transmission lines would have detected abnormal signals, and making adaptive reconfiguration of the system to sectionalize the disturbance and minimize the impact of components failures several hours before the line failed. The look-ahead simulations would have identified the line as having a higher than normal probability of failure. Quickly, cognitive agents (implemented as distributed software and hardware in the infrastructure components and in control centers) would have run failure scenarios on their virtual system models to determine the ideal corrective response. When the high-voltage line actually failed, the sensor network would have detected the voltage fluctuation and communicate the information to reactive agents located at substations. The reactive agents would have executed the pre-determined corrective actions, isolating the high-voltage line and re-routed power to other parts of the grid. No customer in the wider area would even be aware that a catastrophic event had impended, or had seen a few flickers in the light.

Such an approach provides an expanded stability region with larger operational range; as the operating point nears the limit to how much the grid could have adapted (e.g., by automatically rerouting power and/or balancing dropping a small amount of load or generation), rather than cascading failures and large-scale regional system blackouts, the system will be reconfigured to minimize severity/size of outages, to shorten duration of brownouts/blackouts, and to enable rapid/efficient restoration.

This kind of distributed grid control has many advantages if coordination, communication, bandwidth, and security can be assured. This is especially true when the major components are geographically dispersed, as in a large telecommunications, transportation, or computer network. It is almost always preferable to delegate to the local level, as much of the control as is practical.

The simplest kind of distributed control would combine remote sensors and actuators to form regulators (e.g., intelligent electronically controlled secure devices), and adjust their set points or biases with signals from a central location. Such an approach requires a different way of modeling—of thinking about, organizing and designing—the control of a complex, distributed system. Recent research results from a variety of fields, including nonlinear dynamical systems, artificial intelligence, game theory, and software engineering have led to a general theory of complex adaptive systems (CAS). Mathematical and computational techniques originally developed and enhanced for the scientific study of CAS provide new tools for the engineering design of distributed control so that both centralized decision-making and the communication burden it creates can be minimized. The basic approach to analyzing a CAS is to model its components as independent adaptive software and hardware "agents"—partly cooperating and partly competing with each other in their local operations while pursuing global goals set by a minimal supervisory function.

If organized in coordination with the internal structure existing in a complex infrastructure and with the physics specific to the components they control, these agents promise to provide effective local oversight and control without the need of excessive communications, supervision, or initial programming. Indeed, they can be used even if human understanding of the complex system in question is incomplete. These agents exist in every local subsystem-from "horseshoe nail" up to "kingdom"-and perform preprogrammed self-healing actions that require an immediate response. Such simple agents are already embedded in many systems today, such as circuit breakers and fuses as well as diagnostic routines. The observation is that we can definitely account for lose nails and to save the kingdom.

Another key insight came out of analysis of forest fires. In a forest fire the spread of a spark into a conflagration depends on how close together are the trees. If there is just one tree in a barren field and it is hit by lightning, it burns but no big blaze results. But if there are many trees and they are close enough together-which is the usual case with trees because nature is prolific and efficient in using resources-the single lightning strike can result in a forest fire that burns until it reaches a natural barrier such as a rocky ridge, river, or road. If the barrier is narrow enough that a burning tree can fall across it or it includes a burnable flaw such as a wooden bridge, the fire jumps the barrier and burns on. It is the role of first-response wild-land firefighters, such as smokejumpers, to contain a small fire before it spreads by reinforcing an existing barrier or scraping out a defensible fire line barrier around the original blaze.

Similar results hold for failures in electric power grids. For power grids, the "one-tree" situation is a case in which every single electric socket has a dedicated wire connecting it to a dedicated generator. A lightning strike on any wire would take out that one circuit and no more. But like trees in nature, electrical systems are designed for efficient use of resources, which means numerous sockets served by a single circuit and multiple circuits for each generator. A failure anywhere on the system causes additional failures until a barrier-a surge protector or circuit breaker, say-is reached. If the barrier does not function properly or is insufficiently large, the failure bypasses it and continues cascading across the system [16].

These preliminary findings suggest approaches by which the natural barriers in power grids may be made more robust by simple design changes in the configuration of the system, and eventually how small failures might be contained by active smokejumper-like controllers before they grow into large problems. CIN/SI developed, among other things, a new vision for the integrated sensing, communications, and control of the power grid. Some of the pertinent issues are why/how to develop controllers for centralized versus decentralized control and issues involving adaptive operation and robustness to disturbances that include various types of failures.

Modern computer and communications technologies now allow us to think beyond the protection systems and the central control systems to a fully distributed system that places intelligent devices at each component, substation and power plant. This distributed system will enable us to build a truly smart grid.

One of the problems common to the management of central control facilities is the fact that any equipment changes to a substation or power plant must be described and entered manually into the central computer system's database and electrical one-line diagrams. Often this work is performed some time after the equipment is installed and there is thus a permanent set of incorrect data and diagrams in use by the operators. What is needed is the ability to have this information entered automatically when the component is connected to the substation—much as a computer operating system automatically updates itself when a new disk drive or other device is connected.

16 Potential Road Ahead

Electric power systems constitute the fundamental infrastructure of modern society. Often continental in scale, electric power grids and distribution networks reach virtually every home, office, factory, and institution in developed countries and have made remarkable, if remarkably insufficient, penetration in developing countries such as China and India.

Global trends toward interconnectedness, privatization, deregulation, economic development, accessibility of information, and the continued technical trend of rapidly advancing information and telecommunication technologies all suggest that the complexity, interactivity, and interdependence of infrastructure networks will continue to grow.

The existing electricity infrastructure evolved to its technology composition today from the convolution of several major forces, only one of which was technologically based. During the past 15 years, we have systematically scanned science and technology, investment and policy dimensions to gain clearer insight on current science and technology assets when looked at from a consumer-centered future perspective, rather than just incremental contributions to today's electric energy system and services.

The goal of transforming the current infrastructures to self-healing energy delivery, markets, computer and communications networks with unprecedented robustness, reliability, efficiency and quality for customers and our society is ambitious. This will require addressing challenges and developing tools, techniques, and integrated probabilistic risk assessment/impact analysis for wide-area sensing and control for digital-quality infrastructure—sensors, communication and data management, as well as improved state estimation, monitoring and simulation linked to intelligent and robust controllers leading to improved protection and discrete-event control. These follow-on activities will build on the foundations of CIN/SI and current programs that include self-healing systems and real-time dynamic information and emergency management and control.

More specifically, the operation of a modern power system depends on a complex system of sensors and automated and manual controls, all of which are tied together through communication systems. While the direct physical destruction of generators, substations, or power lines, may be the most obvious strategy for causing blackouts, activities that compromise the operation of sensors, communication and control systems by spoofing, jamming, or sending improper commands could also disrupt the system, cause blackouts, and in some cases result in physical damage to key system components. Hacking and cyber attacks are becoming increasingly common.

Most early communication and control systems used in the operation of the power system were carefully isolated from the outside world, and were separated from other systems, such as corporate enterprise computing. However, economic pressures created incentives for utilities to make greater use of commercially available communications and other equipment that was not originally designed with security in mind. Unfortunately, from a security perspective, such interconnections with office and electronic business systems through other layers of communications created vulnerabilities. While this problem is now well understood in the industry and corrective action is being taken, we are still in a transition period during which some control systems have been inadvertently exposed to access from the Internet, intranets, and remote dial-up capabilities that are vulnerable to cyber intrusions.

Many elements of the distributed control systems now in use in power systems are also used in a variety of applications in process control, manufacturing, chemical process control and refineries, transportation, and other critical infrastructure sectors and hence vulnerable to similar modes of attack. Dozens of communication and cyber security intrusions, and penetration red-team attacks have been conducted by DOE, EPRI, electric utilities, commercial security consultants, KEMA, and others. These "attacks" have uncovered a variety of cyber vulnerabilities including unauthorized access, penetration and hijacking of control.

While some of the operations of the system are automatic, ultimately human operators in the system control center make decisions and take actions to control the operation of the system. In addition, to the physical threats to such centers and the communication links that flow in and out of them, one must also be concerned about two other factors: the reliability of the operators within the center, and the possibility that insecure code has been added to one of the programs in a center computer. The threats posed by "insider" threats, as well as the risk of a "Trojan horse" embedded in the software of one or more of the control centers is real, and can only be addressed by careful security measures both within the commercial firms that develop and supply this software, and care security screening of the utility and outside service personnel who perform software maintenance within the center. Today security patches are often not always supplied to end-users, or users are not applying the patches for fear of impacting system performance. Current practice is to apply the upgrades/patches after SCADA vendors thoroughly test and validate patches, sometimes incurring a delay in patch deployment of several months.

As an example, related to numerous major outages, narrowly-programmed protection devices have contributed to worsening the severity and impact of the outage—typically performing a simple on/off logic which locally acts as preprogrammed while destabilizing a larger regional interconnection. With its millions of relays, controls and other components, the parameter settings and structures of the protection devices and controllers in the electricity infrastructure can be a crucial issue. It is analogous to the poem "for want of a horseshoe nail… the kingdom was lost." i.e., relying on an "inexpensive 25 cent chip" and narrow control logic to operate and protect a multi-billion dollar machine.

As a part of enabling a smart self-healing grid, we have developed fast look-ahead modeling and simulation, precursor detection, adaptive protection and coordination methods that minimize the impact on the whole system performance (load dropped as well as robust rapid restoration). There is a need to coordinate the protection actions of such relays and controllers with each other to achieve overall stability; a single controller or relay cannot do all, and they are often tuned for worst cases, therefore control action may become excessive from a system wide perspective. On the other hand, they may be tuned for best case, and then the control action may not be adequate. This calls for a coordinating protection and control—neither agent, using its local signal, can by itself stabilize a system; but with coordination, multiple agents, each using its local signal, can stabilize the overall system.

It is important to note that the key elements and principles of operation for interconnected power systems were established in the 1960s prior to the emergence of extensive computer and communication networks. Computation is now heavily used in all levels of the power network-for planning and optimization, fast local control of equipment, and processing of field data. But coordination across the network happens on a slower time-scale. Some coordination occurs under computer control, but much of it is still based on telephone calls between system operators at the utility control centers, even-or especially!—during emergencies.

Over the last 15 years, our efforts in this area have developed, among other things, a new vision for the integrated sensing, communications, protection and control of the power grid. Some of the pertinent issues are why/how to develop protection and control devices for centralized versus decentralized control and issues involving adaptive operation and robustness to various destabilizers. However, instead of performing in Vivo societal tests which can be disruptive, we have performed extensive "wind-tunnel" simulation testing (in Silico) of devices and policies in the context of the whole system along with prediction of unintended consequences of designs and policies to provide a greater understanding of how policies, economic designs and technology might fit into the continental grid, as well as guidance for their effective deployment and operation.

Advanced technology now under development or under consideration holds the promise of meeting the electricity needs of a robust digital economy. The architecture for this new technology framework is evolving through early research on concepts and the necessary enabling platforms. This architectural framework envisions an integrated, self-healing, electronically controlled electricity supply system of extreme resiliency and responsiveness—one that is fully capable of responding in real time to the billions of decisions made by consumers and their increasingly sophisticated agents. The potential exists to create an electricity system that provides the same efficiency, precision and interconnectivity as the billions of microprocessors that it will power.

17 Next Steps

A new mega-infrastructure is emerging from the convergence of energy (including the electric grid, water, oil and gas pipelines), telecommunications, transportation, Internet and electronic commerce. Furthermore, in the electric power industry and other critical infrastructures, new ways are being sought to improve network efficiency and eliminate congestion problems without seriously diminishing reliability and security.

A balanced, cost-effective approach to investments and use of technology can make a sizable difference in mitigating the risk. As expressed in the July 2001 issue of *Wired* magazine: "The best minds in electricity R&D have a plan: Every node in the power network of the future will be awake, responsive, adaptive, price-smart, eco-sensitive, real-time, flexible, humming—and interconnected with everything else". The technologies include, for example, the concept of self-healing electricity infrastructure, and the methodologies for fast look-ahead simulation and modeling, adaptive intelligent islanding and strategic power infrastructure protection systems are of special interest for improving grid security from terrorist attacks.

How to control a heterogeneous, widely dispersed, yet globally interconnected system is a serious technological problem in any case. It is even more complex and difficult to control it for optimal efficiency and maximum benefit to the ultimate consumers while still allowing all its business components to compete fairly and freely. A similar need exists for other infrastructures, where future advanced systems are predicated on the near perfect functioning of today's electricity, communications, transportation, and financial services.

The increased deployment of feedback and communication implies that loops are being closed where they have never been closed before, across multiple temporal and spatial scales, thereby creating a gold mine of opportunities for control. Control systems are needed to facilitate decision-making under myriad uncertainties, across broad temporal, geographical, and industry scales—from devices to power-system-wide, from fuel sources to consumers, and from utility pricing to demand-response. The various challenges introduced can be posed as a system-of-systems problem, necessitating new control themes, architectures, and algorithms. These architectures and algorithms need to be designed so that they embrace the resident complexity in the grid: large-scale, distributed, hierarchical, stochastic, and uncertain. With information and communication technologies and advanced power electronics providing the infrastructure, these architectures and algorithms will need to provide the smarts, and leverage all advances in communications and computation such as 4G networks, cloud computing, and multi-core processors.

Given economic, societal, and quality-of-life issues and the ever-increasing interdependencies among infrastructures, a key challenge before us is whether the electricity and our interdependent infrastructures will evolve to become the primary support for the 21st century's digital society—smart secure infrastructures with self-healing capabilities—or be left behind as a 20th century industrial relic.

References

1. Amin, S.M.: North America's electricity infrastructure: are we ready for more perfect storms? IEEE Secur. Priv. Mag. **1**(5), 19–25 (2003)
2. Amin, S.M.: Toward self-healing energy infrastructure systems cover feature. IEEE Comput. Appl. Power **14**(1), 20–28 (2001)
3. U.S. Department of Homeland Security. National Infrastructure Protection Plan. http://www.dhs.gov/xlibrary/assets/NIPP_Plan.pdf. Accessed Dec 2012
4. "Sustainable Critical Infrastructure Systems—a Framework for Meeting 21st Century Imperatives," Board on Infrastructure and the Constructed Environment (BICE), Division on Engineering and Physical Sciences, National Research Council of the National Academies, The National Academies Press, Washington, DC (2009)
5. EPRI 2003: Complex Interactive Networks/Systems Initiative: Final Summary Report—Overview and Summary Final Report for Joint EPRI and US DoD University Research Initiative, pp. 155. EPRI, Palo Alto, Dec 2003
6. Amin, S.M., Schewe, P.F.: Preventing Blackouts," Sci. Am. **296**(5), 60–67 (2007)
7. Firestone, "Power Failure Reveals a Creaky System, Energy Experts Believe," The New York Times. http://www.nytimes.com/2003/08/15/nyregion/15GRID.html. Accessed Nov 2012
8. American Society of Civil Engineers, "2009 Report Card for America's Infrastructure," http://www.infrastructurereportcard.org/sites/default/files/RC2009_full_report.pdf. Accessed June 2009
9. "Decision Technology for Rational Selection of Infrastructure Investments: Final Report of ASME Foundation Strategic Proposal Grant Fund," ASME Innovative Technologies Institute, LLC, Washington, DC, July 16, 2009
10. Center for Strategic and International Studies Commission on Public Infrastructure, Guiding Principles for Strengthening America's Infrastructure, 2006
11. Amin, S.M.: National infrastructures as complex interactive networks. In: Samad, T., Weyrauch, J. (eds.) Automation, Control, and Complexity: An Integrated Approach, Chap. 14, pp. 263–286. John Wiley and Sons Ltd., New York (2000)
12. Amin, S.M., Stringer, J.: The electric power grid: today and tomorrow. MRS Bull. 33(4), 399–407 (2008)
13. Amin, S.M.: Powering the 21st century: we can—and must—modernize the grid. IEEE Power Energ. Mag. 3(2), 93–96 (2005)
14. Amin, S.M.: U.S. grid gets less reliable. IEEE Spectr. 48(1), p. 80 (2011)
15. Amin, S.M., Wollenberg, B.F.: Toward a smart grid: power delivery for the 21st century. IEEE Power Energ. Mag. **3**(5), 34–41 (2005)
16. Amin, S.M.: For the good of the grid: toward Increased efficiencies and integration of renewable resources for future electric power networks. IEEE Power Energ. Mag. **6**(6), 48–59 (2008)

Conclusions

Elias Kyriakides and Marios Polycarpou

Abstract Everyday life relies heavily on the reliable operation and intelligent management of large-scale critical infrastructures, such as electric power systems, telecommunication networks, and water distribution networks. The design, monitoring, control and security of such systems are becoming increasingly more challenging as their size, complexity and interactions are steadily growing. Moreover, these critical infrastructures are susceptible to natural disasters, accidents, frequent failures, as well as malicious attacks. There is an urgent need to develop a common framework for modeling the behavior of critical infrastructure systems and for designing algorithms for intelligent monitoring, control and security of such systems. This book provides basic principles as well as new research directions for intelligent monitoring, control, and security of critical infrastructure systems, with an emphasis on electric power systems, telecommunication networks, water distribution systems, and transportation systems.

Keywords Critical infrastructure systems · Monitoring · Control · Security · Fault diagnosis · Interdependencies · System of systems

1 Introduction

Critical infrastructure systems have many common characteristics and requirements, which naturally point to the need for a common methodological framework. They are large-scale, complex, interconnected, distributed, and data-rich systems. They are progressively becoming larger in size and more complex, which makes

E. Kyriakides (✉) · M. Polycarpou
KIOS Research Center for Intelligent Systems and Networks, Department of Electrical and Computer Engineering, University of Cyprus, Nicosia, Cyprus
e-mail: elias@ucy.ac.cy

M. Polycarpou
e-mail: mpolycar@ucy.ac.cy

E. Kyriakides and M. Polycarpou (eds.), *Intelligent Monitoring, Control, and Security of Critical Infrastructure Systems*, Studies in Computational Intelligence 565,
DOI 10.1007/978-3-662-44160-2_13

their management more challenging. Moreover, they are dynamic and their operation is subject to significant uncertainty and time variation. Due to technological advances and deregulation, critical infrastructures are becoming more heterogeneous and distributed, making it more difficult to monitor the overall system and ensure its reliable operation. Their interconnectedness makes critical infrastructures vulnerable to failures, which may be caused by natural disasters, material failures, human error or by intentional attacks. Nevertheless, their interconnectedness also allows a level of operational redundancy for fault tolerance, provided it can be utilized effectively. As technology advances, the interactions and (inter)dependencies in the reliable operation and management between different infrastructures are becoming increasingly more crucial. For example, there are strong links between the operation of power grids and communication networks, as well as between power networks and water networks or between communications and transportation systems. Critical infrastructure systems provide the lifeline that physically ties communities and facilitates quality of life and economic growth. These interdependent systems work together to provide and support the essential services of a modern society. Therefore, there is a crucial and urgent need to investigate under a common framework the reliable operation and interactions between critical infrastructures.

In the last few years there has been significant research on the modeling, monitoring and control of critical infrastructure systems. The vast majority of this work has been application specific. For example, in electric power systems there has been a lot of work on modeling, sensor and actuator instrumentation, simulation and prediction tools, etc. Intelligent electronic devices are starting to become available for monitoring and controlling electric power systems. These devices may include sensing capabilities for real-time measurement, actuators for controlling certain variables, microprocessors for processing information and making decisions based on designed algorithms, and telecommunication units for exchanging information with other electronic devices or possibly with human operators. Such devices, which are sometimes referred to as *distributed or intelligent agents*, provide fundamental tools for intelligent monitoring, control and safety of critical infrastructure systems.

From a data analytics viewpoint, recent technological advances in computing hardware, sensors/actuators, communications and real-time software have provided the capability to generate huge amounts of data in real time. This data is generated continuously every day in a wide range of infrastructure systems, as well as in other applications. The generated data is in various forms (for example, time series measurements, audio, video, etc.) and quite possibly from different geographical locations. As the volume of data generated is rapidly increasing, one of the most important research challenges in the new technology is the development of smart information processing methodologies that can be used to extract meaning and knowledge out of the ever-increasing electronic information that becomes available. Even more important is the capability to utilize the knowledge that is being produced such that one can design software and devices that operate autonomously and cooperatively in some intelligent manner. The ultimate objective is to design intelligent systems or intelligent agents that can make appropriate real-time decisions in the management of large-scale complex systems, such as critical infrastructure systems. It is of outmost significance for the secure and reliable operation of critical infrastructure systems to combine deep knowledge of the application domains with tools from computational intelligence, system theory, optimization, and adaptive systems to develop smart monitoring and control solutions based on the principles of intelligent data interpretation and decision, self-aware/self-improve/self-healing design, scalability, flexibility, modularity and fault tolerance.

The following concepts describe some of the main attributes that a critical infrastructure system of the future is expected to have:

- Distributed Monitoring and Control: As complexity in infrastructure systems increases, there is a need to develop a distributed and decentralized (instead of centralized) monitoring and control framework, where each node in the network is a potential location for a sensor/actuator/controller device, which is referred to as an intelligent agent. The main distinction between a decentralized and a distributed framework, is the capability of some of the agents to communicate with each other in the latter case.
- Intelligent Agents: An intelligent agent is located at a node of the network and is able to gather data, process data, make decisions, possibly change a control variable, and communicate information to other agents. These agents may be in the form of hardware or software. Some intelligent agents are specialized, in the sense that they may be doing a very specific task (such as sensing). A key objective is for the intelligent agents to cooperate with each other based on some global objective.
- Adaptive and Learning Capabilities: Since it is impossible to model exactly all the dynamics of a complex infrastructure system, it is important to devise methods for adaptation and learning of the intelligent agents. These methods may be based on computational intelligence techniques and adaptive control.
- Fault Accommodation and Self-Healing: Using techniques from fault diagnosis and health monitoring, the objective is to develop algorithms that are able to predict or determine a failure as soon as possible, isolate the location of the failure and, if possible, reconfigure the nominal control algorithms to achieve fault recovery and self-healing.
- Intelligent Decision Support Systems: A key factor in the successful implementation of monitoring and control of critical infrastructure systems is the role of human operators. The objective is to design intelligent decision support systems to enhance the decision making capabilities of operators (considering the overwhelming number of alarms that are typically received after a fault and the associated short response time required) and to assist them in normalizing the state of the system at all times. Intelligent decision support systems are capable of processing real-time sensor information, to provide high-level information to human operators in a way that can be easily understood and visualized.

2 Future Directions

On an abstract level, critical infrastructure systems consist of nodes interconnected together with links (arcs) that enable the flow of certain variables from one node to its neighbors. For example, in water systems, the links are the pipes and the flows consist of transport of water, while in communication networks there are communication links that are used to transport packets or (more generally) information. Furthermore, each infrastructure has a number of hardware and/or software agents that are appropriately located in the underlying network (relevant infrastructure) and have sensing capabilities to collect data. If these agents are "intelligent", they also have computational capabilities to process the data and make decisions, have communication capabilities to send/receive information to/from other agents and have actuators to take control actions (autonomously or with human intervention). Moreover, these agents cooperate in order to achieve the desirable team objectives.

The Intelligent Agent behavior and their interactions create certain key research challenges, both from a theoretical perspective, as well as from an application perspective (i.e., as they are used in different critical infrastructure systems). Some of these challenges are described below:

- Distributed Monitoring and Control: An action of one agent may have both a local and a global effect which are not always fully understood. This issue becomes more increasingly difficult as infrastructures grow larger and include a large number of heterogeneous agents.
- Agent Collaboration: In order to achieve the global objectives set for each infrastructure, it is inevitable that the intelligent agents need to collaborate between themselves. How this collaboration should take place is an important question that needs to be addressed.
- Heterogeneous Agent Collaboration: How should agents on different infrastructure networks interact with each other in order to achieve certain objectives? Understanding and facilitating interdependencies between infrastructures is recognized as a key challenge in the years ahead.
- Adaptive and Learning Capabilities: The underlying networks evolve over time and thus the agents should be adaptive and have learning capabilities to appropriately change their behavior. Furthermore, such learning techniques are needed in order to address possible modeling inaccuracies.
- Fault Accommodation and Self-Healing: The underlying infrastructure generally consists of interconnected components which may fail due to usage or due to a deliberate attack. The agents should be able to predict or determine a failure as soon as possible, isolate the location of the failure and reconfigure the distributed control algorithm to achieve fault accommodation and self-healing.
- Architectural Issues: One of the key issues that requires attention is the determination of the overall architecture that connects the local intelligent agents. How is the communication between agents organized? Who communicates with whom? When do they communicate? One of the approaches

that can be considered is the hierarchical architecture, where the information from a number of local controllers is coordinated by regional or global controllers.

Research in critical infrastructures is expected to be at the forefront of engineering research for the next several decades. Some of the key research objectives are:

- To develop innovative intelligent management, monitoring and control models and algorithms for fault tolerant operation of critical infrastructure systems.
- To develop behavioral and functional models for accurately characterizing the operation of critical infrastructure systems under steady state, dynamic, and post-fault conditions.
- To develop interactive models between two or more critical infrastructure systems; for example, modeling the interdependence between communication and electric power networks.
- To develop network design and recovery techniques which can provide solutions for mitigating massive attacks or catastrophic failure scenarios.
- To design inter-infrastructure fault management scenarios (including fault detection, isolation and recovery).
- To develop a theoretical framework that enables the derivation of robustness analysis of large-scale, complex and distributed systems.
- To refine and apply tools and methodologies from computational intelligence and system theory for the management, monitoring, and control of electric power systems.
- To develop network security models to enhance network security, and to develop resource allocation and Quality of Service (QoS) provisioning methodologies for telecommunication networks.
- To investigate Wireless Sensor Network (WSN) based technologies as effective solutions for the problem of distributed, reliable and economic monitoring of critical infrastructure systems.
- To develop optimal sensor and actuator placement methodologies for control, health monitoring and security of critical infrastructure systems.

Some more fundamental questions that deserve attention are data collection and storage, processing, aggregation, and information fusion. During the last few years, Big Data techniques have become more prominent in data analytics, especially for handling large-scale heterogeneous data sets. Such techniques will undoubtedly generate key advances and research interests in the area of intelligent monitoring, control, and security of critical infrastructure systems.